Solid-State Lasers

Springer
New York
Berlin
Heidelberg
Hong Kong
London
Milan
Paris
Tokyo

Physics and Astronomy ONLINE LIBRARY

http://www.springer.de/phys/

Advanced Texts in Physics

This program of advanced texts covers a broad spectrum of topics that are of current and emerging interest in physics. Each book provides a comprehensive and yet accessible introduction to a field at the forefront of modern research. As such, these texts are intended for senior undergraduate and graduate students at the M.S. and Ph.D. levels; however, research scientists seeking an introduction to particular areas of physics will also benefit from the titles in this collection.

Walter Koechner
Michael Bass

Solid-State Lasers

A Graduate Text

With 252 Figures

Springer

Walter Koechner
Fibertek, Inc.
510 Herndon Parkway
Herndon, VA 20170
USA

Michael Bass
School of Optics/CREOL
University of Central Florida
Orlando, FL 32816
USA

Cover illustration: Diode-pumped ND: YAG slab laser with positive branch unstable resonator and variable reflectivity output coupler (adapted from Figure 5.24, page 182).

Library of Congress Cataloging-in-Publication Data
Koechner, Walter, 1937–
    Solid state lasers : a graduate text / Walter Koechner, Michael Bass.
      p. cm.—(Advanced texts in physics)
    Includes bibliographical references and index.
    ISBN 0-387-95590-9 (alk. paper)
    1. Solid-state lasers. I. Bass, Michael, 1939–. II. Title   III. Series.
TA1705 .K633 2003
621.36′61—dc21   2002030568

ISBN 0-387-95590-9   Printed on acid-free paper.

Printed in the United States of America.

9 8 7 6 5 4 3 2 1   SPIN 10893625

www.springer-ny.com

Springer-Verlag New York Berlin Heidelberg
A member of BertelsmannSpringer Science+Business Media GmbH

Preface

This college textbook describes the theory, operating characteristics, and design features of solid-state lasers. The book is intended for students who want to familiarize themselves with solid-state lasers beyond the level of a general textbook.

Although the book is aimed at students who are thinking of entering this fascinating field, it might also be used by practicing scientists and engineers who are changing their technical direction and want to learn more about this particular class of lasers. After studying the material presented in this book, the reader should be able to follow the scientific and technical literature and have an understanding of the basic principles and engineering issues of solid-state lasers, as well as an appreciation of the subtleties, richness of design, and operating possibilities afforded by these systems.

Solid-state lasers and systems represent a one-billion dollar industry, and they are the dominant class of lasers for research, industrial, medical, and military applications. Given the importance of solid-state lasers, a graduate text is required that deals explicitly with these devices.

Following the demonstration of the first laser over 40 years ago, an extraordinary number of different types of lasers have been invented using a wide variety of active media and pump techniques to create an inversion. As a sign of a maturing industry, laser research and engineering has developed into many specialized disciplines depending on the laser medium (solid-state, semiconductor, neutral or ionized gas, liquid) and excitation mechanism (optical pumping, electric current, gas discharge, chemical reaction, electron beam).

The development of solid-state systems represents a multidisciplinary effort and is the result of the interaction of professionals from many branches of science and engineering, such as spectroscopy, solid-state and laser physics, optical design, and electronic and mechanical engineering. Today, solid-state laser systems are very sophisticated devices, and the field has developed so far that it is difficult for a professional to enter it without prior familiarization with the basic concepts and technology of this class of lasers.

For historical reasons, solid-state lasers describe a class of lasers in which active ions in crystal or glass host materials are optically pumped to create a population inversion. Other types of lasers that employ solid-state gain media are semiconductor lasers and optical fiber lasers and amplifiers. However, since these lasers employ very specialized technologies and design principles, they are usually treated separately from conventional bulk solid-state lasers.

The design and performance characteristics of laser diode arrays are discussed in this book because these devices are employed as pump sources for solid-state lasers. Fiber lasers are very similar to conventional solid-state lases as far as the active material and pump source is concerned. However, they are radically different with respect to beam confinement, mode structure, coupling of pump and laser beams, and the design of optical components.

The content and structure of this textbook follow closely the book by Walter Koechner entitled *Solid-State Laser Engineering* which is currently in its 5th edition. In this college text the material has been streamlined by deleting certain engineering and hardware-related details, and more emphasis is placed on a tutorial presentation of the material. Also, each chapter includes tutorial exercises prepared by Professor Michael Bass to help the student reinforce the discussions in the text. A complete solutions manual for instructors is available from textbook@springer-ny.com.

After a historical overview, the books starts with a review of the basic concepts of laser physics (chapter 1), followed by an overview of the different classes and properties of solid-state laser materials (chapter 2). Analytical expressions of the threshold condition, and gain and output of laser oscillators are derived in chapter 3. An oscillator followed by one or more amplifiers is a common architecture in pulsed solid-state laser systems to boost output energy. Energy storage and gain of amplifiers is discussed in chapter 4. Beam divergence and line width of an oscillator are strongly dependent on the spatial and longitudinal mode structure of the resonator. Resonator configuration and characteristics are presented in chapter 5. Different pump source configurations for transferring pump radiation to the active medium are discussed in chapter 6. Thermal gradients set up as a result of heat removal from the active medium have a profound impact on beam quality and output power limitations. Thermal effects and cooling techniques are treated in chapter 7. The output from a laser can be changed temporally or spectrally by Q-switching, mode-locking, and frequency conversion via nonlinear phenomena. These techniques are discussed in the last three chapters.

We would like to thank Judy Eure and Renate Koechner for typing the new material and the editor, Dr. Hans Koelsch, for suggesting a college text on the subject of solid-state lasers. We also thank Prof. D. Hagan for suggestions related to the nonlinear optics exercises and Drs. Bin Chen and Jun Dong and Mrs. Hong Shun and Teyuan Chung for testing the exercises.

Special thanks are due to our wives Renate Koechner and Judith Bass, who have been very patient and supportive throughout this project.

Herndon, Virginia   *Walter Koechner*
Orlando, Florida   *Michael Bass*
September 2002

Contents

Preface   **v**

Introduction   Overview of the History, Performance Characteristics, and Applications of Solid-State Lasers   **1**

Major Milestones in the Development of Solid-State Lasers   1
Typical Performance Parameters and Applications   7

1   Energy Transfer Between Radiation and Atomic Transitions   **12**

1.1   Optical Amplification   12
1.2   Interaction of Radiation with Matter   15
    1.2.1   Blackbody Radiation   15
    1.2.2   Boltzmann's Statistics   16
    1.2.3   Einstein's Coefficients   17
    1.2.4   Phase Coherence of Stimulated Emission   20
1.3   Absorption and Optical Gain   21
    1.3.1   Atomic Lineshapes   21
    1.3.2   Absorption by Stimulated Transitions   25
    1.3.3   Population Inversion   28
1.4   Creation of a Population Inversion   30
    1.4.1   The Three-Level System   31
    1.4.2   The Four-Level System   33
    1.4.3   The Metastable Level   34
1.5   Laser Rate Equations   35
    1.5.1   Three-Level System   36
    1.5.2   Four-Level System   39
Summary   40
References   41
Exercises   41

2   Properties of Solid-State Laser Materials   **44**

2.1   Overview   45
    2.1.1   Host Materials   46
    2.1.2   Active Ions   48
2.2   Ruby   54
2.3   Nd : YAG   57

2.4   Nd : Glass   60
      2.4.1   Laser Properties   60
2.5   Nd : YLF   63
2.6   Nd : YVO$_4$ 65
2.7   Er : Glass   67
2.8   Yb : YAG   68
2.9   Alexandrite   70
2.10 Ti : Sapphire   72
Summary   74
References   75
Exercises   76

3   Laser Oscillator   **78**

3.1   Operation at Threshold   80
3.2   Gain Saturation   84
3.3   Circulating Power   86
3.4   Oscillator Performance Model   88
      3.4.1   Conversion of Input to Output Energy   88
      3.4.2   Laser Output   95
3.5   Relaxation Oscillations 102
3.6   Examples of Laser Oscillators 106
      3.6.1   Lamp-Pumped cw Nd : YAG Laser 107
      3.6.2   Diode Side-Pumped Nd : YAG Laser 111
      3.6.3   End-Pumped Systems 115
Summary 118
References 119
Exercises 119

4   Laser Amplifier   **121**

4.1   Pulse Amplification 122
4.2   Nd : YAG Amplifiers 127
4.3   Nd : Glass Amplifiers 135
4.4   Depopulation Losses 141
      4.4.1   Amplified Spontaneous Emission 141
      4.4.2   Prelasing and Parasitic Modes 144
4.5   Self-Focusing 144
Summary 147
References 147
Exercises 148

5   Optical Resonator   **149**

5.1   Transverse Modes 149
      5.1.1   Intensity Distribution 150
      5.1.2   Characteristics of a Gaussian Beam 154

5.1.3   Resonator Configurations   156
5.1.4   Stability of Laser Resonators   160
5.1.5   Higher Order Modes   161
5.1.6   Diffraction Losses   162
5.1.7   Active Resonator   164
5.1.8   Mode-Selecting Techniques   166
5.2   Longitudinal Modes   169
5.2.1   The Fabry–Perot Interferometer   169
5.2.2   Laser Resonator   172
5.2.3   Longitudinal Mode Control   175
5.3   Unstable Resonators   178
Summary   183
References   183
Exercises   184

6   Optical Pump Systems   **187**

6.1   Pump Sources   187
6.1.1   Flashlamps   187
6.1.2   Continuous Arc Lamps   196
6.1.3   Laser Diodes   198
6.2   Pump Radiation Transfer Methods   213
6.2.1   Side-Pumping with Lamps   214
6.2.2   Side-Pumping with Diodes   220
6.2.3   End-Pumped Lasers   230
6.2.4   Face-Pumped Disks   238
Summary   241
References   242
Exercises   243

7   Thermo-Optic Effects   **245**

7.1   Cylindrical Geometry   248
7.1.1   Temperature Distribution   249
7.1.2   Thermal Stresses   251
7.1.3   Photoelastic Effects   253
7.1.4   Thermal Lensing   255
7.1.5   Stress Birefringence   258
7.1.6   Compensation of Thermally Induced Optical
            Distortions   263
7.2   Slab and Disk Geometries   265
7.2.1   Rectangular-Slab Laser   265
7.2.2   Slab Laser with Zigzag Optical Path   268
7.2.3   Disk Amplifiers   270
7.3   End-Pumped Configurations   271
Summary   276

References 277
Exercises 278

8 Q-Switching   **279**

8.1   Q-Switch Theory 280
       8.1.1   Continuously Pumped, Repetitively
                Q-Switched Systems 284
8.2   Mechanical Devices 288
8.3   Electro-Optical Q-Switches 289
8.4   Acousto-Optic Q-Switches 295
       8.4.1   Device Characteristics 300
8.5   Passive Q-Switch 302
Summary 305
References 306
Exercises 306

9 Mode-Locking   **308**

9.1   Pulse Formation 308
9.2   Passive Mode-Locking 315
       9.2.1   Liquid Dye Saturable Absorber 316
       9.2.2   Kerr Lens Mode-Locking 317
9.3   Active Mode-Locking 322
       9.3.1   AM Modulation 322
       9.3.2   FM Modulation 325
9.4   Picosecond Lasers 326
       9.4.1   AM Mode-Locking 327
       9.4.2   FM Mode-Locking 329
9.5   Femtosecond Lasers 331
       9.5.1   Laser Materials 331
       9.5.2   Resonator Design 332
Summary 336
References 337
Exercises 337

10 Nonlinear Devices   **339**

10.1   Nonlinear Optics 340
       10.1.1   Second-Order Nonlinearities 341
       10.1.2   Third-Order Nonlinearities 343
10.2   Harmonic Generation 345
       10.2.1   Basic Equations of Second-Harmonic Generation . . 345
       10.2.2   Index Matching 348
       10.2.3   Parameters Affecting the Doubling Efficiency 354
       10.2.4   Intracavity Frequency Doubling 358
       10.2.5   Third-Harmonic Generation 360

10.3	Parametric Oscillators		363
	10.3.1	Performance Modeling	365
	10.3.2	Quasi-Phase-Matching	372
10.4	Raman Laser		374
10.5	Optical Phase Conjugation		379
	Summary		384
	References		385
	Exercises		386

A	**Conversion Factors and Constants**	**387**
B	**Definition of Symbols**	**391**
C	**Partial Solutions to the Exercises**	**397**
Index		**405**

Introduction

Overview of the History, Performance Characteristics, and Applications of Solid-State Lasers

Major Milestones in the Development of Solid-State Lasers
Typical Performance Parameters and Applications

In this Introduction we will provide a short overview of the important milestones in the development of solid-state lasers, discuss the range of performance parameters possible with these lasers, and mention major applications. Besides the compactness and benign operating features, it was the enormous flexibility in design and output characteristics which led to the success of solid-state lasers over the last 40 years.

Major Milestones in the Development of Solid-State Lasers

Historically, the search for lasers began as an extension of stimulated amplification techniques employed in the microwave region. Masers, coined from **M**icrowave **A**mplification by **S**timulated **E**mission of **R**adiation, served as sensitive preamplifiers in microwave receivers. In 1954 the first maser was built by C. Townes and utilized the inversion population between two molecular levels of ammonia to amplify radiation at a wavelength around 1.25 cm.

In 1955 an optical excitation scheme for masers was simultaneously proposed by N. Bloembergen, A.M. Prokorov, and N.G. Basov. A few years later, masers were mostly built using optically pumped ruby crystals. In 1958 A. Schawlow and C. Townes proposed extending the maser principle to optical frequencies and the use of a Fabry–Perot resonator for feedback. However, they did not find a suitable material or the means of exciting it to the required degree of population inversion.

This was accomplished by T. Maiman who built the first laser in 1960. It was a pink ruby crystal (sapphire with trivalent chromium impurities), optically pumped by a helical flashlamp that surrounded the cylindrical laser crystal. The parallel ends of the ruby crystal were silvered, with a small hole at one end for observing the radiation. The reflective surfaces comprised the optical resonator. The output wavelength was 694 nm. It was T. Maiman who coined the name "laser," in analogy to maser, as an abbreviation of **L**ight **A**mplification by **S**timulated **E**mission of **R**adiation.

In early ruby laser systems the output consisted of a series of irregular spikes, stretching over the duration of the pump pulse. A key discovery made by R.W. Hellwarth in 1961 was a method called Q-switching for concentrating the output from the ruby laser into a single pulse. The Q-switch is an optical shutter which prevents laser action during the flashlamp pulse, therefore the population inversion can reach large values. If the shutter is suddenly opened, stored energy will be released in a time characterized by a few round trips between the resonator mirrors. Hellwarth initially proposed a Kerr cell, a device which rotates the plane of polarization when voltage is applied. This Q-switch, which consisted of a cell filled with nitrobenzene, required very high voltages for Q-switching; it was soon replaced by spinning one of the resonator mirrors. A further refinement was the insertion of a spinning prism between the fixed mirrors of the resonator.

The earliest application of the laser was in active range-finding by measuring the time of flight of a laser pulse reflected from a target. Investigations in this direction started immediately after the discovery of the ruby laser. Four years later, fully militarized rangefinders containing a flashlamp-pulsed ruby laser with a spinning prism Q-switch went into production. For about 10 years ruby-based rangefinders were manufactured; afterward the ruby laser was replaced by the more efficient neodymium doped yttrium aluminum garnet (Nd : YAG) laser.

Beside the use in range finders, the ruby laser was basically a research tool and, for the next 15 years, ruby lasers became the standard high-power radiation source in the visible region for research at university, government, and industrial laboratories. Applications in an industrial environment were rare, in large part due to the low-pulse repetition rate of the ruby laser (a pulse every few seconds), high cost of the equipment, and the unfamiliarity of the industry with this new radiation source. Some of the specialized applications included drilling holes in diamonds that are used as dies for drawing wires, or spot welding in vacuum through the glass envelope of vacuum tubes. Another application was stress analysis by means of double pulse holography, in which surface deformation due to stress or temperature is measured interferometrically between two pulses.

The discovery of the ruby laser triggered an intensive search for other materials, and in rapid succession laser action in other solids, gases, semiconductors, and liquids was demonstrated. Following the discovery of the ruby laser, the next solid-state material was uranium-doped calcium fluoride which was lased in late 1960. The first solid-state neodymium laser was calcium tungstate doped with neodymium ions. This laser, discovered in 1961, was used in research facilities for a number of years until yttrium aluminate garnet, as a host material for neodymium, was discovered.

In 1961, E. Snitzer demonstrated the first neodymium glass laser. Since Nd : glass could be made in much larger dimensions and with better quality than ruby, it promised to deliver much higher energies. It was quickly realized that high energy, short pulses produced from large Nd : glass lasers possessed the potential to heat matter to thermonuclear temperatures, thus generating energy in small controlled explosions. Large budgets have been devoted to the development and installation of huge Nd : glass laser systems which became the world-wide systems of choice for laser fusion research and weapons simulation. The most powerful of these systems, the NOVA laser, completed in 1985, produced 100 kJ of energy in a 2.5 ns pulse. Systems with energies ten times larger are currently under construction.

Using a ruby laser, P.A. Franken demonstrated second harmonic generation in crystal quartz in 1961. Generation of harmonics is caused by the nonlinear behavior of the refractive index in the presence of a very high electric field strength. The conversion of the fundamental wavelength to the second harmonics was extremely small because the interaction length of the beams was only a few wavelengths and the nonlinearity of quartz is very low.

Soon after these first nonlinear optics experiments were conducted it was realized that efficient nonlinear interactions require a means of achieving phase-velocity matching of the interacting waves over a distance of many wavelengths. Within a year, two basic approaches to achieve efficient harmonic generation were published in the literature. One approach, namely the use of birefringence to offset dispersion, is still the preferred method for most nonlinear processes in use today. Efficient harmonic generation was soon achieved in birefringence compensated potassium dihydrogen phosphate (KDP) crystals, a crystal which is still employed today for the generation of the third harmonic of large Nd : glass lasers. The other method, namely the use of a periodic modulation of the sign of the nonlinear coefficient to restore the optical phase, could only be realized 30 years later. In the early 1990s, lithographic processing techniques enabled the fabrication of quasi-phase-matched small crystals using electric field poling of lithium niobate.

In 1962 the idea of parametric amplification and generation of tunable light was conceived, and a few years later the first experiment demonstrating parametric gain was achieved. Commercial parametric oscillators based on lithium niobate were introduced in 1971. Damage of the nonlinear material and the appearance of tunable dye lasers led to a decline in interest in optical parametric oscillators (OPOs) for almost 20 years. The discovery of damage-resistant nonlinear crystals with large nonlinear coefficients in the early 1990s revived interest in OPOs, and today tunable solid-state lasers covering the wavelength range from the visible to the near-infrared have found widespread applications in spectroscopy, remote sensing, and wherever a tunable radiation source is required.

The possibility of laser action in a semiconductor was explored rather early. Initially, intrinsic semiconductors pumped by an electron beam or by optical radiation were considered. However, at the end of 1962, several groups succeeded in producing pulsed output from gallium–arsenide p–n junctions cooled to cryogenic temperatures. About 10 years later, continuous operation at room temperature was achieved.

The first optical fiber amplifier was demonstrated in 1963 using a 1 m long neodymium-doped glass fiber wrapped around a flashlamp. However, the concept received little attention until the 1980s when low-loss optical fibers became available and the fiber-optic communications industry explored these devices for amplification of signals.

In 1964 the best choice of a host for neodymium ions, namely yttrium aluminum garnet (YAG), was discovered by J. Geusic. Since that time, Nd:YAG remains the most versatile and widely used active material for solid-state lasers. Nd:YAG has a low threshold which permits continuous operation, and the host crystal has good thermal, mechanical, and optical properties and can be grown with relative ease.

An immediate application was the replacement of ruby with Nd:YAG in military rangefinders. Since the system efficiency was about a factor of 10 higher with Nd:YAG as compared to ruby, the weight of storage capacitors and batteries was drastically reduced. This allowed the transition from a tripod-mounted unit, the size of a briefcase, to a hand-held device only slightly larger than a binocular.

Continuously pumped, repetitively Q-switched Nd:YAG lasers were the first solid-state lasers which found applications in a production environment, mainly in the semiconductor industry for resistor trimming, silicon scribing, and marking. The early systems were pumped with tungsten filament lamps and Q-switched with a rotating polygon prism. Reliability was a big issue because lamp-life was short—on the order of 40 hours—and the bearings of the high-speed motors employed in the rotating Q-switches did wear out frequently. The mechanical Q-switches were eventually replaced by acousto-optic Q-switches, and krypton arc lamps replaced tungsten filament lamps.

Up to this point, solid-state lasers were capable of generating very impressive peak powers, but average power was still limited to a few watts or at most a few tens of watts. However, at the end of the 1960s, continuously pumped Nd:YAG lasers with multihundred watts output power became commercially available.

During the first years of laser research, a particular effort was directed toward generation of short pulses from Nd:glass and ruby lasers. With Q-switching, several round trips are required for radiation to build up. Given the length of the resonator and available gain of these early systems, the pulses were on the order of 10 to 20 ns. The next step toward shorter pulses was a technique called cavity dumping, whereby the radiation in the resonator, as it reached its peak, was quickly dumped by a fast Q-switch. Pulses with a duration on the order of one round trip (a few nanoseconds) in the resonator could be generated with this method. In 1965, a technique termed "mode-locking" was invented. Mode-locking is a technique whereby passive loss modulation, with a fast response saturable absorber, or by active loss of frequency modulation, a fixed relationship among the phases of the longitudinal modes is enforced. With either passive or active mode-locking, pulses much shorter than a resonator round trip time can be generated; typically, pulses are on the order of 20 to 100 ps.

By the end of the 1960s, most of the important inventions with regard to solid-state laser technology had been made. Nd:YAG and Nd:glass proved clearly superior over many other solid-state laser materials; short-pulse generation by means of Q-switching and mode-locking, as well as frequency conversion with harmonic generators and parametric oscillators, was well understood. Xenon-filled flashlamps and krypton arc lamps had been developed as pump sources and laser diodes were recognized as an ideal pump source, but due to a lack of suitable devices the technology could not be implemented.

To gain wider acceptance in manufacturing processes, the reliability of the laser systems needed improvement and the operation of the lasers had to be simplified. During the 1970s, efforts concentrated on engineering improvements, such as an increase in component and system lifetime and reliability. The early lasers often worked poorly and had severe reliability problems. At the component level, damage resistant optical coatings and high-quality laser crystals had to be developed; and the lifetime of flash lamps and arc lamps had to be drastically improved. On the system side, the problems requiring solutions were associated with water leaks, corrosion of metal parts by the cooling fluid, deterioration of seals and other parts in the pump cavity due to the ultraviolet radiation of the flashlamps, arcing within the high-voltage section of the laser, and contamination of optical surfaces caused by the environment.

The application of solid-state lasers for military tactical systems proceeded along a clear path since there is no alternative for rangefinders, target illuminators, and designators. At the same time construction of large Nd:glass lasers began at many research facilities. Also solid-state lasers were readily accepted as versatile research tools in many laboratories.

Much more difficult and rather disappointing at first was the acceptance of the solid-state lasers for industrial and medical applications. Despite improvement in systems reliability and performance, it took more than two decades of development and engineering improvements before solid-state lasers moved in any numbers out of the laboratory and onto the production floor or into instruments used in medical procedures. Often applications that showed technical feasibility in the laboratory were not suitable for production because of economic reasons, such as high operating costs or limited processing speeds. Also, other laser systems provided strong competition for a relatively small market. The CO_2 laser proved to be a simpler and more robust system for many industrial and medical applications. Also, the argon ion laser was readily accepted and preferred over solid-state lasers for retinal photocoagulation. The dye laser was the system of choice for tunable laser sources. The entry of solid-state lasers into manufacturing processes started with very specialized applications, either for working with difficult materials, such as titanium, or for difficult machining operations, such as drilling holes in slanted surfaces; for example, in jet fuel nozzles or for precision material removal required in the semiconductor and electronics industry.

In the latter part of the 1970s, and into the 1980s, a number of tunable lasers were discovered, such as alexandrite, titanium-doped sapphire, and chromium-doped fluoride crystals. The most important tunable laser, Ti:sapphire, discovered in the mid-1980s, is tunable between 660 and 980 nm. This laser must be pumped with another laser in the blue–green wavelength region. Alexandrite, first operated in 1979, has a smaller tunable output but can be flashlamp-pumped. Chromium-doped fluoride crystals such as lithium strontium aluminum fluoride and lithium calcium aluminum fluoride are of interest because they can be pumped with laser diodes.

In the late 1980s, the combination of broad band tunable lasers in combination with ultrafast modulation techniques, such as Kerr lens mode-locking, led to the development of mode-locked lasers with pulse widths on the order of femtoseconds. The pulse width limit of a mode-locked laser is inversely proportional to the bandwidth of the laser material. For neodymium-based lasers, the lower limit for the pulse width is a few picoseconds. Laser media with a much larger gain bandwidth, such as Ti : sapphire, can produce much shorter pulses compared to neodymium lasers.

Over the years, the performance of diode lasers has been constantly improved as new laser structures and new material growth and processing techniques were developed. This led to devices with longer lifetimes, lower threshold currents, and higher output powers. In the 1970s, diode lasers capable of continuous operation at room temperature were developed. In the mid-1980s, with the introduction of epitaxial processes and a greatly increased sophistication in the junction structure of GaAs devices, laser diodes became commercially available with output powers of several watts. These devices had sufficient power to render them useful for the pumping of Nd : YAG lasers. The spectral match of the diode laser output with the absorption of neodymium lasers results in a dramatic increase in system efficiency, and a reduction of the thermal load of the solid-state laser material. Military applications and the associated research and development funding provided the basis for exploring this new technology. Since the early laser diodes were very expensive, their use as pump sources could only be justified where diode pumping provided an enabling technology. Therefore the first applications for diode-pumped Nd : YAG lasers were for space and airborne platforms, where compactness and power consumption is of particular importance.

As diode lasers became less expensive, these pump sources were incorporated into smaller commercial solid-state lasers. At this point, laser diode-pumped solid-state lasers began their rapid evolution that continues today. Diode pumping offers significant improvements in overall systems efficiency, reliability, and compactness. In addition, diode pumping has added considerable variety to the design possibilities of solid-state lasers. In many cases laser diode arrays were not just a replacement for flashlamps or arc lamps, but provided means for designing completely new laser configurations.They also led to the exploration of several new laser materials. Radiation from laser diodes can be collimated; this provides great flexibility of designing solid-state lasers with regard to the shape of the laser medium and orientation of the pump beam. In end-pumped lasers, the pump beam and resonator axis are collinear which led to highly efficient lasers with excellent beam quality. In monolithic lasers, the active crystal also provides the resonator structure leading to lasers with high output stability and excellent spatial and temporal beam quality. New laser materials, such as Yb : YAG and Nd : YVO4, that could not be pumped efficiently with flashlamps, are very much suited to laser diode pumping.

In this historical perspective we could sketch only briefly those developments that had a profound impact on the technology of solid-state lasers. Laser emission has been obtained from hundreds of solid-state crystals and glasses. However, most of these lasers are of purely academic interest. There is a big difference between laser research and the commercial laser industry, and there are many reasons why certain lasers did not find their way into the market or disappeared quickly after their introduction. Most of the lasers that did not leave the laboratory were inefficient, low in power, difficult to operate or, simply, less practical to use than other already established systems. Likewise, many pump schemes, laser configurations, and resonator designs did not come into use because of their complexity and commensurate high manufacturing and assembly costs or their difficulty in maintaining performance.

Typical Performance Parameters and Applications

Solid-state lasers provide the most versatile radiation source in terms of output characteristics when compared to other laser systems. A large range of output parameters, such as average and peak power, pulse width, pulse repetition rate, and wavelength, can be obtained with these systems.

Today we find solid-state lasers in industry as tools in many manufacturing processes, in hospitals and in doctors' offices as radiation sources for therapeutic, aesthetic, and surgical procedures, in research facilities as part of the diagnostic instrumentation, and in military systems as rangefinders, target designators, and infrared countermeasure systems. The flexibility of solid-state lasers stems from the fact that:

- The size and shape of the active material can be chosen to achieve a particular performance.
- Different active materials can be selected with different gain, energy storage, and wavelength properties.
- The output energy can be increased by adding amplifiers.
- A large number of passive and active components are available to shape the spectral, temporal, and spatial profile of the output beam.

In this section we will illustrate the flexibility of these systems and indicate the major applications that are based on particular performance characteristics.

Average Output Power. The majority of solid-state lasers available commercially have output powers below 20 W. The systems are continuously pumped, typically equipped with a Q-switch, and often combined with a wavelength converter. Continuously pumped, repetitively Q-switched lasers generate a continuous stream of short pulses at repetition rates between 5 and 100 kHz depending on the material. Since the peak power of each pulse is at least three orders of magnitude above the average power, breakdown of reflective surfaces and subsequent material removal by melting and vaporization is facilitated.

The electronics and electrical industry represents the largest market for applications such as soldering, wire bonding and stripping, scribing of wavers, memory repair, resistor and integrated circuit trimming. In addition, industry-wide, these lasers found uses for marking of parts, precision spot and seam welding, and for general micromachining tasks. In the medical fields solid-state lasers have found applications in ophthomology for vision correction and photocoagulation, skin resurfacing, and as replacements for scalpels in certain surgical procedures. In basic research, solid-state lasers are used in scientific and biomedical instrumentation, Raman and laser-induced breakdown spectroscopy. Application for these lasers are far too broad and diverse to provide a comprehensive listing here.

Higher power solid-state lasers with output powers up to 5 kW are mainly employed in metals working, such as seam and spot welding, cutting, drilling, and surface treatment. In particular, systems with output powers of a few hundred watts have found widespread applications in the manufacturing process. The higher power levels allow for faster processing speed and working with thicker materials.

At the low end of the power scale are very small lasers with output powers typically less than 1 W. These lasers are pumped by diode lasers and have in most cases the resonator mirrors directly coated onto the crystal surfaces. The neodymium-doped crystals are typically only a few millimeters in size. These lasers have an extremely stable, single frequency output and are employed in interferometric instruments, spectroscopic systems, and instruments used in analytical chemistry. They also serve as seed lasers for larger laser systems.

The majority of solid-state lasers with outputs up to 20 W are pumped with diode arrays, whereas systems at the multihundred watt level are for the most part pumped by arc lamps because of the high cost of laser diode arrays, although diode-pumped systems with up to 5 kW of output power are on the market.

Peak Power. Pulsed systems with pulsewidths on the order of 100 μs and energies of several joules are employed in manufacturing processes for hole drilling. The peak power of these systems is on the order of several tens of kilowatts. Substantially higher peak powers are obtained with solid-state lasers that are pulse-pumped and Q-switched. For example, military systems such as rangefinders and target designators have output energies of 10 to 200 mJ and pulsewidths of 10 to 20 ns. Peak power for these systems is on the order of several megawatts. Laser generated plasmas investigated in research facilities require peak powers in the gigawatt regime. Typically, lasers for this work have output energies of several joules and pulsewidths of a few nanoseconds. The highest peak powers from solid-state lasers are generated in huge Nd : glass lasers employed for inertial confinement fusion experiments. The largest of these systems had an energy output around 100 kJ and pulsewidth of 1 ns which resulted in a peak power of 100 TW.

Pulse Width. Solid-state lasers can span the range from continuous operation to pulses as short as one cycle of the laser frequency which is on the order of 1 fs. Long pulses in the milli- and microsecond regime are generated by adjusting the length of the pump pulse. Hole drilling and surface hardening of metals is typically performed with pulses around 100 μs in duration. Continuously pumped Q-switched Nd : YAG lasers generate pulses with pulsewidths on the order of hundreds of nanoseconds.

A reduction in pulsewidth is achieved in lasers that are pulse-pumped and Q-switched. These lasers have pulsewidths from a few nanoseconds to about 20 ns. All military rangefinders and target designators fall into this category. The pulsewidth of pulse-pumped Q-switched lasers is shorter than their continuously pumped counterparts because a higher gain is achieved in pulse-pumped systems. The technique of mode-locking the longitudinal modes provides a means of generating pulses in the picosecond regime with neodymium lasers. Since pulsewidth and gain-bandwidth are inversely related, even shorter pulses are obtained with tunable lasers due to their broad spectrum. For example, with Ti : sapphire lasers, pulses in the femtosecond regime are generated. These short pulses enable researchers, for example, to study dynamic processes that occur during chemical reactions.

Pulse Repetition Rate. At the low end are lasers employed in inertial fusion experiments. In these systems laser pulses are single events with a few experiments conducted each day because the heat generated during each pulse has to be dissipated between shots. Also, some hand-held rangefinders for surveillance purposes are single-shot devices. Most military rangefinders and target designators operate at 20 pulses per second. Welders and drillers, if they are pulse-pumped, generate pulses at a repetition rate of a few hundred pulses per second. Continuously pumped and Q-switched lasers provide a continuous train of pulses between 5 and 100 kHz. A large number of materials-processing applications fall into this mode of operation. Mode-locking generates pulses with repetition rates of several hundred megahertz. These systems are mainly used in photochemistry or in specialized materials processing application. In the latter application, material is removed by ablation that prevents heat from penetrating the surrounding area.

Linewidth. The linewidth of a laser is the result of the gain-bandwidth of the laser material and the number of longitudinal modes oscillating within the resonator. The output of a typical laser is comprised of many randomly fluctuating longitudinal modes, each mode representing a spectral line within the bandwidth of the output beam. The typical linewidth of an Nd : YAG laser is on the order of 10 GHz or 40 pm. Compared to the wavelength, lasers are very narrow-bandwidth radiation sources, and therefore for most applications the linewidth of the laser is not important. Exceptions are applications of the laser in coherent radar systems or in interferometric devices. Also, in lasers that operate at peak powers close to the damage threshold of optical components, it is beneficial to restrict operation to a single longitudinal mode to avoid power spikes as a result of the random superposition of the output from several modes.

Single mode operation of solid-state lasers is most readily achieved with small monolithic devices, having resonators which are so short as to allow only one longitudinal mode to oscillate. Only a few lasers, such as used in interferometers designed for gravitational wave detection, require close to quantum noise-limited performance. Careful temperature and vibration control combined with feedback systems have reduced the bandwidth of these lasers to a few kilohertz.

Spectral Range. A direct approach to tunable output is the use of a tunable laser, such as Ti : sapphire or the alexandrite laser. However, these sources are limited to the spectral region between 600 and 900 nm.The most well-developed and efficient lasers, such as the neodymium-based systems, are essentially fixed wavelength radiation sources with output around 1 μm. Nonlinear crystals employed in harmonic generators will produce second, third, and fourth harmonics, thus providing output in the visible and ultraviolet spectrum. Tunable spectral coverage can be obtained from optical parametric oscillators that convert a portion of the output beam into two beams at longer wavelength. Depending on the region over which tunable output is desired, the optical parametric oscillator can be pumped directly with the fundamental beam of the laser or with one of its harmonics.

The limits of the spectral range for solid-state lasers in the ultraviolet region is reached by quadrupled neodymium lasers at around 266 nm. The longest wavelength at useful power levels is produced around 4 μm by neodymium or erbium lasers operating at 1 or 2 μm, that are shifted to the longer wavelength with optical parametric oscillators. The limits at the short and long wavelengths are determined mainly by a lack of nonlinear crystals with a sufficiently high-damage threshold or nonlinear coefficient.

Many industrial, medical, and military applications require a different wavelength than the fundamental output available from standard lasers. For example, most materials have higher absorption at shorter wavelengths, therefore frequency-doubled neodymium lasers are often preferred over fundamental operation. Also, the smallest spot size diameter that can be achieved from a laser is proportional to the wavelength. The fine structures of integrated circuits and semiconductor devices require operation of the laser at the shortest wavelength possible. Also, by matching the wavelength of a laser to the peak absorption of a specific material, the top layer of a multilayer structure can be removed selectively without damage to the layers underneath.

All Nd : glass lasers employed in inertial confinement fusion experiments are operated at the third harmonic, i.e., 352 nm, because the shorter wavelength is more optimum for pellet compression compared to the fundamental output. Medical applications require solid-state lasers operating in a specific spectral range for control of the absorption depth of the radiation in the skin, tissue, or blood vessels. Frequency agility is required from lasers employed in instruments used for absorption measurements, spectroscopy, sensing devices, analytical chemistry, etc. A fixed or tunable laser in conjunction with harmonic generators and/or an optical parametric oscillator is usually employed to meet these requirements.

Military rangefinders need to operate in a region that does not cause eye damage because most of the time these systems are employed in training exercises. A wavelength of 1.5 μm poses the least eye hazard. This wavelength is obtained from Q-switched erbium lasers or from neodymium lasers that are wavelength-shifted with an optical parametric oscillator or Raman cell. Lasers designed to defeat missile threats, so-called infrared countermeasure lasers, have to operate in the 2 to 4 μm region. Output in this spectral range can be obtained from neodymium or erbium lasers, wavelength-shifted with one or two optical parametric amplifiers.

Spatial Beam Characteristics. Virtually all laser applications benefit from a diffraction-limited beam. Such a beam has the lowest beam divergence and produces the smallest spot if focused by a lens. However, there is a trade-off between output power and beam quality. Lasers in the multihundred or kilowatt output range employed for metal cutting or welding applications have beams that are many times diffraction-limited. On the other hand, lasers that are employed for micromachining applications and semiconductor processing, where a minimum spot size and kerf-width is essential, are mostly operated very close to the diffraction limit.

Future Trends. The replacements of flashlamps and arc lamps with laser diode arrays will continue even for large solid-state lasers because the increase in systems efficiency, beam quality and reliability, is compelling. Also, the push for solid-state lasers, with ever higher average output powers will continue. Concepts for lasers at the 100 kW level are already being developed. Most smaller lasers have output beams that are close to the diffraction limit. A particular challenge is to improve the beam quality of solid-state lasers with output powers in the multihundred or kilowatt regime.

The trend for smaller lasers, certainly for military lasers, is toward systems which do not require liquid cooling. Also, the search continues for new nonlinear crystals with high-damage thresholds and large nonlinear coefficients, particularly for the infrared and ultraviolet regions. Even with diode pumping, solid-state lasers are not particularly efficient radiation sources, converting at best about 10% of electric input into useful output. Further improvements in the efficiency of diode pump sources as a result of refinements in diode structure and processing techniques, coupled with a further optimization of laser materials and designs, could increase the efficiency of solid-state lasers to about 20 or 30%.

1

Energy Transfer Between Radiation and Atomic Transitions

1.1 Optical Amplification

1.2 Interaction of Radiation with Matter

1.3 Absorption and Optical Gain

1.4 Creation of a Population Inversion

1.5 Laser Rate Equations

References

Exercises

In this chapter we shall outline the basic ideas underlying the operation of solid-state lasers. In-depth treatments of laser physics can be found in [1], [2].

1.1 Optical Amplification

To understand the operation of a laser we have to know some of the principles governing the interaction of radiation with matter.

Atomic systems such as atoms, ions, and molecules can exist only in discrete energy states. A change from one energy state to another, called a transition, is associated with either the emission or the absorption of a photon. The wavelength of the absorbed or emitted radiation is given by Bohr's frequency relation

$$E_2 - E_1 = h\nu_{21}, \qquad (1.1)$$

where E_2 and E_1 are two discrete energy levels, ν_{21} is the frequency, and h is Planck's constant. An electromagnetic wave whose frequency ν_{21} corresponds to an energy gap of such an atomic system can interact with it. To the approximation required in this context, a laser medium can be considered an ensemble of very

12

many identical atomic systems. At thermal equilibrium, the lower energy states in the medium are more heavily populated than the higher energy states. A wave interacting with the laser system will raise the atoms (ions, molecules) from lower to higher energy levels and thereby experience absorption.

The operation of a laser requires that the energy equilibrium of a laser material be changed such that more atoms (ions, molecules) populate higher rather than lower energy states. This is achieved by an external pump source that supplies the energy required to transfer atoms (ions, molecules) from a lower energy level to a higher one. The pump energy thereby causes a "population inversion." An electromagnetic wave of appropriate frequency, incident on the "inverted" laser material, will be amplified because the incident photons cause the atoms (ions, molecules) in the higher level to drop to a lower level and thereby emit additional photons. As a result, energy is extracted from the atomic system and supplied to the radiation field. The release of the stored energy by interaction with an electromagnetic wave is based on stimulated or induced emission.

Stated very briefly, when a material is excited in such a way as to provide more atoms (or molecules) in a higher energy level than in some lower level, the material will be capable of amplifying radiation at the frequency corresponding to the energy level difference. The acronym "laser" derives its name from this process: "Light Amplification by Stimulated Emission of Radiation."

A quantum mechanical treatment of the interaction between radiation and matter demonstrates that the stimulated emission is, in fact, completely indistinguishable from the stimulating radiation field. This means that the stimulated radiation has the same directional properties, same polarization, same phase, and same spectral characteristics as the stimulating emission. These facts are responsible for the extremely high degree of coherence which characterizes the emission from lasers. The fundamental nature of the induced or stimulated emission process was already described by Albert Einstein and Max Planck.

Common to all laser amplifiers are at least two elements: a laser medium in which a population inversion among atoms, ions, or molecules can be achieved, and a pump process to supply energy to the system in order to maintain a nonequilibrium state. For a laser oscillator, additionally a feedback mechanism is required for radiation to build up. Typically, two mirrors facing each other provide this feedback.

Whether a population inversion occurs within atoms, ions, or molecules, and whether the pump energy supplied to the medium is in the form of optical radiation, electrical current, kinetic energy due to electron impact in a gas discharge, or an exothermic reaction, depends on the type of laser and the type of active medium, i.e., solid-state, liquid, semiconductor, or gas.

Given below are a few examples of different laser media and excitation methods.

In solid-state lasers, the subject treated in this book, the active medium is a crystal or glass host doped with a relatively small percentage of ions from the rare earth, actinide, or iron groups of the periodic table. The energy levels and associated transition frequencies result from the different quantum energy levels of allowed states of the electrons orbiting the nuclei of atoms. Population inversion occurs within inner, incomplete electron shells of ions embedded in the crystalline or glass host. Excitation is by means of pump radiation from a light source such as a flashlamp, continuous-wave (cw) arc lamp, or diode laser. Pump radiation absorbed by the active ions raises the electrons to higher energy levels.

In a liquid or dye laser, the active medium is an organic dye in a liquid solvent. Intense optical pumping creates a population inversion of electron transitions within the dye molecule.

In semiconductor lasers, also referred to as injection lasers, radiation is emitted as a consequence of carrier injection in a forward-biased semiconductor $p–n$ junction. Since the junction defines a diode, the laser is commonly called a diode laser. Energy levels in a semiconductor are defined by the conduction band and the valence band separated by a band gap. The $p–n$ junction is made of p-type semiconductor material which accepts electrons (or produces holes or positive carriers) and n-type material which is a donor of excess electrons (or negative carriers). The n-type material with a large electron density in the conduction band is brought into intimate contact with the p-type material with a large hole density in the valence band. In a forward biased $p–n$ junction, current will flow, electrons in the n-type material are injected into the p-type region, while positive holes from the p-region are injected into the n-type region. When an electron meets a positive hole, they recombine, the electron transitions from the conduction band to the valence band, emitting a photon equal to the band gap energy. From a laser point of view, injected electrons in the p-type region (so-called minority carriers) and holes injected into the n-region represent a population inversion. Diode lasers are used as pump sources for solid-state lasers; therefore their design and performance characteristics will be discussed later in this book.

In gases, the various particles have the mobility to excite higher energy levels by collision. In most common gas lasers, inversion is achieved by passing current through the gas to create a discharge. The active species can be a neutral atom such as in the helium–neon laser or an ion such as in the argon ion laser. The most typical of the neutral atomic gas lasers is the He–Ne laser. When an He atom, excited to a higher state by electric discharge, collides with an Ne atom in the ground state, the excitation energy of the He is transferred to Ne, and as a result the distribution of Ne atoms at the higher energy state increases. Lasing action is the result of electron transitions in the neutral atom Ne.

In molecular gas lasers the transition between the energy levels of a molecule is exploited for laser action. For example, in a CO_2 laser, the multiatom molecule CO_2 exhibits energy levels that arise from the vibrational and rotational motions of the molecule as a whole. Since CO_2 is a triatomic linear molecule, it has a symmetrical stretching and two bending vibration modes. In the vibrational–rotational molecular lasers, such as CO_2, the motion of the molecule rather than electronic transitions are responsible for laser action. Pump energy to raise the molecules to higher energy levels is provided by the electrical energy of a gas discharge.

In some molecular lasers, a population inversion is achieved on electronic transitions within the molecule, rather than on vibrational–rotational energy levels. An example is the excimer laser. An excimer, which is a contraction of an excited dimer, is a molecule consisting of two atoms that can exist only in the excited state. Population inversion involves transitions between different electronic states of the diatomic molecule. An example of this family of lasers is the xenon fluoride laser.

In the lasers discussed so far, electrical energy is used to drive the pump mechanism which creates a population inversion. In chemical lasers a population inversion is directly produced by an exothermic reaction of two gases. In the helium fluoride and deuterium fluoride laser, the gases enter a chamber through a set of nozzles. As soon as they are mixed together they react and a fraction of the chemical energy released in the exothermic reaction goes into excitation of the molecules to form vibrationally excited HF or DF.

1.2   Interaction of Radiation with Matter

Many of the properties of a laser may be readily discussed in terms of the absorption and emission processes which take place when an atomic system interacts with a radiation field. In the first decade of the last century Planck described the spectral distribution of thermal radiation, and in the second decade Einstein, by combining Planck's law and Boltzmann statistics, formulated the concept of stimulated emission. Einstein's discovery of stimulated emission provided essentially all of the theory necessary to describe the physical principle of the laser.

1.2.1   Blackbody Radiation

When electromagnetic radiation in an isothermal enclosure, or cavity, is in thermal equilibrium at temperature T, the distribution of radiation density $\varrho(v)\,dv$, contained in a bandwidth dv, is given by Planck's law

$$\varrho(v)\,dv = \frac{8\pi v^2\,dv}{c^3}\,\frac{hv}{e^{hv/kT}-1}, \tag{1.2}$$

where $\varrho(v)$ is the radiation density per unit frequency [J s/cm^3], k is Boltzmann's constant, and c is the velocity of light. The spectral distribution of thermal radiation vanishes at $v = 0$ and $v \to \infty$, and has a peak which depends on the temperature.

The factor

$$\frac{8\pi v^2}{c^3} = p_{\mathrm{n}} \tag{1.3}$$

in (1.2) gives the density of radiation modes per unit volume and unit frequency interval. The factor p_{n} can also be interpreted as the number of degrees of freedom associated with a radiation field, per unit volume, per unit frequency interval. The expression for the mode density p_n [modes s/cm^3] plays an important role in connecting the spontaneous and the induced transition probabilities.

For a uniform, isotropic radiation field, the following relationship is valid

$$W = \frac{\varrho(v)c}{4}, \tag{1.4}$$

where W is the blackbody radiation [W/cm^2] which will be emitted from an opening in the cavity of the blackbody. Many solids radiate like a blackbody. Therefore, the radiation emitted from the surface of a solid can be calculated from (1.4).

According to the Stefan–Boltzmann equation, the total blackbody radiation is

$$W = \sigma T^4, \tag{1.5}$$

where $\sigma = 5.68 \times 10^{-12}$ W/cm^2 K^4. The emitted radiation W has a maximum which is obtained from Wien's displacement law

$$\frac{\lambda_{max}}{\mu m} = \frac{2893}{T/K}. \tag{1.6}$$

For example, a blackbody at a temperature of 5200 K has its radiation peak at 5564 Å, which is about the center of the visible spectrum.

1.2.2 Boltzmann's Statistics

According to a basic principle of statistical mechanics, when a large collection of similar atoms is in thermal equilibrium at temperature T, the relative populations of any two energy levels E_1 and E_2, such as the ones shown in Fig. 1.1, must be related by the Boltzmann ratio

$$\frac{N_2}{N_1} = \exp\left(\frac{-(E_2 - E_1)}{kT}\right), \tag{1.7}$$

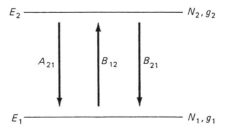

FIGURE 1.1. Two energy levels with population N_1, N_2 and degeneracies g_1, g_2, respectively.

where N_1 and N_2 are the number of atoms in the energy levels E_1 and E_2, respectively. For energy gaps large enough that $E_2 - E_1 = h\nu_{21} \gg kT$, the ratio is close to zero, and there will be very few atoms in the upper energy level at thermal equilibrium. The thermal energy kT at room temperature ($T \approx 300\,\text{K}$) corresponds to an energy gap $h\nu$ with $\nu \approx 6 \times 10^{12}\,\text{Hz}$, which is equivalent in wavelength to $\lambda \approx 50\,\mu\text{m}$. Therefore, for any energy gap whose transition frequency ν_{21} lies in the near-infrared or visible regions, the Boltzmann exponent will be very small at normal temperatures. The number of atoms in any upper level will then be very small compared to the lower levels. For example, in ruby the ground level E_1 and the upper laser level E_2 are separated by an energy gap corresponding to a wavelength of $\lambda \approx 0.69\,\mu\text{m}$. Since $h = 6.6 \times 10^{-34}\,\text{W\,s}^2$, then $E_2 - E_1 = h\nu = 2.86 \times 10^{-19}\,\text{W\,s}$. With $k = 1.38 \times 10^{-23}\,\text{W\,s\,K}$ and $T = 300\,\text{K}$, it follows that $N_2/N_1 \approx \exp(-69)$. Therefore at thermal equilibrium virtually all the ions will be in the ground level.

Equation (1.7) is valid for atomic systems having only nondegenerate levels. If there are g_i different states of the atom corresponding to the energy E_i, then g_i is defined as the degeneracy of the ith energy level.

We recall that atomic systems, such as atoms, ions, and molecules, can exist only in certain stationary states, each of which corresponds to a definite value of energy and thus specifies an energy level. When two or more states have the same energy, the respective level is called degenerate, and the number of states with the same energy is the multiplicity of the level. All states of the same energy level will be equally populated, therefore the number of atoms in levels 1 and 2 is $N_1 = g_1 N_1'$ and $N_2 = g_2 N_2'$, where N_1' and N_2' refer to the population of any of the states in levels 1 and 2, respectively. It follows then from (1.7) that the populations of the energy levels 1 and 2 are related by the formula

$$\frac{N_2}{N_1} = \frac{g_2}{g_1} \frac{N_2'}{N_1'} = \frac{g_2}{g_1} \exp\left(\frac{-(E_2 - E_1)}{kT}\right). \tag{1.8}$$

At absolute zero temperature, Boltzmann's statistics predict that all atoms will be in the ground state. Thermal equilibrium at any temperature requires that a state with a lower energy be more densely populated than a state with a higher energy. Therefore N_2/N_1 is always less than unity for $E_2 > E_1$ and $T > 0$. This means that optical amplification is not possible in thermal equilibrium.

1.2.3 Einstein's Coefficients

We can most conveniently introduce the concept of Einstein's A and B coefficients by loosely following Einstein's original derivation. To simplify the discussion, let us consider an idealized material with just two nondegenerate energy levels, 1 and 2, having populations of N_1 and N_2, respectively. The total number of atoms in these two levels is assumed to be constant

$$N_1 + N_2 = N_{\text{tot}}. \tag{1.9}$$

Radiative transfer between the two energy levels, which differ by $E_2 - E_1 = h\nu_{21}$, is allowed. The atom can transfer from state E_2 to the ground state E_1 by emitting energy; conversely, transition from state E_1 to E_2 is possible by absorbing energy. The energy removed or added to the atom appears as quanta of $h\nu_{21}$. We can identify three types of interaction between electromagnetic radiation and a simple two-level atomic system.

Absorption. If a quasi-monochromatic electromagnetic wave of frequency ν_{21} passes through an atomic system with energy gap $h\nu_{21}$, then the population of the lower level will be depleted at a rate proportional both to the radiation density $\varrho(\nu)$ and to the population N_1 of that level

$$\frac{\partial N_1}{\partial t} = -B_{12}\varrho(\nu)N_1, \tag{1.10}$$

where B_{12} is a constant of proportionality with dimensions cm^3/s^2 J.

The product $B_{12}\varrho(\nu)$ can be interpreted as the probability per unit frequency that transitions are induced by the effect of the field.

Spontaneous Emission. After an atom has been raised to the upper level by absorption, the population of the upper level 2 decays spontaneously to the lower level 1 at a rate proportional to the upper level population

$$\frac{\partial N_2}{\partial t} = -A_{21}N_2, \tag{1.11}$$

where A_{21} is a constant of proportionality with dimension s^{-1}. The quantity A_{21}, being a characteristic of the pair of energy levels in question, is called the spontaneous transition probability because this coefficient gives the probability that an atom in level 2 will spontaneously change to a lower level 1 within a unit of time.

Spontaneous emission is a statistical function of space and time. With a large number of spontaneously emitting atoms there is no phase relationship between the individual emission processes; the quanta emitted are incoherent. Spontaneous emission is characterized by the lifetime of the electron in the excited state, after which it will spontaneously return to the lower state and radiate away the energy. This can occur without the presence of an electromagnetic field.

Equation (1.11) has a solution

$$N_2(t) = N_2(0)\,\exp\!\left(\frac{-t}{\tau_{21}}\right), \tag{1.12}$$

where τ_{21} is the lifetime for spontaneous radiation of level 2. This radiation lifetime is equal to the reciprocal of the Einstein coefficient,

$$\tau_{21} = A_{21}^{-1}. \tag{1.13}$$

In general, the reciprocal of the transition probability of a process is called its lifetime.

Stimulated Emission. Emission takes place not only spontaneously but also under stimulation by electromagnetic radiation of the appropriate frequency. In this case, the atom gives up a quantum to the radiation field by "induced emission" according to

$$\frac{\partial N_2}{\partial t} = -B_{21}\varrho(\nu_{21})N_2, \tag{1.14}$$

where B_{21} again is a constant of proportionality.

Radiation emitted from an atomic system in the presence of external radiation consists of two parts. That part whose intensity is proportional to A_{21} is the spontaneous radiation; its phase is independent of that of the external radiation. The part whose intensity is proportional to $\varrho(\nu)B_{21}$ is the stimulated radiation; its phase is the same as that of the stimulating external radiation. The probability of induced transition is proportional to the energy density of external radiation in contrast to spontaneous emission.

The quantum that is emitted to the field by the induced emission is coherent with it. The useful parameter for laser action is the B_{21} coefficient; the A_{21} coefficient represents a loss term and introduces into the system photons that are not phase-related to the incident photon flux of the electric field. Thus the spontaneous process represents a noise source in a laser.

If we combine absorption, spontaneous, and stimulated emission, as expressed by (1.10), (1.11), and (1.14) we can write, for the change of the upper- and lower-level populations in our two-level model,

$$\frac{\partial N_1}{\partial t} = -\frac{\partial N_2}{\partial t} = B_{21}\varrho(\nu)N_2 - B_{12}\varrho(\nu)N_1 + A_{21}N_2. \tag{1.15}$$

The relation

$$\frac{\partial N_1}{\partial t} = -\frac{\partial N_2}{\partial t} \tag{1.16}$$

follows from (1.9).

In thermal equilibrium, the number of transitions per unit time from E_1 to E_2 must be equal to the number of transitions from E_2 to E_1. Certainly, in thermal equilibrium

$$\frac{\partial N_1}{\partial t} = \frac{\partial N_2}{\partial t} = 0. \tag{1.17}$$

Therefore we can write

$$\underset{\substack{\text{Spontaneous} \\ \text{emission}}}{N_2 A_{21}} + \underset{\substack{\text{Stimulated} \\ \text{emission}}}{N_2\varrho(\nu)B_{21}} = \underset{\text{Absorption}}{N_1\varrho(\nu)B_{12}}. \tag{1.18}$$

Using the Boltzmann equation (1.8) for the ratio N_2/N_1, we then write the above expression as

$$\varrho(\nu_{21}) = \frac{(A_{21}/B_{21})}{(g_1/g_2)(B_{12}/B_{21}) \exp(h\nu_{21}/kT) - 1}. \tag{1.19}$$

Comparing this expression with the blackbody radiation law (1.2), we see that

$$\frac{A_{21}}{B_{21}} = \frac{8\pi \nu^2 h\nu}{c^3} \quad \text{and} \quad B_{21} = \frac{g_1 B_{12}}{g_2}. \tag{1.20}$$

The relations between the A's and B's are known as Einstein's relations. The factor $8\pi \nu^2/c^3$ in (1.20) is the mode density p_n given by (1.3).

In solids the speed of light is $c = c_0/n$, where n is the index of refraction and c_0 is the speed of light in vacuum.

For a simple system with no degeneracy, that is, one in which $g_1 = g_2$, we see that $B_{21} = B_{12}$. Thus, the Einstein coefficients for stimulated emission and absorption are equal. If the two levels have unequal degeneracy, the probability for stimulated absorption is no longer the same as that for stimulated emission.

1.2.4 Phase Coherence of Stimulated Emission

The stimulated emission provides a phase-coherent amplification mechanism for an applied signal. The signal extracts from the atoms a response that is directly proportional to, and phase-coherent with, the electric field of the stimulating signal. Thus the amplification process is phase-preserving. The stimulated emission is, in fact, completely indistinguishable from the stimulating radiation field. This means that the stimulated emission has the same directional properties, same polarization, same phase, and same spectral characteristics as the stimulating emission. These facts are responsible for the extremely high degree of coherence that characterizes the emission from lasers. The proof of this fact is beyond the scope of this elementary introduction, and requires a quantum mechanical treatment of the interaction between radiation and matter. However, the concept of induced transition, or the interaction between a signal and an atomic system, can be demonstrated, qualitatively, with the aid of the classical electron-oscillator model.

Electromagnetic radiation interacts with matter through the electric charges in the substance. Consider an electron that is elastically bound to a nucleus. One can think of electrons and ions held together by spring-type bonds which are capable of vibrating around equilibrium positions. An applied electric field will cause a relative displacement between the electron and nucleus from their equilibrium position. They will execute an oscillatory motion about their equilibrium position. Therefore, the model exhibits an oscillatory or resonant behavior and a response to an applied field. Since the nucleus is so much heavier than the electron, we assume that only the electron moves. The most important model for understanding the interaction of light and matter is that of the harmonic oscillator. We take as our model a single electron, assumed to be bound to its equilibrium position by a linear restoring force. We may visualize the electron as a point of mass suspended by springs. Classical electromagnetic theory asserts that any oscillating electric charge will act as a miniature antenna or dipole and will continuously radiate away electromagnetic energy to its surroundings.

1.3   Absorption and Optical Gain

In this section we will develop the quantitative relations that govern the absorption and amplification processes in substances. This requires that we increase the realism of our mathematical model by introducing the concept of atomic lineshapes. Therefore, the important features and the physical processes that lead to different atomic lineshapes will be considered first.

1.3.1   Atomic Lineshapes

In deriving Einstein's coefficients we have assumed a monochromatic wave with frequency v_{21} acting on a two-level system with an infinitely sharp energy gap hv_{21}. We will now consider the interaction between an atomic system having a finite transition linewidth Δv and a signal with a bandwidth dv.

Before we can obtain an expression for the transition rate for this case, it is necessary to introduce the concept of the atomic lineshape function $g(v, v_0)$. The distribution $g(v, v_0)$, centered at v_0, is the equilibrium shape of the linewidth-broadened transitions. Suppose that N_2 is the total number of ions in the upper energy level considered previously. The spectral distribution of ions per unit frequency is then

$$N(v) = g(v, v_0)N_2. \tag{1.21}$$

If we integrate both sides over all frequencies we have, to obtain N_2 as a result,

$$\int_0^\infty N(v)\, dv = N_2 \int_0^\infty g(v, v_0)\, dv = N_2. \tag{1.22}$$

Therefore, the lineshape function must be normalized to unity

$$\int_0^\infty g(v, v_0)\, dv = 1. \tag{1.23}$$

If we know the function $g(v, v_0)$, we can calculate the number of atoms $N(v)\, dv$ in level 1 which are capable of absorbing in the frequency range v to $v + dv$, or the number of atoms in level 2 which are capable of emitting in the same range.

From (1.21) we have

$$N(v)\, dv = g(v, v_0)\, dv\, N_2. \tag{1.24}$$

From the foregoing it follows that $g(v, v_0)$ can be defined as the probability of emission or absorption per unit frequency. Therefore $g(v)\, dv$ is the probability that a given transition will result in an emission (or absorption) of a photon with energy between $h\nu$ and $h(\nu + d\nu)$. The probability that a transition will occur between $\nu = 0$ and $\nu = \infty$ has to be 1.

It is clear from the definition of $g(\nu, \nu_0)$ that we can, for example, rewrite (1.11) in the form

$$-\frac{\partial N_2}{\partial t} = A_{21} N_2 g(\nu, \nu_0) \, d\nu, \qquad (1.25)$$

where N_2 is the total number of atoms in level 2 and $\partial N_2 / \partial t$ is the number of photons spontaneously emitted per second between ν and $\nu + d\nu$.

The linewidth and lineshape of an atomic transition depends on the cause of line broadening. Optical frequency transitions in gases can be broadened by lifetime, collision, or Doppler broadening, whereas transitions in solids can be broadened by lifetime, dipolar or thermal broadening, or by random inhomogeneities. All these linewidth-broadening mechanisms lead to two distinctly different atomic lineshapes, the homogeneously and the inhomogeneously broadened line.

The Homogeneously Broadened Line

The essential feature of a homogeneously broadened atomic transition is that every atom has the same atomic lineshape and frequency response, so that a signal applied to the transition has exactly the same effect on all atoms in the collection. This means that within the linewidth of the energy level each atom has the same probability function for a transition.

Differences between homogeneously and inhomogeneously broadened transitions show up in the saturation behavior of these transitions. This has a major effect on the laser operation. The important point about a homogeneous lineshape is that the transition will saturate uniformly under the influence of a sufficiently strong signal applied anywhere within the atomic linewidth.

Mechanisms which result in a homogeneously broadened line are lifetime broadening, collision broadening, dipolar broadening, and thermal broadening.

Lifetime Broadening. This type of broadening is caused by the decay mechanisms of the atomic system. Spontaneous emission or fluorescence has a radiative lifetime. Broadening of the atomic transition due to this process is related to the fluorescence lifetime τ_{21} by $\Delta\omega_a \tau_{21} = 1$, where ω_a is the bandwidth.

Actually, physical situations in which the lineshape and linewidth are determined by the spontaneous emission process itself are vanishingly rare. Since the natural or intrinsic linewidth of an atomic line is extremely small, it is the linewidth that would be observed from atoms at rest without interaction with one another.

Collision Broadening. Collision of radiating particles (atoms or molecules) with one another and the consequent interruption of the radiative process in a random manner leads to broadening. Since an atomic collision interrupts either the emission or the absorption of radiation, the long wave train that otherwise would be present becomes truncated. The atom restarts its motion after the collision with a completely random initial phase. After the collision, the process is restarted without memory of the phase of the radiation prior to the collision. The result of frequent collisions is the presence of many truncated radiative or absorptive processes.

Since the spectrum of a wave train is inversely proportional to the length of the train, the linewidth of the radiation in the presence of collision is greater than that of an individual uninterrupted process.

Collision broadening is observed in gas lasers operated at higher pressures, hence the name pressure broadening. At higher pressures, collisions between gas atoms limit their radiative lifetime. Collision broadening, therefore, is quite similar to lifetime broadening, in that the collisions interrupt the initial state of the atoms.

Dipolar Broadening. Dipolar broadening arises from interactions between the magnetic or electric dipolar fields of neighboring atoms. This interaction leads to results very similar to collision broadening, including a linewidth that increases with increasing density of atoms. Since dipolar broadening represents a kind of coupling between atoms, so that excitation applied to one atom is distributed or shared with other atoms, dipolar broadening is a homogeneous broadening mechanism.

Thermal Broadening. Thermal broadening is brought about by the effect of the thermal lattice vibrations on the atomic transition. The thermal vibrations of the lattice surrounding the active ions modulate the resonance frequency of each ion at a very high frequency. This frequency modulation represents a coupling mechanism between the ions, therefore a homogeneous linewidth is obtained. Thermal broadening is the mechanism responsible for the linewidth of the ruby laser and Nd : YAG laser.

The lineshape of homogeneous broadening mechanisms lead to a Lorentzian lineshape for atomic response. For the normalized Lorentz distribution, the equation

$$g(v) = \left(\frac{\Delta v}{2\pi} \right) \left[(v - v_0)^2 + \left(\frac{\Delta v}{2} \right)^2 \right]^{-1} \tag{1.26}$$

is valid. Here, v_0 is the center frequency and Δv is the width between the half-power points of the curve. The factor $\Delta v/2\pi$ assures normalization of the area under the curve according to (1.23). The peak value for the Lorentz curve is

$$g(v_0) = \frac{2}{\pi \Delta v}. \tag{1.27}$$

The Inhomogeneously Broadened Line

Mechanisms which cause inhomogeneous broadening tend to displace the center frequencies of individual atoms, thereby broadening the overall response of a collection without broadening the response of individual atoms. Different atoms have slightly different resonance frequencies on the same transition, for example, owing to Doppler shifts. As a result, the overall response of the collection is broadened. An applied signal at a given frequency within the overall linewidth interacts strongly only with those atoms whose shifted resonance frequencies lie close to the signal frequency. The applied signal does not have the same effect on all the atoms in an inhomogeneously broadened collection.

Since, in an inhomogeneously broadened line, interaction occurs only with those atoms whose resonance frequencies lie close to the applied signal frequency, a strong signal will eventually deplete the upper-laser level in a very narrow frequency interval. The signal will eventually "burn a hole" in the atomic absorption curve. Examples of inhomogeneous frequency-shifting mechanisms include Doppler broadening and broadening due to crystal inhomogeneities.

Doppler Broadening. The apparent resonance frequencies of atoms undergoing random motions in a gas are shifted randomly so that the overall frequency response of the collection of atoms is broadened. A particular atom moving with a velocity component v relative to an observer in the z-direction will radiate at a frequency measured by the observer as $v_0(1 + v/c)$. When these velocities are averaged, the resulting lineshape is Gaussian. Doppler broadening is one form of inhomogeneous broadening since each atom emits a different frequency rather than one atom having a probability distribution for emitting any frequency within the linewidth. In the actual physical situation, the Doppler line is best visualized as a packet of homogeneous lines of width Δv_n, which superimpose to give the observed Doppler shape. The He–Ne laser has a Doppler-broadened linewidth. Most visible and near-infrared gas laser transitions are inhomogeneously broadened by Doppler effects.

Line Broadening Due to Crystal Inhomogeneities. Solid-state lasers may be inhomogeneously broadened by crystalline defects. This happens only at low temperatures where the lattice vibrations are small. Random variations of dislocations, lattice strains, and so forth, may cause small shifts in the exact energy level spacings and transition frequencies from ion to ion. Like Doppler broadening, these variations do not broaden the response on an individual atom, but they do cause the exact resonance frequencies of different atoms to be slightly different. Thus random crystal imperfection can be a source of inhomogeneous broadening in a solid-state laser crystal.

A good example of an inhomogeneously broadened line occurs in the fluorescence of neodymium-doped glass. As a result of the so-called glassy state, there are variations, from rare earth site to rare earth site, in the relative atomic positions occupied by the surrounding lattice ions. This gives rise to a random distribution of static crystalline fields acting on the rare earth ions. Since the line shifts corresponding to such crystal-field variations are larger, generally speaking, than the width contributed by other factors associated with the transition, an inhomogeneous line results.

The inhomogeneous-broadened linewidth can be represented by a Gaussian frequency distribution. For the normalized distribution, the equation

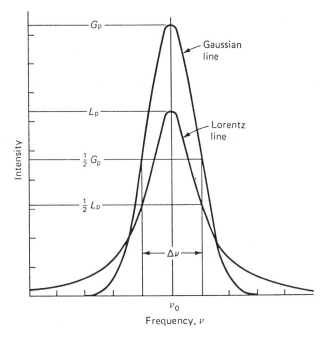

FIGURE 1.2. Gaussian and Lorentz lines of common linewidth (G_p and L_p are the peak intensities).

$$g(v) = \frac{2}{\Delta v} \left(\frac{\ln 2}{\pi} \right)^{1/2} \exp\left[-\left(\frac{v - v_0}{\Delta v/2} \right)^2 \ln 2 \right] \qquad (1.28)$$

is valid, where v_0 is the frequency at the center of the line and Δv is the linewidth at which the amplitude falls to one-half. The peak value of the normalized Gaussian curve is

$$g(v_0) = \frac{2}{\Delta v} \left(\frac{\ln 2}{\pi} \right)^{1/2}. \qquad (1.29)$$

In Fig. 1.2 the normalized Gaussian and Lorentz lines are plotted for a common linewidth.

1.3.2 Absorption by Stimulated Transitions

We assume a quasi-collimated beam of energy density $\varrho(v)$ incident on a thin absorbing sample of thickness dx; as before, we consider the case of an optical system that operates between only two energy levels as illustrated schematically in Fig. 1.1. The populations of the two levels are N_1 and N_2, respectively. Level 1 is the ground level and level 2 is the excited level. We consider absorption of radiation in the material and emission from the stimulated processes but neglect the spontaneous emission. From (1.15) and (1.20) we obtain

$$-\frac{\partial N_1}{\partial t} = \varrho(v)B_{21}\left(\frac{g_2}{g_1}N_1 - N_2\right). \tag{1.30}$$

As we recall, this relation was obtained by considering infinitely sharp energy levels separated by hv_{21} and a monochromatic wave of frequency v_{21}.

We will now consider the interaction between two linewidth-broadened energy levels with an energy separation centered at v_0, and a half-width of Δv characterized by $g(v, v_0)$ and a signal with center frequency v_s and bandwidth dv. The situation is shown schematically in Fig. 1.3. The spectral width of the signal is narrow, as compared to the linewidth-broadened transition. If N_1 and N_2 are the total number of atoms in levels 1 and 2, then the number of atoms capable of interacting with a radiation of frequency v_s and bandwidth dv are

$$\left(\frac{g_2}{g_1}N_1 - N_2\right)g(v_s, v_0)\,dv. \tag{1.31}$$

The net change of atoms in energy level 1 can be expressed in terms of energy density $\varrho(v)\,dv$ by multiplying both sides of (1.30) with photon energy hv and dividing by the volume V. We will further express the populations N_1 and N_2 as population densities n_1 and n_2.

Equation (1.30) now becomes

$$-\frac{\partial}{\partial t}[\varrho(v_s)\,dv] = \varrho(v_s)\,dv\,B_{21}hvg(v_s, v_0)\left(\frac{g_2}{g_1}n_1 - n_2\right). \tag{1.32}$$

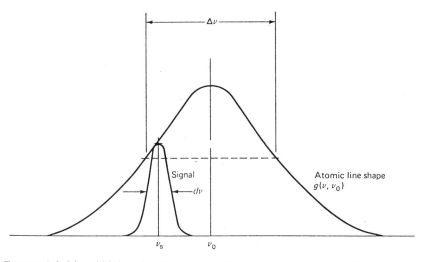

FIGURE 1.3. Linewidth-broadened atomic transition line centered at v_0 and narrow-band signal centered at v_s.

This equation gives the net rate of absorbed energy in the frequency interval dv centered around v_s. In an actual laser system the wavelength of the emitted radiation, corresponding to the signal bandwidth dv in our model, is very narrow as compared to the natural linewidth of the material. Ruby, for example, has a fluorescence linewidth of 5 Å, whereas the linewidth of the laser output is typically 0.1 to 0.01 Å. The operation of a laser, therefore, can be fairly accurately characterized as the interaction of linewidth-broadened energy levels with a monochromatic wave. The photon density of a monochromatic radiation of frequency v_0 can then be represented by a delta function $\delta(v - v_0)$. After integrating (1.32) in the interval dv, we obtain, for a monochromatic signal of frequency v_s and a linewidth-broadened transition,

$$-\frac{\partial \varrho(v_s)}{\partial t} = \varrho(v_s) B_{21} h v_s g(v_s, v_0) \left(\frac{g_2}{g_1} n_1 - n_2 \right). \tag{1.33}$$

The signal will travel through the material of thickness dx in time $dt = dx/c = (n_0/c_0)\, dx$. Then, as the wave advances from x to $x + dx$, the decrease of energy in the beam is

$$-\frac{\partial \varrho(v_s)}{\partial x} = h v_s \varrho(v_s) g(v_s, v_0) B_{21} \left(\frac{g_2}{g_1} n_1 - n_2 \right) \frac{1}{c}. \tag{1.34}$$

Integration of (1.34) gives

$$\frac{\varrho(v_s)}{\varrho_0(v_s)} = \exp\left[-h v_s g(v_s, v_0) B_{21} \left(\frac{g_2}{g_1} n_1 - n_2 \right) \frac{x}{c} \right]. \tag{1.35}$$

If we introduce an absorption coefficient $\alpha_0(v_s)$,

$$\alpha_0(v_s) = \left(\frac{g_2}{g_1} n_1 - n_2 \right) \sigma_{21}(v_s), \tag{1.36}$$

where

$$\sigma_{21}(v_s) = \frac{h v_s g(v_s, v_0) B_{21}}{c}. \tag{1.37}$$

Then we can write (1.35) as

$$\varrho(v_s) = \varrho_0(v_s) \exp[-\alpha_0(v_s) x]. \tag{1.38}$$

Equation (1.38) is the well-known exponential absorption equation for the thermal equilibrium condition $n_1 g_2/g_1 > n_2$. The energy of the radiation decreases exponentially with the depth of penetration into the substance. The maximum possible absorption occurs when all atoms exist in the ground state n_1. For equal population of the energy states $n_1 = (g_1/g_2)n_2$, the absorption is eliminated and the material is transparent. The parameter σ_{21} is the cross section for the radiative transition $2 \rightarrow 1$. The cross section for stimulated emission σ is related to the absorption cross section σ_{12} by the ratio of the level degeneracies,

$$\frac{\sigma}{\sigma_{12}} = \frac{g_1}{g_2}. \tag{1.39}$$

The cross section is a very useful parameter to which we will refer in the following chapters. If we replace B_{21} by the Einstein relation (1.20), we obtain σ_{21} in a form which we will find most useful:

$$\sigma(\nu_s) = \frac{A_{21}\lambda_0^2}{8\pi n_0^2} g(\nu_s, \nu_0). \tag{1.40}$$

As we will see later, the gain for the radiation building up in a laser resonator will be highest at the center of the atomic transitions. Therefore, in lasers we are mostly dealing with stimulated transitions which occur at the center of the linewidth.

If we assume $\nu \approx \nu_s \approx \nu_0$, we obtain, for the spectral stimulated emission cross section at the center of the atomic transition for a Lorentzian lineshape,

$$\sigma = \frac{A_{21}\lambda_0^2}{4\pi^2 n_0^2 \Delta \nu}, \tag{1.41}$$

and for a Gaussian lineshape,

$$\sigma = \frac{A_{21}\lambda_0^2}{4\pi n_0^2 \Delta \nu} \left(\frac{\ln 2}{\pi}\right)^{1/2}. \tag{1.42}$$

Here we have introduced into (1.40) the peak values of the lineshape function, as given in (1.27) and (1.29) for the Lorentzian and Gaussian curves, respectively. For example, in the case of the R_1 line of ruby, where $\lambda_0 = 6.94 \times 10^{-5}$ cm, $n_0 = 1.76$, $\tau_{21} = (1/A_{21}) = 3$ ms, and $\Delta \nu = 11$ cm^{-1} one finds, according to (1.41), $\sigma = 4.0 \times 10^{-20}$ cm^2. In comparing this value with the data provided in Table 2.2 (page 58), we have to distinguish between the spectroscopic cross section and the effective stimulated emission cross section. (This will be discussed in Section 2.3.1 for the case of Nd : YAG.) The effective stimulated emission cross section is the spectroscopic cross section times the occupancy of the upper laser level relative to the entire manifold population. In ruby, the upper laser level is split into two sublevels, therefore the effective stimulated emission cross section is about half of the value calculated from (1.41).

1.3.3 Population Inversion

According to the Boltzmann distribution (1.7), in a collection of atoms at thermal equilibrium there are always fewer atoms in a higher-lying level E_2 than in a lower level E_1. Therefore the population difference $N_1 - N_2$ is always positive, which means that the absorption coefficient $\alpha_0(\nu_s)$ in (1.36) is positive and the incident radiation is absorbed (Fig. 1.4).

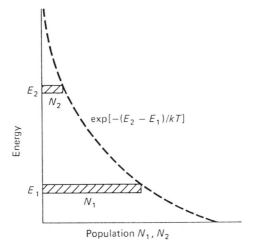

FIGURE 1.4. Relative populations in two energy levels as given by the Boltzmann relation for thermal equilibrium.

Suppose that it were possible to achieve a temporary situation such that there are more atoms in an upper energy level than in a lower energy level. The normally positive population difference on that transition then becomes negative, and the normal stimulated absorption, as seen from an applied signal on that transition, is correspondingly changed to stimulated emission or amplification of the applied signal. That is, the applied signal gains energy as it interacts with the atoms and hence is amplified. The energy for this signal amplification is supplied by the atoms involved in the interaction process. This situation is characterized by a negative absorption coefficient $\alpha(\nu_s)$ according to (1.36). From (1.34) it follows that $\partial\varrho(\nu)/\partial x > 0$.

The essential condition for amplification is that there are more atoms in an upper energy level than in a lower energy level; i.e., for amplification,

$$N_2 > N_1 \quad \text{if} \quad E_2 > E_1, \tag{1.43}$$

as illustrated in Fig. 1.5. The resulting negative sign of the population difference $(N_2 - g_2 N_1/g_1)$ on that transition is called a population inversion. Population inversion is clearly an abnormal situation; it is never observed at thermal equilibrium. The point at which the population of both states is equal is called the "inversion threshold."

Stimulated absorption and emission processes always occur side by side independently of the population distribution among the levels. So long as the population of the higher energy level is smaller than that of the lower energy level, the number of absorption transitions is larger than that of the emission transitions, so that there is an overall attenuation of the radiation. When the numbers of atoms in both states are equal, the number of emissions becomes equal to the number

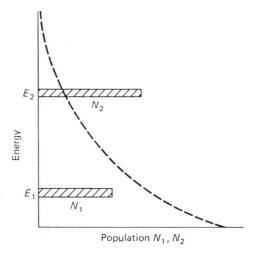

FIGURE 1.5. Inverted population difference required for optical amplification.

of absorptions; the material is then transparent to the incident radiation. As soon as the population of the higher level becomes larger than that of the lower level, emission processes predominate and the radiation is enhanced collectively during passage through the material. To produce an inversion requires a source of energy to populate a specified energy level; we call this energy the pump energy.

In Section 1.4 we will discuss the type of energy level structure an atomic system must possess in order to make it possible to generate an inversion. Techniques by which the ions of a solid-state laser can be raised or pumped into upper energy levels are discussed in Section 6.1. Depending on the atomic system involved, an inverted population condition may be obtainable only on a transient basis, yielding pulsed laser action; or it may be possible to maintain the population inversion on a steady-state basis, yielding cw laser action.

The total amount of energy which is supplied by the atoms to the light wave is

$$E = \Delta N h \nu, \tag{1.44}$$

where ΔN is the total number of atoms which are caused to drop from the upper to the lower energy level during the time the signal is applied. If laser action is to be maintained, the pumping process must continually replenish the supply of upper-state atoms. The size of the inverted population difference is reduced not only by the amplification process but also by spontaneous emission that always tends to return the energy level populations to their thermal equilibrium values.

1.4 Creation of a Population Inversion

We are concerned in this section with how the necessary population inversion for laser action is obtained in solid-state lasers. We can gain considerable understanding on how laser devices are pumped and how their population densities are inverted by studying some simplified but fairly realistic models.

The discussion up to this point has been based on a hypothetical 2 ↔ 1 transition and has not been concerned with how the levels 2 and 1 fit into the energy level scheme of the atom. This detached point of view must be abandoned when one tries to understand how laser action takes place in a solid-state medium. As already noted, the operation of the laser depends on a material with narrow energy levels between which electrons can make transitions. Usually these levels are due to impurity ions in a host crystal. The pumping and laser processes in real laser systems typically involve a very large number of energy levels, with complex excitation processes and cascaded relaxation processes among all these levels. Operation of an actual laser material is properly described only by a many-level energy diagram. The main features can be understood, however, through the familiar three- or four-level idealizations of Figs. 1.6 and 1.7. More detailed energy level diagrams of some of the most important solid-state laser materials are presented in Chapter 2.

1.4.1 The Three-Level System

Figure 1.6 shows a diagram that can be used to explain the operation of an optically pumped three-level laser, such as ruby. Initially, all ions of the laser material are in the lowest level 1. Excitation is supplied to the solid by radiation of frequencies that produce absorption into the broad band 3. Thus, the pump light raises ions from the ground state to the pump band, level 3. In general, the "pumping" band, level 3, is actually made up of a number of bands, so that the optical pumping can be accomplished over a broad spectral range. Most of the excited ions are transferred by fast radiationless transitions into the intermediate sharp level 2. In this process the energy lost by the electron is transferred to the lattice.

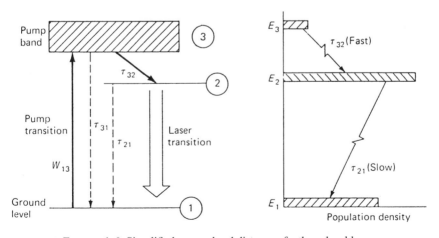

FIGURE 1.6. Simplified energy level diagram of a three-level laser.

Finally, the electron returns to the ground level by the emission of a photon. It is this last transition that is responsible for the laser action. If pumping intensity is below laser threshold, atoms in level 2 predominantly return to the ground state by spontaneous emission. Ordinary fluorescence acts as a drain on the population of level 2. After the pump radiation is extinguished, level 2 is emptied by fluorescence at a rate that varies from material to material. In ruby, at room temperature, the lifetime of level 2 is 3 ms. When the pump intensity is above laser threshold, the decay from the fluorescent level consists of stimulated as well as spontaneous radiation; the stimulated radiation produces the laser output beam. Since the terminal level of the laser transition is the highly populated ground state, a very high population must be reached in the E_2 level before the $2 \rightarrow 1$ transition is inverted.

It is necessary, in general, that the rate of radiationless transfer from the uppermost level to the level at which the laser action begins is fast compared with the other spontaneous transition rates in a three-level laser. Therefore, the lifetime of the E_2 state should be large in comparison with the relaxation time of the $3 \rightarrow 2$ transition, that is,

$$\tau_{21} \gg \tau_{32}. \tag{1.45}$$

The number of ions N_3 in level E_3 is then negligible compared with the number of ions in the other two states, i.e., $N_3 \ll N_1, N_2$. Therefore,

$$N_1 + N_2 \approx N_{\text{tot}}. \tag{1.46}$$

A vital aspect of the three-level system is that the ions are in effect pumped directly from level 1 into the metastable level 2 with only a momentary pause as they pass through level 3. With these conditions, we can calculate as if only two levels were present. In order that an equal population is achieved between the E_2 and E_1 levels, one-half of all ions must be excited to the E_2 level:

$$N_2 = N_1 = \frac{N_{\text{tot}}}{2}. \tag{1.47}$$

In order to maintain a specified amplification, the population of the second level must be larger than that of the first level. In most cases that are of practical importance, however, the necessary inversion $(N_2 - N_1)$ is small compared with the total number of all ions. The pump power necessary for maintaining this inversion is also small compared with the power necessary for achieving equal population of the levels.

The disadvantage of a three-level system is that more than half of the ions in the ground state must be raised to the metastable level E_2. There are thus many ions present to contribute to the spontaneous emission. Moreover, each of the ions which participate in the pump cycle transfer energy into the lattice from the $E_3 \rightarrow E_2$ transition. This transition is normally radiationless, the energy being carried into the lattice by phonons.

1.4.2   The Four-Level System

The four-level laser system, which is characteristic of the rare earth ions in glass
or crystalline host materials, is illustrated in Fig. 1.7. Note that a characteristic of
the three-level laser material is that the laser transition takes place between the
excited laser level 2 and the final ground state 1, the lowest energy level of the
system. This leads to low efficiency. The four-level system avoids this disadvantage. The pump transition extends again from the ground state (now level E_0) to
a wide absorption band E_3. As in the case of the three-level system, the ions so
excited will proceed rapidly to the sharply defined level E_2. The laser transition,
however, proceeds now to a fourth, terminal level E_1, which is situated above the
ground state E_0. From here the ion undergoes a rapid nonradiative transition to
the ground level. In a true four-level system, the terminal laser level E_1 will be
empty. To qualify as a four-level system a material must possess a relaxation time
between the terminal laser level and the ground level which is fast compared to
the fluorescence lifetime, that is, $\tau_{10} \ll \tau_{21}$. In addition, the terminal laser level
must be far above the ground state so that its thermal population is small. The
equilibrium population of the terminal laser level 1 is determined by the relation

$$\frac{N_1}{N_0} = \exp\left(\frac{-\Delta E}{kT}\right), \tag{1.48}$$

where ΔE is the energy separation between level 1 and the ground state, and T is
the operating temperature of the laser material. If $\Delta E \gg kT$, then $N_1/N_0 \ll 1$,
and the intermediate level will always be relatively empty. In some laser materials
the energy gap between the lower laser level and the ground state is relatively
small and, therefore, they must be cooled to function as four-level lasers. In a four-
level system an inversion of the $2 \rightarrow 1$ transition can occur even with vanishingly
small pump power, and the high pump rate, necessary to maintain equilibrium

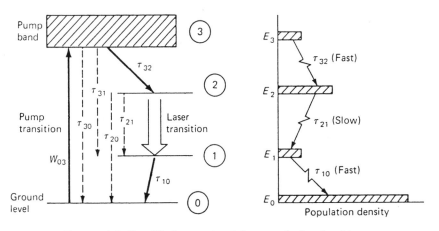

FIGURE 1.7. Simplified energy level diagram of a four-level laser.

population in the aforementioned three-level system, is no longer needed. In the most favorable case, the relaxation times of the $3 \rightarrow 2$ and $1 \rightarrow 0$ transitions in the four-level system are short compared with the spontaneous emission lifetime of the laser transition τ_{21}. Hence we can also carry out the calculations as if only the E_1 and E_2 states were populated.

1.4.3   The Metastable Level

After this brief introduction to the energy level structure of solid-state lasers we can ask the question, "What energy level scheme must a solid possess to make it a useful laser?" As we have seen in the previous discussion, the existence of a metastable level is of paramount importance for laser action to occur. The relatively long lifetime of the metastable level provides a mechanism by which inverted population can be achieved. Most transitions of ions show rapid non-radiative decay, because the coupling of the internal atomic oscillations to the surrounding lattice is strong. Radiative decay processes can occur readily, but most have short lifetimes and broad linewidths. Only a few transitions of selected ions in solids turn out to be decoupled from the lattice vibrations. These transitions have a radiative decay that leads to relatively long lifetimes.

In typical laser systems with energy levels, such as illustrated by Figs. 1.6 and 1.7, the $3 \rightarrow 2$ transition frequencies, as well as the $1 \rightarrow 0$ transition frequencies, all fall within the frequency range of the vibration spectrum of the host crystal lattice. Therefore, all these transitions can relax extremely rapidly by direct nonradiative decay, that is, by emitting a phonon to the lattice vibrations, with $\tau_{32}, \tau_{10} \approx 10^{-8}$ to 10^{-11} s. However, the larger $3 \rightarrow 0$, $3 \rightarrow 1$, $2 \rightarrow 0$, and $2 \rightarrow 1$ energy gaps in these ions often correspond to transition frequencies that are higher than the highest possible vibration frequency of the crystal lattice. Such transitions cannot relax via simple single-phonon spontaneous emission since the lattice simply cannot accept phonons at those high frequencies. These transitions must then relax either by radiative (photon) emission or by multiple-phonon processes. Since both these processes are relatively weak compared to direct single-phonon relaxation, the high-frequency transitions will have much slower relaxation rates ($\tau_{21} \approx 10^{-5}$ to 10^{-3} s in many cases). Therefore, the various levels lumped into level 3 will all relax mostly into level 2 while level 2 itself is metastable and long-lived because there are no other levels located close below it into which it can decay directly.

The existence of metastable levels follows from quantum mechanical considerations that will not be discussed here. However, for completeness we will at least explain the term "forbidden transition." As we have seen in Section 1.2.4, the mechanism by which energy exchange takes place between an atom and the electromagnetic fields is the dipole radiation. As a consequence of quantum-mechanical considerations and the ensuing selection rules, transfer between certain states cannot occur due to forbidden transitions. The term "forbidden" means that a transition among the states concerned does not take place as a result of the interaction of the electric dipole moment of the atom with the radiation field. As a result of the selection rules, an atom may get into an excited state from which it will have difficulty returning to the ground state. A state from which all dipole transitions to lower energy states are forbidden is metastable; an atom entering such a state will generally remain in that state much longer than it would in an ordinary excited state from which escape is comparatively easy.

In the absence of a metastable level, the ions which become excited by pump radiation and are transferred to a higher energy level will return either directly to the ground state by spontaneous radiation or by cascading down on intermediate levels, or they may release energy by phonon interaction with the lattice. In order for the population to increase at the metastable laser level, several other conditions have to be met. Let us consider the more general case of a four-level system illustrated in Fig. 1.7. (Note that a three-level system can be thought of as a special case of a four-level scheme where levels 1 and 0 coincide.) Pumping takes place between two levels and laser action takes place between two other levels. Energy from the pump band is transferred to the upper laser level by fast radiative transitions. Energy is removed from the lower laser level again by fast radiationless transitions.

For electrons in the pump band at level 3 to transfer to level 2 rather than return directly to the ground state, it is required that $\tau_{30} \gg \tau_{32}$. For population to build up, relaxation out of the lower level 1 has to be fast, $\tau_{21} \gg \tau_{10}$. Thus, as a first conclusion, we may say that if the right relaxation time ratio exists between any two levels (such as 3 and 2) in an energy level system, a population inversion should be possible. If so, then obtaining a large enough inversion for successful laser operation becomes primarily a matter of the right pumping method. The optical pumping method is generally feasible only in laser materials that combine a narrow laser emission line with a broad absorption transition, so that a broadband intense light source can be used as the pump source. An exception is a solid-state laser that is pumped by another laser, such as a diode laser for example. In this case the requirement for a broad absorption range for the pump band can be relaxed.

Having achieved population inversion in a material by correct combination of relaxation times and the existence of broad pump bands, the linewidth of the laser transition becomes very important. In Chapter 2 we will see that the optical gain for a given population inversion is inversely proportional to linewidth. Therefore, the metastable level should have a sufficiently narrow linewidth.

1.5 Laser Rate Equations

The dynamic behavior of a laser can be described with reasonable precision by a set of coupled rate equations. In their simplest forms, a pair of simultaneous differential equations describe the population inversion and the radiation density within a spatially uniform laser medium. We will describe the system in terms of the energy-level diagrams shown in Figs. 1.6 and 1.7. As we have seen in the preceding discussions, two energy levels are of prime importance in laser action:

the excited upper laser level E_2 and the lower laser level E_1. Thus, for many analyses of laser action, an approximation of the three- and four-level systems by a two-level representation is very useful.

The rate-equation approach used in this section involves a number of simplifying assumptions; in using a single set of rate equations we are ignoring longitudinal and radial variations of the radiation within the laser medium. In spite of these limitations, the simple rate-equation approach remains a useful tool and, properly used, provides a great deal of insight into the behavior of real solid-state laser devices. We will derive from the rate equations the threshold condition for laser actions, and obtain a first-order approximation of the relaxation oscillations in a solid-state laser. Furthermore, in Chapter 4 we will use the rate equations to calculate the gain in a laser amplifier.

In general, the rate equations are useful in predicting the gross features of the laser output, such as average and peak power, Q-switched pulse-envelope shape, threshold condition, and so forth. On the other hand, many details of the nature of the laser emission are inaccessible from the point of view of a simple rate equation. These include detailed descriptions of the spectral, temporal, and spatial distributions of the laser emission. Fortunately, these details can often be accounted for independently.

In applying the rate equations to the various aspects of laser operation, we will find it more convenient to express the probability for stimulated emission $\varrho(\nu)B_{21}$ by the photon density ϕ and the stimulated emission cross section σ.

With (1.37) we can express the Einstein coefficient for stimulated emission B_{21} in terms of the stimulated emission cross section $\sigma(\nu)$,

$$B_{21} = \frac{c}{h\nu g(\nu)}\sigma(\nu), \qquad (1.49)$$

where $c = c_0/n_0$ is the speed of light in the medium. The energy density per unit frequency $\varrho(\nu)$ is expressed in terms of the lineshape factor $g(\nu)$, the energy $h\nu$, and the photon density ϕ [photons/cm^3] by

$$\varrho(\nu) = h\nu g(\nu)\phi. \qquad (1.50)$$

From (1.49) and (1.50) we obtain

$$B_{21}\varrho(\nu) = c\sigma(\nu)\phi. \qquad (1.51)$$

1.5.1 Three-Level System

In order to approximate the three-level system with a two-level scheme, we assume that the transition from the pump band to the upper laser level is so fast that $N_3 \approx 0$. Therefore pumping does not affect the other processes at all, except to allow a mechanism of populating the upper level and thereby obtaining population inversion ($N_2 > N_1$).

Looking at Fig. 1.6, this assumption requires that the relaxation time ratio τ_{32}/τ_{21} be very small. In solid-state lasers $\tau_{32}/\tau_{21} \approx 0$ is a good approximation. Spontaneous losses from the pump band to the ground state can be expressed by the quantum efficiency η_Q. This parameter, defined as

$$\eta_Q = \left(1 + \frac{\tau_{32}}{\tau_{31}}\right)^{-1} \leq 1, \tag{1.52}$$

specifies what fraction of the total atoms excited to level 3 drop from there to level 2, thus becoming potentially useful for laser action. A small η_Q obviously requires a correspondingly larger pump power.

The changes in the electron population densities in a three-level system, based on the assumption that essentially all of the laser ions are in either levels 1 or 2, are

$$\frac{\partial n_1}{\partial t} = \left(n_2 - \frac{g_2}{g_1}n_1\right)c\phi\sigma + \frac{n_2}{\tau_{21}} - W_p n_1 \tag{1.53}$$

and

$$\frac{\partial n_2}{\partial t} = -\frac{\partial n_1}{\partial t}, \tag{1.54}$$

since

$$n_{\text{tot}} = n_1 + n_2, \tag{1.55}$$

where W_p is the pumping rate [s^{-1}].

The terms of the right-hand side of (1.53) express the net stimulated emission, the spontaneous emission, and the optical pumping.

The time variation of the population in both levels due to absorption, spontaneous, and stimulated emission is obtained from (1.15). Note that the populations N_1 and N_2 are now expressed in terms of population densities n_1 and n_2. To take into account the effect of pumping, we have added the term $W_p n_1$, which can be thought of as the rate of supply of ions to the metastable level 2. More precisely, $W_p n_1$ is the number of ions transferred from the ground level to the upper laser level per unit time per unit volume. The pump rate W_p is related to the pump parameter W_{13} in Fig. 1.6 by

$$W_p = \eta_Q W_{13}. \tag{1.56}$$

The negative sign in front of $W_p n_1$ in (1.53) indicates that the pump mechanism removes atoms from the ground level 1 and increases the population of level 2.

If we now define the inversion population density by

$$n = n_2 - \frac{g_2 n_1}{g_1}, \tag{1.57}$$

we can combine (1.53), (1.54), and (1.57) to obtain

$$\frac{\partial n}{\partial t} = -\gamma n \phi \sigma c - \frac{(n + n_{tot})(\gamma - 1)}{\tau_f} + W_p(n_{tot} - n), \tag{1.58}$$

where

$$\gamma = 1 + \frac{g_2}{g_1} \quad \text{and} \quad \tau_f = \tau_{21}. \tag{1.59}$$

In obtaining (1.58) we have used the relations

$$n_1 = \frac{n_{tot} - n}{1 + g_2/g_1} \quad \text{and} \quad n_2 = \frac{n + (g_2/g_1)n_{tot}}{1 + g_2/g_1}. \tag{1.60}$$

Another equation, usually regarded together with (1.58), describes the rate of change of the photon density within the laser resonator,

$$\frac{\partial \phi}{\partial t} = c\phi\sigma n - \frac{\phi}{\tau_c} + S, \tag{1.61}$$

where τ_c is the decay time for photons in the optical resonator and S is the rate at which spontaneous emission is added to the laser emission.

If we consider for the moment only the first term on the right, which is the increase of the photon density by stimulated emission, then (1.61) is identical to (1.33). However, for the time variation of the photon density in the laser resonator we must also take into account the decrease of radiation due to losses in the system and the increase of radiation due to a small amount of spontaneous emission that is added to the laser emission. Although very small, this term must be included because it provides the source of radiation which initiates laser emission.

An important consideration for initiation of laser oscillation is the total number p of resonant modes possible in the laser resonator volume V since in general only a few of these modes are initiated into oscillations. This number is given by the familiar expression (1.3),

$$p = 8\pi v^2 \frac{\Delta v V}{c^3}, \tag{1.62}$$

where v is the laser optical frequency and Δv is the bandwidth of spontaneous emission. Let p_L be the number of modes of the laser output. Then S can be expressed as the rate at which spontaneous emission contributes to stimulated emission, namely,

$$S = \frac{p_L n_2}{p \tau_{21}}. \tag{1.63}$$

The reader is referred to Chapter 3 for a more detailed description of the factor τ_c which appears in (1.61). For now we only need to know that τ_c represents all the losses in an optical resonator of a laser oscillator. Since τ_c has the dimension of time, the losses are expressed in terms of a relaxation time. The decay of the photon population in the cavity results from transmission and absorption at the end mirrors, "spillover" diffraction loss due to the finite apertures of the mirrors, scattering and absorptive losses in the laser material itself, etc. In the absence of the amplifying mechanism, (1.61) becomes

$$\frac{\partial \phi}{\partial t} = -\frac{\phi}{\tau_c}, \tag{1.64}$$

the solution of which is $\phi(\tau) = \phi_0 \exp(-t/\tau_c)$.

The importance of (1.61) should be emphasized by noting that the right-hand side of this equation describes the net gain per transit of an electromagnetic wave passing through a laser material.

1.5.2 Four-Level System

We will assume again that the transition from the pump band into the upper laser level occurs very rapidly. Therefore the population of the pump band is negligible, that is, $n_3 \approx 0$. With this assumption the rate of change of the two laser levels in a four-level system is

$$\frac{dn_2}{dt} = W_p n_g - \left(n_2 - \frac{g_2}{g_1} n_1 \right) \sigma \phi c - \left(\frac{n_2}{\tau_{21}} + \frac{n_2}{\tau_{20}} \right), \tag{1.65}$$

$$\frac{dn_1}{dt} = \left(n_2 - \frac{g_2}{g_1} n_1 \right) \sigma \phi c + \frac{n_2}{\tau_{21}} - \frac{n_1}{\tau_{10}}, \tag{1.66}$$

$$n_{tot} = n_g + n_1 + n_2 \tag{1.67}$$

where n_g is the population density of the ground level 0. From (1.65) it follows that the upper laser level population in a four-level system increases because of pumping and decreases because of stimulated emission and spontaneous emissions into levels 1 and 0. The lower level population increases because of stimulated and spontaneous emission and decreases by a radiationless relaxation process into the ground level. This process is characterized by the time constant τ_{10}. In an ideal four-level system the terminal level empties infinitely fast to the ground level. If we let $\tau_{10} \approx 0$, then it follows from (1.66) that $n_1 = 0$. In this case the entire population is divided between the ground level 0 and the upper level of the laser transition. The system appears to be pumping from a large source that is independent of the lower laser level. With $\tau_{10} = 0$ and $n_1 = 0$, we obtain the following rate equation for the ideal four-level system

$$n = n_2 \tag{1.68}$$

and

$$n_{tot} = n_g + n_2 \approx n_g \qquad \text{since} \quad n_2 \ll n_g. \tag{1.69}$$

Therefore, instead of (1.58), we have

$$\frac{\partial n_2}{\partial t} = -n_2 \sigma \phi c - \frac{n_2}{\tau_f} + W_p(n_g - n_2). \tag{1.70}$$

The fluorescence decay time τ_f of the upper laser level is given by

$$\frac{1}{\tau_f} = \frac{1}{\tau_{21}} + \frac{1}{\tau_{20}}. \tag{1.71}$$

In the equation for the rate of change of the upper laser level we have again taken into account the fact that not all ions pumped to level 3 will end up at the upper laser level. It is

$$W_p = \eta_Q W_{03}, \tag{1.72}$$

where the quantum efficiency η_Q depends on the branching ratios that are the relative relaxation rates for the ions along the various possible downward paths,

$$\eta_Q = \left(1 + \frac{\tau_{32}}{\tau_{31}} + \frac{\tau_{32}}{\tau_{30}}\right)^{-1} \leq 1. \tag{1.73}$$

Strictly speaking the definition of η_Q given in (1.73) is the pump quantum efficiency that addresses the fact that some of the absorbed pump photons will decay to levels other than the upper laser level. The general definition of η_Q as given in Chapters 3 and 7 also includes losses from the upper laser level as a result of non-radiative decay. These losses caused by quenching processes reduce the inversion available for laser action.

Summary

In thermal equilibrium lower energy states of ions or atoms are more heavily populated than higher energy states according to Boltzmann's statistics. In order for stimulated emission rather than absorption to occur, the population between two energy states has to be inverted, such that the higher energy level is more heavily populated compared to the lower level. Energy to achieve this population inversion is supplied by a pump source.

In a three-level laser the ground state of the electronic transition is also the lower laser level. At thermal equilibrium the majority of ions are in this level. Thus, at least half of the ions at the ground level must be transferred to the upper laser level before laser action is possible.

In a four-level laser, the lower laser level lies at an energy level that is above the ground state. Since the number of thermally excited ions in the lower laser level is small, the population can be easily inverted by pumping a relatively small number of ions into the upper laser level. Therefore it requires less energy to generate a population inversion in a four-level laser and the threshold will be lower compared to a three-level system.

For both systems there must be a fast process that transfers excited ions from the pump level to the upper laser level. Futhermore in a four-level system, the lower laser level has to drain rapidly to the ground state to prevent laser action to cease.

The rate equation applicable to three- and four-level systems can be expressed by a single pair of equations, namely, (1.58) and (1.61), where $\gamma = 1 + g_2/g_1$ for a three-level system and $\gamma = 1$ for a four-level system. The factor γ can be thought of as an "inversion reduction factor" since it corresponds to the net reduction in the population inversion after the emission of a single photon. In a four-level system, see (1.70), we have $\gamma = 1$ since the population inversion density is only reduced by one for each photon emitted. For a three-level system (see (1.58)), we have $\gamma = 2$ if we assume no degeneracy, that is, $g_2/g_1 = 1$. This reflects the fact that in this case the population inversion is reduced by two for each stimulated emission of a photon because the emitting photon is not only lost to the upper laser level, but also increases the lower laser level by one.

References

[1] A. Yariv: *Quantum Electronics*, 4th ed. (Wiley, New York), 1991.
[2] A.E. Siegman: *Lasers* (University Science Books, Mill Valley, CA), 1986.

Exercises

1. A pretty good classical model of matter results when you think of an atom as having an electron attached to a massive nucleus by a spring so that it can undergo damped harmonic motion. Electrons, being charged particles, are driven to move under these constraints by applied electric fields. Since light is an electromagnetic wave phenomenon having an electric field transverse to the direction of propagation, we can try to understand light–matter interactions by asking how such a field causes the matter to react.

   Solve for the motion of the driven, damped harmonic oscillator shown below, where the driving electric field propagating in the z-direction and polarized in the y-direction is $\mathbf{E} = E_0 \exp(i(\omega t - kz))\mathbf{a}_y$ where ω is the frequency of the driving field, k is its wave vector, and $\mathbf{a}_y$ is a unit vector in the y-direction.

   (a) Find the resonant frequency for the electron by solving for the motion of the electron when it is not driven. When there is no damping, the resonant frequency is called the natural frequency of the oscillator.

   (b) What is the effect of damping on the motion (mathematically and in words)?

   (c) A material interacts with an electromagnetic wave by becoming polarized and that polarization then radiates an electromagnetic wave. The polarization of a material is proportional to the sum of the dipole moments of

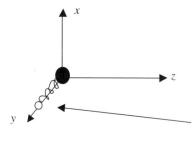

An electron of charge $-e$ and mass m attached to a massive nucleus fixed at the origin and constrained to move along the y axis. Its motion is damped with a damping constant γ.

the atoms in the material. Recall that the dipole moment is just the charge on the electron times its displacement from equilibrium. From the general solution to the driven, damped harmonic oscillator we are discussing, identify the terms related to Einstein's A and B coefficients and explain why you made each identification. In other words, show that Einstein's model is actually a classical model of light–matter interactions.

(*Hint:* The equation for a damped harmonic oscillator is

$$\frac{d^2y}{dt^2} + \frac{\gamma}{m}\frac{dy}{dt} + \frac{k}{m}y = 0$$

and, when a force, F, drives it we have

$$\frac{d^2y}{dt^2} + \frac{\gamma}{m}\frac{dy}{dt} + \frac{k}{m}y = \frac{F}{m}.$$

You solve the first homogeneous equation and find the natural motion of the oscillator. Then you solve the driven oscillator by assuming a solution at the frequency of the driving field. The general solution is the sum of the solutions.)

2. Equation (1.20) gives a relation between the stimulated and spontaneous emission coefficients. Recall the implication of spontaneous emission on the excited state population and explain how this relation affects the possibility of making short wavelength lasers.

3. Consider two different materials each having an emission line centered at frequency ν_0 with the same line width, $\Delta\nu$ (full width at half-maximum). You know that one is homogeneously broadened and the other is inhomogeneously broadened. A friend suggests that to tell which is which you measure the line strengths at say, three times $\Delta\nu$. Which will be stronger and by how much?

4. Is it possible to achieve an inverted population in a purely two-level system (assume equal degeneracy)? If so, then find the threshold for lasing in such a system. If not, describe the interaction of a beam of light having a frequency corresponding to the energy difference between the two levels when its photon density gets very high? When this happens, what is the population of each level?

5. Under what circumstances and at what temperature will the rate of transition due to thermal background radiation (proportional to the photon density times the Einstein B coefficient) be equal to the rate of spontaneous transition (proportional to the Einstein A coefficient) for a two-level system?

2

Properties of Solid-State Laser Materials

2.1 Overview

2.2 Ruby

2.3 Nd : YAG

2.4 Nd : Glass

2.5 Nd : YLF

2.6 Nd : YVO$_4$

2.7 Er : Glass

2.8 Yb : YAG

2.9 Alexandrite

2.10 Ti : Sapphire

References

Exercises

Materials for laser operation must possess sharp fluorescent lines, strong absorption bands, and a reasonably high quantum efficiency for the fluorescent transition of interest. These characteristics are generally shown by solids (crystals or glass) that incorporate in small amounts elements in which optical transitions can occur between states of inner, incomplete electron shells. Thus the transition metals, the rare earth (lanthanide) series, and the actinide series are of interest in this connection. The sharp fluorescence lines in the spectra of crystals doped with these elements result from the fact that the electrons involved in transitions in the optical regime are shielded by the outer shells from the surrounding crystal lattice. The corresponding transitions are similar to those of the free ions. In addition to a sharp fluorescence emission line, a laser material should possess pump bands within the emission spectrum of readily available pump sources such as arc lamps and laser diode arrays.

The three principal elements leading to gain in a solid-state laser are:

- *The host material* with its macroscopic mechanical, thermal and optical properties, and its unique microscopic lattice properties.
- *The activator/sensitizer ions* with their distinctive charge states and free-ion electronic configurations.
- *The optical pump source* with its particular geometry, spectral irradiance, and temporal characteristic.

These elements are interactive and must be selected self-consistently to achieve a given system performance.

In this chapter we consider the properties of various host materials and activator/sensitizer combinations.

2.1 Overview

The conditions for laser action at optical frequencies were first described by Schawlow and Townes [1] in 1958. The first demonstration of laser action by Maiman [2] was achieved in 1960 using ruby ($Cr^{3+} : Al_2O_3$), a crystalline solid system. The next step in the development of solid-state lasers was the operation of trivalent uranium in CaF_2 and divalent samarium in CaF_2.

Laser action in neodymium-doped glass and the first continuously operating crystal laser using $Nd^{3+} : CaWO_4$ were reported in 1961. Since then laser action has been achieved from trivalent rare earths (Nd^{3+}, Er^{3+}, Ho^{3+}, Ce^{3+}, Tm^{3+}, Pr^{3+}, Gd^{3+}, Eu^{3+}, Yb^{3+}), divalent rare earths (Sm^{2+}, Dy^{2+}, Tm^{2+}), transition metals (Cr^{3+}, Ni^{2+}, Co^{2+}, Ti^{3+}, V^{2+}), and the actinide ion U^{3+} embedded in various host materials. Optically pumped laser action has been demonstrated in hundreds of ion-host crystal combinations covering a spectral range from the visible to the mid-infrared.

The exceptionally favorable characteristics of the trivalent neodymium ion for laser action were recognized at a relatively early stage in the search for solid-state laser materials. Thus, Nd^{3+} was known to exhibit a satisfactorily long fluorescence lifetime and narrow fluorescence linewidths in crystals with ordered structures, and to possess a terminal state for the laser transition sufficiently high above the ground state so that cw operation at room temperature was readily feasible. Therefore, this ion was incorporated as a dopant in a variety of host materials, that is, glass, $CaWO_4$, $CaMoO_4$, CaF_2, LaF_3, and so forth, in an effort to make use of its great potential. However, most of these early hosts displayed undesirable shortcomings, either from the standpoint of their intrinsic physical properties or because of the way in which they interacted with the Nd^{3+} ions. Finally, yttrium aluminum garnet (YAG) was explored by Geusic et al. [3] as a host for Nd^{3+}, and its superiority to other host materials was quickly demonstrated.

2.1.1   Host Materials

Solid-state host materials may be broadly grouped into crystalline solids and glasses. The host must have good optical, mechanical, and thermal properties to withstand the severe operating conditions of practical lasers. Desirable properties include hardness, chemical inertness, absence of internal strain and refractive index variations, resistance to radiation-induced color centers, and ease of fabrication.

Several interactions between the host crystal and the additive ion restrict the number of useful material combinations. These include size disparity, valence, and spectroscopic properties. Ideally, the size and valence of the additive ion should match that of the host ion it replaces.

In selecting a crystal suitable for a laser ion host one must consider the following key criteria:

(a) The crystal must possess favorable optical properties. Variations in the index of refraction lead to inhomogeneous propagation of light through the crystal which results in poor beam quality.

(b) The crystal must possess mechanical and thermal properties that will permit high-average-power operation. The most important parameters are thermal conductivity, hardness, and fracture strength.

(c) The crystal must have lattice sites that can accept the dopant ions and that have local crystal fields of symmetry and strength needed to induce the desired spectroscopic properties. In general, ions placed in a crystal host should have long radiative lifetimes with emission cross sections near 10^{-20} cm^2.

(d) It must be possible to scale the growth of the impurity-doped crystal, while maintaining high optical quality and high yield.

Glasses

Glasses form an important class of host materials for some of the rare earths, particularly Nd^{3+}. The outstanding practical advantage compared to crystalline materials is the tremendous size capability for high-energy applications. Rods up to 1 m in length and over 10 cm in diameter, and disks up to 90 cm in diameter and several centimeters thick have been produced. The optical quality is excellent, and glass, of course, is easily fabricated and takes a good optical finish. Laser ions placed in glass generally show a larger fluorescent linewidth than in crystals as a result of the lack of a unique and well-defined crystalline field surrounding the individual active atoms. Therefore, the laser thresholds for glass lasers have been found to run higher than their crystalline counterparts. Also, glass has a much lower thermal conductivity than most crystalline hosts. The latter factor leads to a large thermally induced birefringence and optical distortion in glass laser rods when they are operated at high average powers.

For bulk solid-state lasers, glass doped with Nd^{3+} or Er^{3+} is important and will be discussed in Sections 2.4 and 2.7. In fiber oscillators and amplifiers, Er^{3+}- and Yb^{3+}-doped glass provides the active medium.

Oxides

Sapphire. The first laser material to be discovered (ruby laser) employed sapphire as a host. The Al_2O_3 (sapphire) host is hard, with high thermal conductivity, and transition metals can readily be incorporated substitutionally for the Al. The Al site is too small for rare earths, and it is not possible to incorporate appreciable concentrations of these impurities into sapphire. Besides ruby which is still used today, Ti-doped sapphire has gained significance as a tunable-laser material. The properties of ruby and Ti-sapphire will be discussed in Sections 2.2 and 2.10.

Garnets. Some of the most useful laser hosts are the synthetic garnets: yttrium aluminum garnet, $Y_3Al_5O_{12}$ (YAG); gadolinium gallium garnet, $Gd_3Ga_5O_{12}$ (GGG), and gadolinium scandium aluminum garnet $Gd_3Sc_2Al_3O_{12}$ (GSGG). These garnets have many properties that are desirable in a laser host material. They are stable, hard, optically isotropic, and have good thermal conductivities, which permits laser operation at high average power levels.

In particular, yttrium aluminum garnet doped with neodymium (Nd : YAG) has achieved a position of dominance among solid-state laser materials. YAG is a very hard, isotropic crystal, which can be grown and fabricated in a manner that yields rods of high optical quality. At the present time, it is the best commercially available crystalline laser host for Nd^{3+}, offering low threshold and high gain. The properties of Nd : YAG are discussed in more detail in Section 2.3. Besides Nd^{3+}, the host crystal YAG has been doped with Tm^{3+}, Er^{3+}, Ho^{3+}, and Yb^{3+}. The Yb : YAG laser material will be discussed in Section 2.8 and Er : YAG, Tm : YAG, and Ho : YAG crystals are briefly covered under the appropriate active ion.

In recent years, Nd : GSGG co-doped with Cr^{3+} has been employed in a number of laser systems. Cr^{3+} considerably increases the absorption of flashlamp radiation and transfers the energy very efficiently to Nd. A comparison with Nd : YAG reveals a lower thermal conductivity and a lower heat capacity for GSGG but, in general, the materials parameters are fairly close. The stimulated emission cross section, and therefore the gain of Cr : Nd : GSGG is about half of that in YAG. This can be an advantage for Q-switch operation since more energy can be stored before amplified spontaneous emission (ASE) starts to deplete the metastable level.

Vanadates

Nd^{3+}-doped yttrium orthovanadate (YVO_4) has shown a relatively low threshold at pulsed operation. However, early studies of this crystal were hampered by severe crystal growth problems, and as a result YVO_4 was discarded as a host. With the emergence of diode pumping, Nd : YVO_4 has become an important solid-state laser material (Section 2.6), because it has very attractive features, such as a large stimulated emission cross section and a high absorption of the pump wavelength, and the growth problem has been overcome for the small crystals required with this pump source.

Fluorides

Doping of fluorides with trivalent rare earth ions requires charge compensation which complicates the crystal growth process. The important representative of this crystal family is yttrium fluoride ($YLiF_4$), a uniaxial crystal. $YLiF_4$ is transparent to 150 nm. Therefore, high-current-density xenon flashlamps which emit strongly in the blue and near-ultraviolet can be used as pump sources without damage to the material. The most common dopant of YLF is Nd^{3+}. Nd : YLF offers a reduction in thermal lensing and birefringence combined with improved energy storage relative to Nd : YAG. The thermomechanical properties of Nd : YLF, however, are not as good as those of Nd : YAG. Details of Nd : YLF are discussed in Section 2.5.

2.1.2   Active Ions

Before proceeding to a discussion of the active laser ions, we will review briefly the nomenclature of atomic energy levels.

Different energy states of electrons are expressed by different quantum numbers. The electrons in an atom are characterized by a principal quantum number n, an orbital angular momentum l, the orientation of the angular momentum vector m, and a spin quantum number s. A tradition from the early days of line-series allocation has established the following method of designating individual electronic orbits: a number followed by a letter symbolizes the principal quantum number n and the angular momentum number l, respectively. The letters s, p, d, f stand for $l = 0, 1, 2, 3$, respectively. For example, a $3d$ electron is in an orbit with $n = 3$ and $l = 2$.

To designate an atomic energy term one uses by convention capital letters with a system of subscripts and superscripts. The symbol characterizing the term is of the form $^{2S+1}L_J$, where the orbital quantum numbers $L = 0, 1, 2, 3, 4, 5$, are expressed by the capital letters S, P, D, F, G, H, I. A superscript to the left of the letter indicates the value $(2S + 1)$, that is, the multiplicity of the term due to possible orientation of the resultant spin S. Thus a one-electron system $(S = \frac{1}{2})$ has a multiplicity 2. L and S can combine to a total angular momentum J, indicated by a subscript to the right of the letter. Thus the symbol $^2P_{3/2}$ shows an energy level with an orbital quantum number $L = 1$, a spin of $S = \frac{1}{2}$, and a total angular momentum of $J = \frac{3}{2}$. The complete term description must include the configuration of the excited electron, which precedes the letter symbol. Thus the ground state of lithium has the symbol $2s\,^2S_{1/2}$.

When an atom contains many electrons, the electrons that form a closed shell may be disregarded and the energy differences associated with transitions in the atom may be calculated by considering only the electrons outside the closed shell.

In describing the state of a multielectron atom, the orbital angular momenta and the spin angular momenta are added separately. The sum of the orbital angular momenta are designated by the letter L, and the total spin is characterized by S. The total angular momentum J of the atom may then be obtained by vector addition of L and S. The collection of energy states with common values of J, L, and S is called a term.

In the following section, a qualitative description is given of some of the prominent features of the most important rare earth, actinide, and transition metal ions.

Rare Earth Ions

The rare earth ions are natural candidates to serve as active ions in solid-state laser materials because they exhibit a wealth of sharp fluorescent transitions representing almost every region of the visible and near-infrared portions of the electromagnetic spectrum. It is a characteristic of these lines that they are very sharp, even in the presence of the strong local fields of crystals, as a result of the shielding effect of the outer electrons.

The ground state electronic configuration of the rare earth atom consists of a core that is identical to xenon, plus additional electrons in higher orbits. In xenon, the shells with quantum numbers $n = 1, 2, 3$ are completely filled. The shell $n = 4$ has its s, p, and d subshells filled, whereas the $4f$ subshell capable of accommodating 14 electrons is completely empty. However, the $n = 5$ shell has acquired its first eight electrons which fill the $5s$ and $5p$ orbits. The electronic configuration for xenon is

$$1s^2 2s^2 2p^6 3s^2 3p^6 3d^{10} 4s^2 4p^6 4d^{10} 5s^2 5p^6 .$$

Elements beyond xenon, which has the atomic number 54, have this electronic structure and, in addition, have electrons in the $4f$, $5d$, $6s$, etc. orbits. Cesium, barium, and lanthanum are the elements between xenon and the rare earths. Cesium has one, and barium has two $6s$ electrons, and lanthanum has in addition one electron in the $5d$ orbit. Rare earth elements begin with the filling of the inner vacant $4f$ orbits. For example, the first rare earth element cerium has only one electron in the f-orbit:

$$\text{Ce:} \quad \ldots 4f 5s^2 5p^6 5d 6s^2$$

and the important rare earth neodymium has four electrons in the f orbit

$$\text{Nd:} \quad \ldots 4f^4 5s^2 5p^6 6s^2.$$

Since the first nine shells and subshells up to $4d^{10}$ are completely filled, only the outer electron configuration is indicated.

In crystals, rare earth ions are normally trivalent, but under appropriate conditions the valence state can also be divalent. When a trivalent ion is formed the atom gives up its outermost $6s$ electrons, the atom loses also its $5d$ electron if it has one, otherwise one of the $4f$ electrons is lost. For example, trivalent cesium has the electronic configuration

$$\text{Ce}^{3+}: \quad \ldots 4f 5s^2 5p^6$$

and trivalent neodymium has the configuration

$$Nd^{3+}: \quad \ldots 4f^3 5s^2 5p^6.$$

As one can see, the trivalent ions of rare earths have a simpler configuration than the corresponding atoms. Ions from the rare earths differ in electronic structure only by the number of electrons in the $4f$ shell as illustrated in Table 2.1. When a divalent rare earth ion is formed, the atom gives up its outermost $6s$ electrons.

The fluorescence spectra of rare earth ions arise from electronic transitions between levels of the partially filled $4f$ shell. Electrons present in the $4f$ shell can be raised by light absorption into unoccupied $4f$ levels. The $4f$ states are well-shielded by the filled $5s$ and $5p$ outer shells. As a result, emission lines are relatively narrow and the energy level structure varies only slightly from one host to another. The effect of the crystal field is normally treated as a perturbation on the free-ion levels. The perturbation is small compared to spin-orbit and electrostatic interactions among the $4f$ electrons. The primary change in the energy levels is a splitting of each of the free-ion levels in many closely spaced levels caused by the Stark effect of the crystal field. In crystals the free-ion levels are then referred to as manifolds. For example, Figure 2.2 provides a nice illustration of the splitting of the Nd^{3+} manifolds into sublevels as a result of the YAG crystal field.

Neodymium. Nd^{3+} was the first of the trivalent rare earth ions to be used in a laser, and it remains by far the most important element in this group. Stimulated emission has been obtained with this ion incorporated in at least 100 different host materials, and a higher power level has been obtained from neodymium lasers than from any other four-level material. The principal host materials are YAG and glass. In these hosts stimulated emission is obtained at a number of frequencies

TABLE 2.1. Electronic configuration of trivalent rare earths.

Element number	Trivalent rare earth	Number of $4f$ electrons	Ground state
58	Cerium, Ce^{3+}	1	$^2F_{5/2}$
59	Praseodymium, Pr^{3+}	2	$^3H_4$
60	Neodymium, Nd^{3+}	3	$^4I_{9/2}$
61	Promethium, Pm^{3+}	4	$^5I_4$
62	Samarium, Sm^{3+}	5	$^6H_{5/2}$
63	Europium, Eu^{3+}	6	$^7F_0$
64	Gadolinium, Gd^{3+}	7	$^8S_{7/2}$
65	Terbium, Tb^{3+}	8	$^7F_6$
66	Dysprosium, Dy^{3+}	9	$^6H_{15/2}$
67	Holmium, Ho^{3+}	10	$^5I_8$
68	Erbium, Er^{3+}	11	$^4I_{15/2}$
69	Thulium, Tm^{3+}	12	$^3H_6$
70	Ytterbium, Yb^{3+}	13	$^2F_{7/2}$
71	Lutetium, Lu^{3+}	14	$^1S_0$

within three different groups of transitions centered at 0.9, 1.06, and 1.35 μm. Radiation at these wavelengths results from $^4F_{3/2} \rightarrow {}^4I_{9/2}, {}^4I_{11/2}$, and $^4I_{13/2}$ transitions, respectively.

The nomenclature of the energy levels may be illustrated by a discussion of the Nd^{3+} ion. This ion has three electrons in the $4f$ subshell. In the ground state their orbits are so aligned that the orbital angular momentum adds up to $3 + 2 + 1 = 6$ atomic units. The total angular momentum $L = 6$ is expressed by the letter I. The spins of the three electrons are aligned parallel to each other, providing an additional $\frac{3}{2}$ units of angular momentum, which, when added antiparallel to the orbital angular momentum, gives a total angular momentum of $6 - \frac{3}{2} = \frac{9}{2}$ units. According to the quantum rules for the addition of angular momenta, the vector sum of an orbital angular momentum of 6 and a spin angular momentum of $\frac{3}{2}$ may result in the following four values of the total angular momentum: $\frac{9}{2}, \frac{11}{2}, \frac{13}{2}$, and $\frac{15}{2}$. The levels corresponding to these values are $^4I_{9/2}, {}^4I_{11/2}, {}^4I_{13/2}$, and $^4I_{15/2}$. The first of these, which has the lowest energy, is the ground state; the others are among the first few excited levels of Nd^{3+}. These levels are distinguished by the orientation of the spins with respect to the resultant orbital angular momentum. Other excited levels are obtained when another combination of the orbital angular momenta is chosen. The most important neodymium laser materials, namely Nd : YAG, Nd : glass, Nd : YLF, and Nd : YVO$_4$ are described later in this chapter.

Erbium. Laser action in erbium has been demonstrated in a variety of garnets, fluorides, and glasses. Erbium-laser performance is not very impressive in terms of efficiency or energy output. However, erbium has attracted attention because of two particular wavelengths of interest. A crystal, such as YAG, highly doped with erbium produces an output around 2.9 μm, and erbium-doped phosphate glass generates an output at 1.54 μm. Both of these wavelengths are absorbed by water, which leads to interesting medical applications in the case of the 2.9 μm lasers, and to eye safe military rangefinders in the case of the shorter wavelength.

The wavelength at 1.54 μm arises from a transition between the $^4I_{13/2}$ state and the $^4I_{15/2}$ ground state of Er^{3+}. At room temperature all levels of the terminal $^4I_{15/2}$ manifold are populated to some degree; thus this transition forms a three-level laser scheme with a correspondingly high threshold. Er : glass is discussed in Section 2.7. The wavelength at 2.9 μm stems from a laser transition from the $^4I_{11/2}$ to the $^4I_{13/2}$ state. The lower laser level of this transition is considerably above the ground state, similar to the 1.06 μm transition in Nd^{3+}. Therefore this transition is four-level in nature. However, complicated energy transfer mechanisms between the levels and a lower laser level which has a longer lifetime compared to the upper laser level has prevented cw operation at this transition. But with highly concentrated pump beams at 963 nm from quasi-cw laser diodes average output powers of several watts have been obtained from 50% Er : YAG.

Holmium. Laser action in Ho^{3+} has been reported in many different host materials. Because the terminal level is only about 250 cm^{-1} above ground level, the lower laser level has a relatively high thermal population at room temperature.

While Ho : YAG and Ho : YLF have proven to be efficient lasers, operation has been limited in most cases to cryogenic temperatures, which will depopulate the lower laser level. Previous efforts in flashlamp-pumped 2 μm lasers have concentrated on Er : Tm : Ho-doped YAG and YLF.

It was discovered that chromium-sensitized Tm : Ho : YAG offers several advantages over the erbium-sensitized materials. In a Cr : Tm : Ho : YAG laser, a very efficient energy transfer process between the co-dopants takes place. Cr^{3+} acts to efficiently absorb flashlamp energy, which is then transferred to thulium with a transfer quantum efficiency approaching 2 (two photons in the infrared for each pump photon). From thulium the energy is efficiently transferred to holmium. Lasing occurs at the $^5I_7$–$^5I_8$ holmium transition at a wavelength of 2080 nm. Laser diode pumping of a Tm : Ho : YAG laser via an absorption line in Tm^{3+} at 780 nm is also possible. Chromium doping is not necessary in this case.

Thulium. Efficient flashlamp and laser diode-pumped laser operation has been achieved in Tm^{3+} : YAG and Tm^{3+} : YLF co-doped either with Cr^{3+} or Ho^{3+}. The output wavelength for the $^3F_4$–$^3H_6$ transition is 2.01 μm. The thulium ion has an absorption at 785 nm which is useful for diode pumping. With diode pumping Tm : YAG lasers have been designed with output powers in excess of 100 W. Cr-doping provides for efficient absorption of the flashlamp radiation. Flashlamp-pumped Cr : Tm : YAG can achieve tunable output between 1.945 and 1.965 μm.

Thulium and thulium-sensitized holmium lasers have outputs in the 2 μm region, a wavelength of interest for coherent radar systems, remote sensing, and medical applications. The possibility of pumping thulium-doped crystals, with readily available powerful GaAlAs laser diodes at 785 nm, has stimulated interest in these materials.

Praseodymium, Gadolinium, Europium, Ytterbium, and Cerium. Laser action in these triply ionized rare earths has been reported; however, only marginal performance was obtained in hosts containing these ions with the exception of ytterbium. Diode-pumped Yb : YAG has become an important laser and is described in Section 2.8.

Samarium, Dysprosium, and Thulium. The divalent rare earths Sm^{2+}, Dy^{2+}, and Tm^{2+} have an additional electron in the $4f$ shell, which lowers the energy of the $5d$ configuration. Consequently, the allowed $4f$–$5d$ absorption bands fall in the visible region of the spectrum. These bands are particularly suitable for pumping the laser systems. Tm^{2+}, Dy^{2+}, and Sm^{2+} have been operated as lasers, all in a CaF_2 host. For laser operation, these crystals must be refrigerated to at least 77 K.

Actinide Ions

The actinides are similar to the rare earths in having the $5f$ electrons partially shielded by $6s$ and $6p$ electrons. Most of the actinide elements are radioactive, and only uranium in CaF_2 has been successfully used in a laser. The host was doped with 0.05% uranium. Laser action occurred at 2.6 μm between a metastable level and a terminal level some 515 cm^{-1} above the ground state.

Transition Metals

Important members of the transition metal group include chromium and titanium. The chromium atom has 18 electrons in orbitals which make up the filled core. The outer configuration includes six electrons in the d and s orbitals:

$$\text{Cr:} \quad \ldots 3d^5 4s.$$

The trivalent chromium ion has given up three electrons from these outer two sets of orbitals

$$\text{Cr}^{3+}: \quad \ldots 3d^3,$$

leaving three electrons in the $3d$ orbitals. These electrons determine the optical properties of the ion. Since they are unshielded by outer shells of electrons, in contrast to rare earth ions, the optical properties of transition metal ions are strongly influenced by the host crystal field.

The titanium atom has the same electronic configuration as chromium but only four electrons in the d and s orbitals

$$\text{Ti:} \quad \ldots 3d^2 4s^2.$$

The trivalent titanium has only a single electron in the d orbital

$$\text{Ti}^{3+}: \quad \ldots 3d.$$

Transition metal–host crystal combinations which have resulted in important lasers include ruby ($\text{Cr}^{3+} : \text{Al}_2\text{O}_3$), alexandrite ($\text{Cr}^{3+} : \text{BeAl}_2\text{O}_4$), and Ti : sapphire ($\text{Ti}^{3+} : \text{Al}_2\text{O}_3$) lasers, which are discussed in separate sections.

In addition, two other lasers based on chromium-doped crystals, Cr : LiSAF ($\text{Cr}^{3+} : \text{LiSrAlF}_6$) and Cr : forsterite ($\text{Cr}^{4+} : \text{Mg}_2\text{SiO}_4$) need to be mentioned. Cr : LiSAF has found applications as a flashlamp or diode-pumped laser source with tunable output from 780 to 920 nm. The broadband emissions of Cr : LiSAF makes this crystal attractive for the generation and amplification of femtosecond mode-locked pulses. Of particular interest is a diode-pumped, all solid-state tunable source for femtosecond pulse generation.

Systems have been developed which range from small diode-pumped Cr : LiSAF mode-locked oscillators to very large flashlamp-pumped amplifier stages. Both pulsed and cw laser operation has been achieved in $\text{Cr}^{4+} : \text{Mg}_2\text{SiO}_4$ with Nd : YAG lasers at 1.06 μm and 532 nm as pump sources. The tuning range covers the spectral region from 1167 to 1345 nm. A distinguishing feature of laser actions in Cr : Mg_2SiO_4 is that the lasing ion is not trivalent chromium (Cr^{3+}) as in the case with other chromium-based lasers, but the active ion in this crystal is tetravalent chromium (Cr^{4+}) which substitutes for silicon (Si^{4+}) in a tetradedral site.

In the following section we will describe some of the features of the most prominent laser materials. The ruby laser, still in use today, provides a good illustration of a three-level system. Nd : YAG, because of its high gain and good thermal and mechanical properties, is by far the most important solid-state laser for scientific, medical, industrial, and military applications. Nd : glass is important for laser fusion drivers because it can be produced in large sizes. Nd : YLF is a good material for a number of applications; the crystal exhibits lower thermal birefringence and has a higher energy storage capability (due to its lower gain coefficient) compared to Nd : YAG. Nd : YVO_4 has become a very attractive material for small diode-pumped lasers because of its large emission cross section and strong absorption at 809 nm. Er : glass is of special importance because the laser output at 1.55 μm is in the eye-safe region of the spectrum. The Yb : YAG laser has been developed for high-power applications requiring output in the kW range.

The last two lasers described in this chapter, alexandrite and Ti : sapphire, are both broadly tunable lasers. Tunability of the emission in solid-state lasers is achieved when the stimulated emission of photons is intimately coupled to the emission of vibrational quanta (phonons) in a crystal lattice. In these "vibronic" lasers, the total energy of the lasing transition is fixed but can be partitioned between photons and phonons in a continuous fashion. The result is broad wavelength tunability of the laser output. In other words, the existence of tunable solid-state lasers is due to the subtle interplay between the Coulomb field of the lasing ion, the crystal field of the host lattice, and electron–phonon coupling permitting broadband absorption and emission. Therefore, the gain in vibronic lasers depends on transitions between coupled vibrational and electronic states; that is, a phonon is either emitted or absorbed with each electronic transition. In alexandrite and Ti : sapphire tunability is achieved by the 3d electron of the transition metal ions Cr^{3+} and Ti^{3+}.

Overviews of laser host materials and active ions, as well as compilations of useful laser and materials parameters, can be found in a number of handbooks and summary articles [4]–[10].

2.2 Ruby

The ruby laser, $Cr^{3+} : Al_2O_3$, is a three-level system, which makes it necessary to pump the crystal at high intensity to achieve at least inversion. Consequently, the overall system efficiency is very low, which is the main reason for its limited use today. Ruby chemically consists of sapphire (Al_2O_3) in which a small percentage of the Al^{3+} has been replaced by Cr^{3+}. This is done by adding small amounts of Cr_2O_3 to the melt of highly purified Al_2O_3. Sapphire has the best thermomechanical properties of all solid-state laser materials. The crystal is very hard, has high thermal conductivity, and can be grown to very high quality by the Czochralski method. The Cr^{3+} ion has three d electrons in its unfilled shell; the ground state of the free ion is described by the spectroscopic symbol ^4A. The amount of doping is nominally 0.05 wt.% Cr_2O_3.

A simplified energy level diagram of ruby is given in Fig. 2.1. In ruby lasers, population inversion with respect to the so-called ^2E level is obtained by opti-

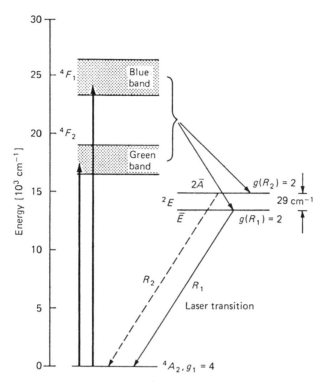

FIGURE 2.1. Important energy levels of Cr^{3+} in ruby (separation of $2E$ levels not to scale).

cally pumping Cr^{3+} ions from the $^4A_2$ ground state to the broad pump bands $^4F_2$ and $^4F_1$. The lifetime at the pump bands, which are each about 100 nm wide, located in the green $(18,000\,cm^{-1})$ and in the violet $(25,000\,cm^{-1})$, is extremely short, with the ions returning to a metastable state ^2E that has a fluorescent lifetime of 3 ms. This metastable level is split into two sublevels with a separation of $\Delta E = 29\,cm^{-1}$. The upper sublevel is the $2\overline{A}$ and the lower one the $\overline{E}$ sublevel. The two transitions $(\overline{E} \rightarrow {}^4A_2$ and $2\overline{A} \rightarrow {}^4A_2)$ are referred to as the R_1 and R_2 lines, which are located at the end of the visible region at 694.3 and 692.9 nm, respectively. The width of these fluorescent lines is $\Delta\nu = 11\,cm^{-1}$ at room temperature and they are homogeneously broadened because of the interaction of the Cr^{3+} ions with lattice vibrations. The two R lines are connected to each other by a very fast nonradiative decay, thermalization of the population occurs, and this results in the lower-lying $\overline{E}$ level being the more heavily populated.

At thermal equilibrium the difference in population between the $\overline{E}$ and $2\overline{A}$ level is determined by the Boltzmann factor $\kappa = \exp(\Delta E/kT)$. At room temperature the Boltzmann factor is $\kappa = 0.87$. The R_1 line attains laser threshold before the R_2 line because of the higher inversion. Once laser action commences in the R_1 line, the $\overline{E}$ level becomes depleted and population transfer from the nearby $2\overline{A}$ level proceeds at such a fast rate that the threshold level is never reached for the R_2 line. Therefore the entire initial population of the two states decays through R_1 emission. If we compare the energy-level diagram of ruby with our simplified scheme of a three-level system in Figure 1.6, the levels $^4F_1$ and $^4F_2$ jointly constitute level 3, whereas ^2E and $^4A_2$ states represent levels 2 and 1, respectively. We can write

$$n_2 = n_2(R_1) + n_2(R_2) \tag{2.1}$$

for the metastable level and n_1 for the ground level. Threshold and gain in ruby depends only on the population of level $n_2(R_1)$. However, in relating gain and threshold to the population of the ground level n_1 or to the total number of Cr^{3+} ions n_{tot}, one has to take the population of $n_2(R_2)$ into account. In ruby all levels are degenerate, that is,

$$g(n_1) = 4, \qquad g(R_1) = g(R_2) = 2. \tag{2.2}$$

Because of the higher degeneracy of the ground state, amplification occurs when the R_1 level is at least one-half as densely populated as the ground state. Since

$$n_2(R_1) + n_2(R_2) + n_1 = n_{tot} , \tag{2.3}$$

we have the following population at threshold (300 K)

$$n_2(R_1) = \frac{n_{tot}}{3 + \kappa} = 0.26 n_{tot},$$

$$n_2(R_2) = \frac{n_{tot}}{3 + \kappa} = 0.22 n_{tot}, \tag{2.4}$$

$$n_1 = \frac{2 n_{tot}}{3 + \kappa} = 0.52 n_{tot}.$$

Thus we must have just under one-half of the atoms in the two upper levels in order to reach threshold. With a Cr^{3+} concentration of $n_{tot} = 1.58 \times 10^{19}$ cm^3 and a photon energy of $h\nu = 2.86 \times 10^{-19}$ W s, we obtain for the energy density at the metastable level required to achieve transparency in ruby according to (2.4),

$$J_{th} = \left(\frac{\kappa + 1}{\kappa + 3} \right) n_{tot} h\nu = 2.18 \, \text{J/cm}^3. \tag{2.5}$$

The small-signal gain coefficient in ruby is

$$g_0 = \sigma \left[n_2(R_1) - \frac{g(R_1) n_1}{g(n_1)} \right], \tag{2.6}$$

where σ is the stimulated emission cross section of the R_1 line and $n_2(R_1)$ is the population density of the E level.

Since $\sigma = \sigma_{12}g(n_1)/g(R_1)$, the gain coefficient can be expressed as

$$g_0 = \alpha_0 \left(\frac{2n_2}{n_{\text{tot}}} - 1 \right), \tag{2.7}$$

where $\alpha_0 = \sigma_{12}n_{\text{tot}}$ is the absorption coefficient of ruby. Ruby has an absorption cross section of $\sigma_{12} = 1.22 \times 10^{-20}$ cm^2 and an absorption coefficient of $\alpha_0 = 0.2$ cm^{-1}. With all the chromium ions in the ground state ($n_2 = 0$) the gain of the unpumped ruby crystal is $g_0 = -\alpha_0$. With all the ions in the upper levels ($n_2 = n_{\text{tot}}$) the maximum gain coefficent is $g_0 = \alpha_0$.

The example of ruby illustrates very clearly the drawbacks of a three-level system.

(a) Compared to a four-level laser, three-level lasers such as ruby and Er : glass, and quasi-three-level lasers such as Yb : YAG require high-pump intensities because a certain amount of pump power is required just to achieve transparency in the material.

(b) Since three- and quasi-three-level lasers absorb at the laser wavelength unless pumped to inversion, any portion of the laser crystal which is shielded from the pump light, for example, by the rod holder, will present a high absorption loss.

Despite these obvious drawbacks, three-level lasers have other redeeming features, such as desirable wavelength or a particular efficient pump method, which makes these lasers attractive. For example, ruby lases in the visible region, it has broad absorption bands at a wavelength region particularly well-suited for flash-lamp pumping, and it has a long fluorescent lifetime and a reasonably large stimulated emission cross section.

2.3   Nd : YAG

The Nd : YAG laser is by far the most commonly used type of solid-state laser. Neodymium-doped yttrium aluminum garnet (Nd : YAG) possesses a combination of properties uniquely favorable for laser operation. The YAG host is hard, of good optical quality, and has a high thermal conductivity. Furthermore, the cubic structure of YAG favors a narrow fluorescent linewidth, which results in high gain and low threshold for laser operation. In Nd : YAG, trivalent neodymium substitutes for trivalent yttrium, so charge compensation is not required.

Physical Properties

Pure $Y_3Al_5O_{12}$ is a colorless, optically isotropic crystal that possesses a cubic structure characteristic of garnets. In Nd : YAG about 1% of Y^{3+} is substituted by Nd^{3+}. The radii of the two rare earth ions differ by about 3%. Therefore, with the addition of large amounts of neodymium, strained crystals are obtained—indicating that either the solubility limit of neodymium is exceeded or that the

TABLE 2.2. Physical and optical properties of Nd : YAG.

Chemical formula	$Nd : Y_3Al_5O_{12}$
Weight % Nd	0.725
Atomic % Nd	1.0
Nd atoms/cm^3	1.38×10^{20}
Melting point	$1970°$ C
Knoop hardness	1215
Density	$4.56 \, g/cm^3$
Rupture stress	$1.3–2.6 \times 10^6 \, kg/cm^2$
Modulus of elasticity	$3 \times 10^6 \, kg/cm^2$
Thermal expansion coefficient	
[100] orientation	$8.2 \times 10^{-6} \, °C^{-1}, 0–250°$ C
[110] orientation	$7.7 \times 10^{-6} \, °C^{-1}, 10–250°$ C
[111] orientation	$7.8 \times 10^{-6} \, °C^{-1}, 0–250°$ C
Linewidth	120 GHz
Stimulated emission cross section	
$R_2 - Y_3$	$\sigma = 6.5 \times 10^{-19} \, cm^2$
$^4F_{3/2} - {}^4I_{11/2}$	$\sigma = 2.8 \times 10^{-19} \, cm^2$
Fluorescence lifetime	$230 \, \mu s$
Photon energy at $1.06 \, \mu m$	$h\nu = 1.86 \times 10^{-19} \, J$
Index of refraction	1.82 (at $1.0 \, \mu m$)

lattice of YAG is seriously distorted by the inclusion of neodymium. Some of the important physical properties of YAG are listed in Table 2.2 together with optical and laser parameters.

Laser Properties

The Nd : YAG laser is a four-level system as depicted by a simplified energy level diagram in Fig. 2.2. The laser transition, having a wavelength of 1064.1 nm, originates from the R_2 component of the $^4F_{3/2}$ level and terminates at the Y_3 component of the $^4I_{11/2}$ level. At room temperature only 40% of the $^4F_{3/2}$ population is at level R_2; the remaining 60% are at the lower sublevel R_1 according to Boltzmann's law. Lasing takes place only by R_2 ions whereby the R_2 level population is replenished from R_1 by thermal transitions. The ground level of Nd : YAG is the $^4I_{9/2}$ level. There are a number of relatively broad energy levels, which together may be viewed as comprising pump level 3. Of the main pump bands shown, the 0.81 and 0.75 μm bands are the strongest. The terminal laser level is 2111 cm^{-1} above the ground state, and thus the population is a factor of $\exp(\Delta E/kT) \approx \exp(-10)$ of the ground-state density. Since the terminal level is not populated thermally, the threshold condition is easy to obtain.

The upper laser level, $^4F_{3/2}$, has a fluorescence efficiency greater than 99.5% and a fluorescence lifetime of 230 μs. The branching ratio of emission from $^4F_{3/2}$ is as follows: $^4F_{3/2} \to {}^4I_{9/2} = 0.25$, $^4F_{3/2} \to {}^4I_{11/2} = 0.60$, $^4F_{3/2} \to {}^4I_{13/2} = 0.14$, and $^4F_{3/2} \to {}^4I_{15/2} < 0.01$. This means that almost all the ions transferred from the ground level to the pump bands end up at the upper laser level, and 60% of the ions at the upper laser level cause fluorescence output at the $^4I_{11/2}$ manifold.

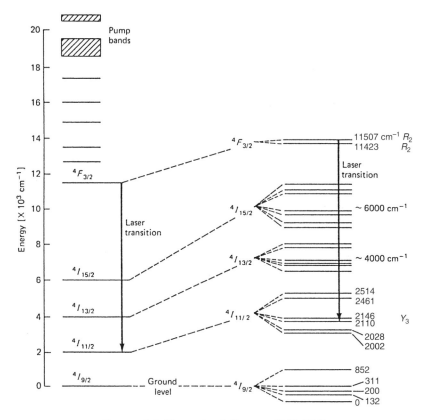

FIGURE 2.2. Energy level diagram of Nd:YAG.

At room temperature the main $1.06 \mu m$ line in Nd:YAG is homogeneously broadened by thermally activated lattice vibrations. The spectroscopic cross section for the individual transition between Stark sublevels has been measured to be $\sigma(R_2 - Y_3) = 6.5 \times 10^{-19} \, \text{cm}^2$. At a temperature of 295 K, the Maxwell–Boltzmann fraction in the upper Stark sublevel is 0.427, implying an effective cross section for Nd:YAG of $\sigma^*(^4F_{3/2} - ^4I_{11/2}) = 2.8 \times 10^{-19} \, \text{cm}^2$. The effective stimulated-emission cross section is the spectroscopic cross section times the occupancy of the upper laser level relative to the entire $^4F_{3/2}$ manifold population.

Figure 2.3 shows the fluorescence spectrum of Nd^{3+} in YAG near the region of the laser output with the corresponding energy levels for the various transitions. The absorption of Nd:YAG in the range 0.3 to $0.9 \mu m$ is given in Fig. 2.4(a). In Fig. 2.4(b) the absorption spectrum is expanded around the wavelength of 808 nm, which is important for laser diode pumping. Thermal properties of Nd:YAG are summarized in Table 2.3. Under normal operating conditions the Nd:YAG laser oscillates at room temperature on the strongest $^4F_{3/2} \rightarrow ^4I_{11/2}$ transition at $1.0641 \mu m$. It is possible, however, to obtain oscillation at other wavelengths by inserting etalons or dispersive prisms in the resonator, by utilizing a specially

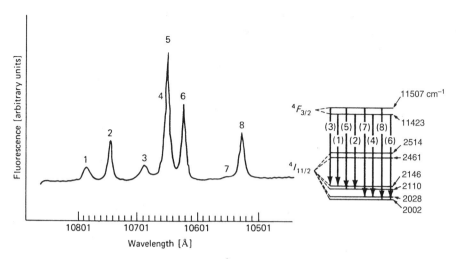

FIGURE 2.3. Fluorescence spectrum of Nd^{3+} in YAG at 300 K in the region of 1.06 μm [10], [11].

designed resonant reflector as an output mirror, or by employing highly selective dielectrically coated mirrors. These elements suppress laser oscillation at the 1.06 μm wavelength and provide optimum conditions at the wavelength desired. With this technique laser systems have been built which utilize the 946 and 1330 nm transitions.

2.4   Nd : Glass

There are a number of characteristics that distinguish glass from other solid-state laser host materials. Its properties are isotropic. It can be doped at very high concentrations with excellent uniformity, and it can be made in large pieces of diffraction-limited optical quality. In addition, glass lasers have been made, in a variety of shapes and sizes, from fibers a few micrometers in diameter to rods 2 m long and 7.5 cm in diameter and disks up to 90 cm in diameter and 5 cm thick.

The most common commercial optical glasses are oxide glasses, principally silicates and phosphates, that is, SiO_2 and P_2O_5 based. Table 2.4 summarizes some important physical and optical properties of commercially available silicate and phosphate glasses. The gain cross sections at 1053 nm of available phosphates range from 3.0 × 10^{-20} to 4.2 × 10^{-20} cm^2 and are generally larger than the 1064 nm cross sections of silicate glasses. Silicate and phosphate glasses have fluorescent decay times of around 300 μs at doping levels of 2 × 10^{20} Nd atoms/cm^3.

2.4.1   Laser Properties

There are two important differences between glass and crystal lasers. First, the thermal conductivity of glass is considerably lower than that of most crystal hosts.

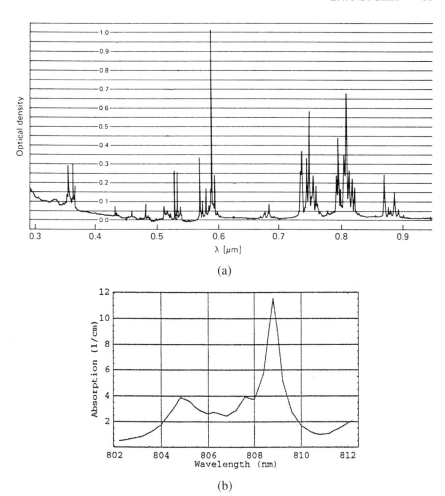

FIGURE 2.4. (a) Absorption spectrum of Nd : YAG from 0.3 to 0.9 μm and (b) expanded scale around 808 nm.

TABLE 2.3. Thermal properties of Nd : YAG.

Property	Units	300 K	200 K	100 K
Thermal conductivity	W cm^{-1} K^{-1}	0.14	0.21	0.58
Specific heat	W s g^{-1} K^{-1}	0.59	0.43	0.13
Thermal diffusivity	cm^2 s^{-1}	0.046	0.10	0.92
Thermal expansion	K^{-1} $\times$ 10^{-6}	7.5	5.8	4.25
$\partial n/\partial T$	K^{-1}	7.3 $\times$ 10^{-6}	—	—

TABLE 2.4. Physical and optical properties of Nd-doped glasses.

Glass Type	Q–246	Q–88	LHG–5	LHG–8	LG–670	LG–760
	Silicate	Phosphate	Phosphate	Phosphate	Silicate	Phosphate
Spectroscopic properties	(Kigre)	(Kigre)	(Hoya)	(Hoya)	(Schott)	(Schott)
Peak wavelength [nm]	1062	1054	1054	1054	1061	1054
Cross section [$\times 10^{-20}$ cm^2]	2.9	4.0	4.1	4.2	2.7	4.3
Fluorescence lifetime [μs]	340	330	290	315	330	330
Linewidth FWHM [nm]	27.7	21.9	18.6	20.1	27.8	19.5
Density [gm/cm^3]	2.55	2.71	2.68	2.83	2.54	2.60
Index of refraction [Nd]	1.568	1.545	1.539	1.528	1.561	1.503
Nonlinear coefficient						
[10^{-16} cm^2/W]	3.74	2.98	3.48	3.10	3.78	2.90
dn/dt (20°–40° C [10^{-6}/° C]	2.9	−0.5	8.6	−5.3	2.9	−6.8
Thermal coefficient of optical						
path (20°–40° C) [10^{-6}/° C]	+8.0	+2.7	+4.6	+0.6	8.0	—
Transformation point [° C]	518	367	455	485	468	—
Thermal expansion coefficient						
(20°–40° [10^{-7}/° C]	90	104	86	127	92.6	138
Thermal conductivity						
[W/m° C]	1.30	0.84	1.19	—	1.35	0.67
Specific heat [J/g° C]	0.93	0.81	0.71	0.75	0.92	0.57
Knoop hardness	600	418	497	321	497	—
Young's modulus [kg/mm^2]	8570	7123	6910	5109	6249	—
Poisson's ratio	0.24	0.24	0.237	0.258	0.24	0.27

Second, the emission lines of ions in glasses are inherently broader than in crystals. A wider line increases the laser threshold value of amplification. Nevertheless, this broadening has an advantage. A broader line offers the possibility of obtaining and amplifying shorter light pulses and, in addition, it permits the storage of larger amounts of energy in the amplifying medium for the same linear amplification coefficient. Thus, glass and crystalline lasers complement each other. For continuous or very high repetition-rate operation, crystalline materials provide higher gain and greater thermal conductivity. Glasses are more suitable for high-energy pulsed operation because of their large size, flexibility in their physical parameters, and the broadened fluorescent line.

Unlike many crystals, the concentration of the active ions can be very high in glass. The practical limit is determined by the fact that the fluorescence lifetime, and therefore the efficiency of stimulated emission, decreases with higher concentrations. In silicate glass, this decrease becomes noticeable at a concentration of 5% Nd_2O_3.

Figure 2.5 shows a simplified energy level diagram of Nd : glass. The Nd^{3+} ion in glass represents a four-level system. The upper laser level indicated in Fig. 2.5 is the lower-lying component of the $^4F_{3/2}$ pair with a several hundred microsecond spontaneous emission lifetime. The terminal laser level is the lower-lying level of the pair in the $^4I_{11/2}$ multiplet. The $^4I_{11/2}$ group empties spontaneously by a radiationless phonon transition to the $^4I_{9/2}$ ground state about 1950 cm^{-1} below.

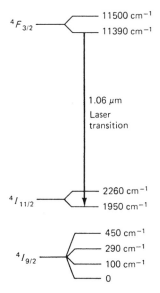

$^4F_{3/2}$ — 11500 cm^{-1}
— 11390 cm^{-1}

1.06 μm
Laser
transition

— 2260 cm^{-1}
$^4I_{11/2}$ — 1950 cm^{-1}

— 450 cm^{-1}
— 290 cm^{-1}
$^4I_{9/2}$ — 100 cm^{-1}
— 0

FIGURE 2.5. Partial energy level diagram of Nd^{3+} in glass.

Owing to the large separation of the terminal laser level from the ground state, even at elevated temperatures there is no significant terminal-state population and, therefore, no degradation of laser performance. In addition, the fluorescent linewidth of the neodymium ion in glass is quite insensitive to temperature variation; only a 10% reduction is observed in going from room temperature to liquid nitrogen temperature. As a result of these two characteristics, it is possible to operate a neodymium-doped glass laser with little change in performance over a temperature range of $-100°$ to $+100°$ C.

Figure 2.6 shows the pump bands of Nd : glass. In comparing Fig. 2.6 with Fig. 2.4, one notes that the locations of the absorption peaks in Nd : YAG and Nd : glass are about the same; however, in Nd : glass the peaks are much wider and have less fine a structure compared to Nd : YAG.

2.5   Nd : YLF

During the last 10 years, the crystal quality of the scheelite-structured host lithium yttrium fluoride (YLF) has been dramatically improved, and the material has gained a firm foothold for a number of applications.

The material has advantages for diode laser pumping since the fluorescence lifetime in Nd : YLF is twice as long as in Nd : YAG. Laser diodes are power limited; therefore, a larger pump time afforded by the longer fluorescence time provides for twice the energy storage from the same number of diode lasers.

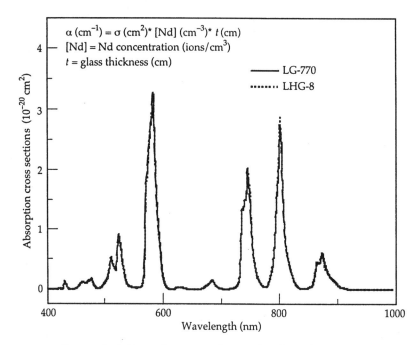

FIGURE 2.6. Absorption spectra for two phosphate glasses [12].

Figure 2.7 shows a simplified energy level diagram of Nd : YLF. Depending on the polarization, two lines each are obtained around 1.05 and 1.3 μm. For example, with an intracavity polarizer one can select either the 1047 nm (extraordinary) or 1053 nm (ordinary) transition. The same can be done for the two 1.3 μm transitions; however, in addition lasing at the 1.05 μm lines has to be suppressed. All lines originate on the same Stark split $^4F_{3/2}$ upper level.

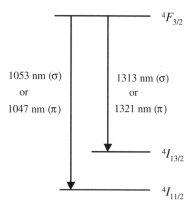

FIGURE 2.7. Simplified energy level diagram of Nd : YLF.

TABLE 2.5. Properties of Nd-doped lithium
yttrium fluoride (YLF).

Lasing wavelength [nm]	1053 (σ)
	1047 (π)
Index of refraction	$n_0 = 1.4481$
$\lambda = 1.06\,\mu$m	$n_e = 1.4704$
Fluorescent lifetime	$480\,\mu$s
Stimulated emission	1.8×10^{-19} (π)
Cross section [cm^2]	1.2×10^{-19} (σ)
Density [g/cm^3]	3.99 (undoped)
Hardness [Mohs]	4–5
Elastic modulus [N/m^2]	7.5×10^{10}
Strength [N/m^2]	3.3×10^7
Poisson's ratio	0.33
Thermal conductivity [W/cm-K]	0.06
Thermal expansion coefficient [$^\circ$ C^{-1}]	a-axis: 13×10^{-6}
	c-axis: 8×10^{-6}
Melting point [$^\circ$ C]	825

Material properties of Nd : YLF are listed in Table 2.5. The relatively large thermal conductivity allows efficient heat extraction, and its natural birefringence overwhelms thermally induced birefringence eliminating the thermal depolarization problems of optically isotropic hosts like YAG. The cross section for YLF is about a factor of 2 lower than YAG. For certain lasers requiring moderate Q-switch energies, the lower gain offers advantages in system architecture compared to the higher gain material Nd : YAG.

The energy storage in Q-switched operation of neodymium oscillators and amplifiers is constrained by the onset of parasitic oscillations. To first order, the energy storage limit of two materials is inversely proportional to the ratio of the stimulated emission cross section. Therefore, higher storage densities are obtained in the lower cross section material Nd : YLF as compared to Nd : YAG.

2.6   Nd : YVO$_4$

Neodymium-doped yttrium vanadate has several spectroscopic properties that are particularly relevant to laser diode pumping. The two outstanding features are a large stimulated emission cross section that is five times higher than Nd : YAG, and a strong broadband absorption at 809 nm.

The vanadate crystal is naturally birefringent and laser output is linearly polarized along the extraordinary π-direction. The polarized output has the advantage that it avoids undesirable thermally induced birefringence.

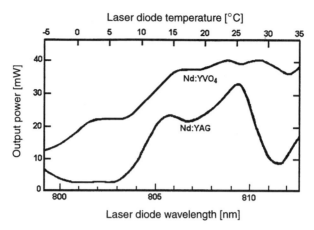

FIGURE 2.8. Output from a Nd: YVO$_4$ and Nd: YAG laser as a function of diode pump temperature and wavelength [13].

Pump absorption in this uniaxial crystal is also polarization dependent. The strongest absorption occurs for pump light polarized in the same direction as the laser radiation. The absorption coefficient is about four times higher compared to Nd: YAG in the π-direction. The sublevels at the $^4F_{5/2}$ pump band are more resolved in Nd: YAG, whereas in Nd: YVO$_4$ Stark splitting is smaller and the multiple transitions are more compacted. The result is a broader and less spiky absorption profile of Nd: YVO$_4$ compared to Nd: YAG around the pump wavelength of 809 nm. Figure 2.8 gives an indication of the broader and smoother absorption profile of this material as compared to Nd: YAG. From this data it also follows that Nd: YVO$_4$ laser performance is more tolerant to diode temperature variations because of the large pump bandwidth. If one defines this bandwidth as the wavelength range where at least 75% of the pump radiation is absorbed in a 5 mm thick crystal, then one obtains for Nd: YVO$_4$ a value of 15.7 nm, and 2.5 nm for Nd: YAG.

Table 2.6 summarizes important material parameters for Nd: YVO$_4$. The material does have several drawbacks, the principal one being a shorter excited state lifetime than Nd: YAG. As expressed in (3.65) the pump input power to reach

TABLE 2.6. Material parameters for Nd : YVO$_4$.

Laser cross section	15.6×10^{-19} cm^2
Laser wavelength	1064.3 nm
Linewidth	0.8 nm
Fluorescence lifetime	100 μs
Peak pump wavelength	808.5 nm
Peak absorption coefficient	37 (π polarization)
at 808 nm [cm^{-1}]	10 (σ polarization)
Nd doping	1% (atomic Nd)

threshold for cw operation depends on the product of $\sigma \tau_f$. Therefore, the large cross section σ of Nd : YVO$_4$ is partially offset by its shorter fluorescence lifetime τ_f. Fluorescence lifetime is also a measure of the energy storage capability in Q-switched operation. Large energy storage requires a long fluorescence lifetime. As far as thermal conductivity is concerned, it is only half as high as Nd : YAG and somewhat lower than Nd : YLF.

The properties of Nd : YVO$_4$ can best be exploited in an end-pumped configuration, and a number of commercial laser systems are based on fiber-coupled diode arrays pumping a small vanadate crystal. Actually, Nd : YVO$_4$ is the material of choice for cw end-pumped lasers in the 5 W output region. These systems are often also internally frequency doubled to provide output at 532 nm. In end-pumped systems the pump beam is usually highly focused, and it is difficult to maintain a small beam waist over a distance of more than a few millimeters. In this case a material such as Nd : YVO$_4$, which has a high absorption coefficient combined with high gain, is very advantageous.

2.7 Er : Glass

Erbium has been made to lase at 1.54 μm in both silicate and phosphate glasses. Owing to the three-level behavior of erbium, and the weak absorption of pump radiation, codoping with other rare earth ions is necessary to obtain satisfactory system efficiency. In fact, erbium must be sensitized with ytterbium if the material is to lase at all at room temperature. Pump radiation is absorbed by ytterbium, which has an absorption band between 0.9 and 1 μm, and transferred to the erbium ions. This transfer is illustrated in Fig. 2.9. In the co-doped erbium, Yb : glass radiation from a flashlamp or laser diode pumps the strong $^2F_{7/2} - {}^2F_{5/2}$ transition in Yb^{3+}. Because of the good overlap between the upper states of Yb^{3+} and Er^{3+}, the excited Yb^{3+} ions transfer energy to the $^4I_{11/2}$ level of Er^{3+}. The erbium ions then relax to the upper laser level $^4I_{13/2}$ and lasing occurs down to the $^4I_{15/2}$ man-

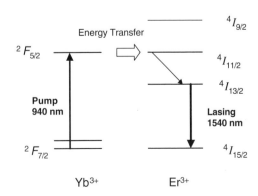

FIGURE 2.9. Energy level diagram in codoped Er, Yb : glass.

ifold. For flashlamp pumping, the efficiency of Er : glass can be further improved by adding Cr^{3+} in addition to Yb^{3+}. The Cr^{3+} ions absorb flashlamp radiation in two broad bands centered at 450 and 640 μm, and emit radiation in a broad band centered at 760 nm. This allows energy to be transferred from Cr^{3+} to Yb^{3+} and Er^{3+}. With the development of strained layer InGaAs laser diodes, which have emission between 0.9 and 1 μm, it is now possible to pump the strong Yb^{3+} transition $^2F_{7/2} - {}^2F_{5/2}$. Diode pumping at 940 nm has dramatically improved the efficiency and average power achievable from Er : glass lasers.

2.8   Yb : YAG

Although Yb : YAG has been known for decades, interest was not very high in this material since it lacks any pump bands in the visible spectrum. The crystal has only a single absorption feature around 942 nm. With conventional pump sources such as flashlamps, the threshold is very high, which eliminated this material from any serious consideration. With the emergence of powerful InGaAs laser diodes that emit at 942 nm, the picture has changed completely and Yb : YAG has become an important laser material [15]. The relevant energy level diagram of Yb : YAG is very simple and consists of the $^2F_{7/2}$ lower level and $^2F_{5/2}$ excited state manifolds separated by about 10,000 cm^{-1}. The laser wavelength is at 1.03 μm, a representative emission spectrum for a 5.5 at.% doped sample is shown in Fig. 2.10.

The laser transition $^2F_{5/2}$ to $^2F_{7/2}$ has a terminal level 612 cm^{-1} above the ground state (see Fig. 2.11). The thermal energy at room temperature is 200 cm^{-1}; therefore the terminal state is thermally populated making Yb : YAG a quasi-three-level system. By comparison, the terminal laser level in Nd : YAG is about 2000 cm^{-1} above the ground state. Being a quasi-three-level laser, Yb : YAG ab-

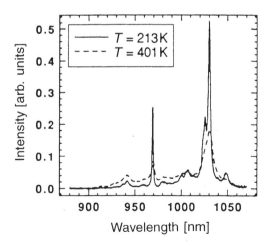

FIGURE 2.10. Emission spectra of Yb : YAG with 5.5 at.% doping [14].

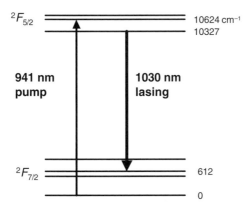

FIGURE 2.11. Relevant energy levels in Yb : YAG.

sorbs at 1.03 μm unless it is pumped to inversion. At room temperature the thermal population of the lower laser level is about 5.5%. The absorbed pump power per volume needed to achieve and maintain transparency at the laser wavelength is $I = f_a n_t h \nu_p / \tau_f$, where f_a is the fraction of the total ion density n_t occupying the lower laser level, $h \nu_p$ is the energy per pump photon, and τ_f is the lifetime of the upper level. With $f_a = 0.055$, $n_t = 1.38 \times 10^{20}$ cm^{-3} at 1% doping, $h \nu_p = 2.11 \times 10^{-19}$ J, and $\tau_f = 0.95$ ms, the absorbed pump power needed to reach inversion is 1.7 kW/cm^3. Of course, a higher power density is required to overcome optical losses and reach laser threshold, and for an efficient operation the laser has to be pumped about five to six times above threshold. Typically, in this laser, small volumes of material are pumped on the order of 10 kW/cm^3.

Yb : YAG performance is strongly dependent on temperature and can be improved by cooling the crystal, which reduces the thermal population and increases the stimulated emission cross section from 2.1×10^{-20} cm^2 at room temperature to twice this value at 220 K.

In most of the large systems built to date the temperature of the crystal is maintained between $-10°$ C and $20°$ C depending on the particular design and mode of operation. To maintain the laser crystal at such a temperature under intensive pump radiation requires, as a minimum, some kind of refrigeration.

Despite these obvious drawbacks, Yb : YAG has a number of redeeming features that motivated the development of very powerful systems.

Pumping of Yb : YAG with an InGaAs pump source produces the smallest amount of crystal heating compared to any other major laser system. Actually the pump radiation in this material generates only about one-third of the heat compared to Nd : YAG. The fractional thermal loading, that is, the ratio of heat generated to absorbed energy, is around 11% for Yb : YAG pumped at 943 nm and 32% for Nd : YAG pumped at 808 nm. This substantially reduced thermal dissipation is the result of a very small energy difference between the photons of the pump and laser radiation. This quantum defect or Stokes shift is 9% in Yb : YAG versus 24% in Nd : YAG. The thermal load generated in a laser medium is of

TABLE 2.7. Pertinent material parameters for Yb : YAG.

Laser wavelength	1030 nm
Radiative lifetime at room temperature	951 μs
Peak emision cross section	2.1×10^{-20} cm^2
Peak absorption wavelength	942 nm
Pump bandwidth at 942 nm	18 nm
Doping density (1% at.)	1.38×10^{20} cm^3

primary concern for high-power applications. The reduced thermal heat load can potentially lead to higher-average-power systems with better beam quality than possible with Nd : YAG. Other unique advantages of Yb : YAG are an absorption bandwidth of 18 nm for diode pumping, which is about 10 times broader than the 808 nm absorption in Nd : YAG. This significantly relaxes temperature control needed for the diode pump source. Yb : YAG has a long lifetime of 951 μs, which reduces the number of quasi-cw pump diodes required for a given energy per pulse output. As will be discussed in Chapter 6, since laser diodes are power limited, a long fluorescence time permits the application of a long pump pulse that, in turn, generates a high-energy output. Important materials parameters of Yb : YAG are listed in Table 2.7.

The presence of only one excited-state manifold eliminates problems associated with excited-state absorption and up-conversion processes. Despite the quasi-three-level behavior of Yb : YAG, the obvious advantages of very low fractional heating, broad absorption at the InGaAs wavelength, long lifetime combined with high conductivity, and tensile strength of the host crystal have raised the prospect of high-power generation with good beam quality.

The major interest lies in large systems at the kilowatt level for industrial applications.

2.9   Alexandrite

Alexandrite (BeAl$_2$O$_4$: Cr^{3+}) is the common name for chromium-doped chrysoberyl. The crystal is grown in large boules by the Czochralski method much like ruby and YAG. Alexandrite was the first tunable laser considered for practical use. Alexandrite, and Ti : sapphire, discussed in the next section, belong to the family of vibronic lasers. These lasers derive their name from the fact that laser transitions take place between coupled vibrational and electronic states. In this case, the emission of a photon is accompanied by emission of a phonon. Tunability in alexandrite is achieved by the band of vibrational levels indicated in Fig. 2.12, which are the result of a strong coupling between the Cr^{3+} ion and the lattic vibrations.

Alexandrite lases at room temperature with flashlamp pumping throughout the range 701 to 818 nm. The alexandrite absorption bands are very similar to those of ruby and span the region from about 380 to 630 nm with peaks occurring at

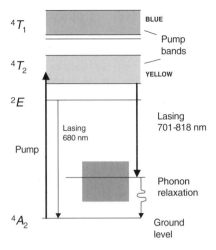

FIGURE 2.12. Simplified energy level diagram for chromium ions in alexandrite.

410 and 590 nm. Figure 2.13 shows the absorption bands of alexandrite. The laser gain cross section increases from 7×10^{-21} cm^2 at 300 K to 2×10^{-20} cm^2 at 475 K, which results in improved laser performance at elevated temperature.

The basic physics of the four-level alexandrite laser can be discussed with reference to the energy level diagram (Fig. 2.12). The $^4A_2$ level is the ground state, and $^4T_2$ is the absorption state continuum. Vibronic lasing is due to emission from the $^4T_2$ state to excited vibronic states within $^4A_2$. Subsequent phonon emission returns the system to equilibrium. The laser wavelength depends on which vibrationally excited terminal level acts as the transition terminus; any energy not released by the laser photon will then be carried off by a vibrational phonon, leaving the chromium ion at its ground state.

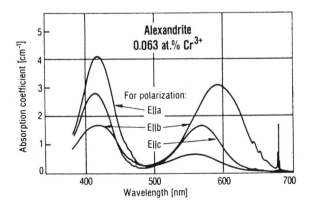

FIGURE 2.13. Absorption spectrum of alexandrite [Allied Corp. Data Sheet].

Alexandrite can lase both as a four-level vibronic laser and as a three-level system analogous to ruby. In the latter mode of operation the laser transition is from the ^2E state, which is coupled to $^4T_2$, down to the ground state $^4A_2$. As a three-level laser, alexandrite has a high threshold, fixed output wavelength (680.4 nm at room temperature), and relatively low efficiency. Obviously, the primary interest of alexandrite lies in its vibronic nature.

2.10   Ti : Sapphire

Since laser action was first reported, the Ti : Al_2O_3 laser has been the subject of extensive investigations and today it is the most widely used tunable solid-state laser. The Ti : sapphire laser combines a broad tuning range of about 400 nm with a relatively large gain cross section that is half of Nd : YAG at the peak of its tuning range. The energy level structure of the Ti^{3+} ion is unique among transition-metal laser ions in that there are no d state energy levels above the upper laser level. The simple energy-level structure ($3d^1$ configuration) eliminates the possibility of excited-state absorption of the laser radiation, an effect which has limited the tuning range and reduced the efficiency of other transition-metal-doped lasers.

In this material, a Ti^{3+} ion is substituted for an Al^{3+} ion in Al_2O_3. Laser crystals, grown by the Czochralski method, consist of sapphire doped with 0.1% Ti^{3+}

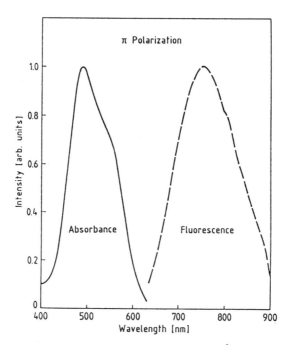

FIGURE 2.14. Absorption and fluorescence spectra of the Ti^{3+} ion in Al_2O_3 (sapphire) [16].

TABLE 2.8. Laser parameters of Ti : Al_2O_3.

Index of refraction	1.76
Fluorescent lifetime	$3.2 \mu s$
Fluorescent linewidth (FWHM)	180 nm
Peak emission wavelength	790 nm
Peak stimulated emission cross section	
parallel to c-axis	$\sigma_\parallel \sim 4.1 \times 10^{-19} cm^2$
perpendicular to c-axis	$\sigma_\perp \sim 2.0 \times 10^{-19} cm^2$
Stimulated emission cross section	
at $0.795 \mu m$ (parallel to c-axis)	$\sigma_\parallel = 2.8 \times 10^{-19} cm^2$
Quantum efficiency of	
converting a $0.53 \mu m$ pump photon	
into an inverted site	$\eta_Q \approx 1$
Saturation fluence at $0.795 \mu m$	$E_s = 0.9 \, J/cm^2$

by weight. Crystals of Ti : Al_2O_3 exhibit a broad absorption band, located in the blue–green region of the visible spectrum with a peak around 490 nm. The great interest in this material arises from the broad vibronic fluorescence band that allows tunable laser output between 670–1070 nm, with the peak of the gain curve around 800 nm.

The absorption and fluorescence spectra for Ti : Al_2O_3 are shown in Fig. 2.14. The broad, widely separated absorption and fluorescence bands are caused by the strong coupling between the ion and host lattice and are the key to broadly tunable laser operation. The laser parameters of Ti : Al_2O_3 are listed in Table 2.8.

Besides having favorable spectroscopic and lasing properties, one other advantage of Ti : Al_2O_3 is the material properties of the sapphire host itself, namely very high thermal conductivity, exceptional chemical inertness, and mechanical rigidity. Ti : sapphire lasers have been pumped with a number of sources such as argon and copper vapor lasers, frequency doubled Nd : YAG and Nd : YLF lasers, as well as flashlamps. Flashlamp pumping is very difficult to achieve in Ti : sapphire because a very high pump flux is required. The reason for that is the short fluorescence lifetime of $3.2 \mu m$, which results in a small product of stimulated emission cross section times fluorescence lifetime ($\sigma \tau_f$). The population inversion in a laser required to achieve threshold is inversely proportional to $\sigma \tau_f$, as will be shown in Section 3.1.

Commercial Ti : sapphire lasers are pumped by argon lasers or frequency-doubled Nd : YAG or Nd : YLF lasers. Figure 2.15 displays the output of a Ti : sapphire laser pumped at 1 kHz by a frequency-doubled Nd : YLF laser which had an average output of 1.7 W at 527 nm. A very important application of Ti : sapphire lasers is the generation and amplification of femtosecond mode-locked pulses. Kerr lens mode-locking and chirped pulse amplification with Ti : sapphire lasers is discussed in Chapter 9.

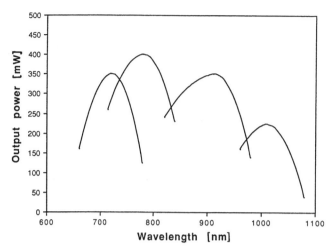

FIGURE 2.15. Tuning range of a Ti : sapphire laser pumped by a Nd : YLF laser at 1 kHz [17].

Summary

In this chapter an overview is given of the laser-related properties of rare earth, transition metal, and actinide ions embedded in solid hosts. The characteristics of the most important and highly developed laser materials are discussed in detail. The active element of a solid-state laser is a crystalline or glass host doped with a relatively small percentage of ions either from the rare earth, transition metals, or actinide group of the periodic table. In terms of the density of active ions, solid-state laser materials are intermediate to gaseous and semiconductor lasers. Laser action has been produced in many hundreds of active ion-host combinations. However, only a small fraction of laser materials have been found to be useful in actual laser systems.

Besides having pump bands that overlap the output spectra of flash lamps or diode lasers, desirable features of laser ions are a large stimulated emission cross section, long lifetime, and high quantum efficiency. A large stimulated emission cross section leads to high gain and low threshold operation, a long lifetime determines whether the laser can be operated cw or must be pulsed, and the quantum efficiency is a measure of the competition between radiative and nonradiative decay. The host material must possess favorable optical, thermal, and mechanical properties and lend itself to an economical growth process.

Important active ions include trivalent Nd, Er, and Yb from the rare earth group and Cr and Ti from the transition metals. Since the $^4F_{3/2} - {}^4I_{11/2}$ transition of Nd^{3+} has the largest cross section of all active ions, it has exhibited laser operation in the largest number of host crystals and glasses. Optical pumping of the $^4F_{3/2}$ level is efficient because of the large number of closely spaced levels at higher energy levels that creates absorption in the visible and near infrared.

As a host, YAG has particular favorable optical, thermal, and mechanical properties; this makes Nd:YAG the most widely used ion-host combination for solid state lasers. The principal advantages offered by a glass over a crystalline host are the flexibility in size and shape and the excellent optical quality. The major disadvantage of glass is a low thermal conductivity.

Erbium-doped glass is attractive because the output wavelength of this material is of interest for fiberoptic amplifiers and oscillators and for military applications of lasers. Compared to other materials, ytterbium-doped YAG crystals have a much smaller energy difference between pump and laser wavelength.This results in reduced heating of the crystals and makes Yb : YAG attractive for high power applications. Ti : sapphire and alexandrite are important tunable lasers in the near infrared. Nd : YVO$_4$ has a large stimulated emission cross section which makes it a low-threshold and high-gain laser. The longer lifetime of Nd : YLF as compared to Nd : YAG is beneficial for diode pumping since diode lasers are power limited and the longer pump pulse duration afforded by the larger upper state lifetime increases the energy storage capability of the material.

References

[1] A.L. Schawlow and C.H. Townes: *Phys. Rev.* **112**, 1940 (1958).

[2] T.H. Maiman: *Nature* **187**, 493 (1960).

[3] J.E. Geusic, H.M. Marcos, and L.G. Van Uitert: *Appl. Phys. Lett.* **4**, 182 (1964).

[4] S.E. Stokowski: Glass lasers, in *Handbook of Laser Science and Technology* (edited by M.J. Weber) (CRC Press, Boca Raton, FL), 1982, pp. 215–264.

[5] P.F. Moulton: Paramagnetic ion lasers, in *Handbook of Laser Science and Technology* (edited by M.J. Weber) (CRC Press, Boca Raton, FL), 1986, Vol. 1, pp. 21–295.

[6] L.G. DeShazer, S.C. Rund, and B.A. Wechsler: Laser crystals, in *Handbook of Laser Science and Technology* (edited by M.J. Weber) (CRC Press, Boca Raton, FL), 1987, Vol. 5, pp. 281–338.

[7] J.C. Walling: Tunable paramagnetic-ion solid-state lasers, in *Tunable Lasers*, 2nd edition (edited by L.F. Mollenauer, J.C. White, C.R. Pollock), *Topics Appl. Phys.*, Vol. 59 (Springer, Berlin), 1992, Chap. 9.

[8] R.C. Powell: *Physics of Solid-State Laser Materials* (Springer, New York), 1998.

[9] Special issue on solid-state lasers, IEEE J. Quantum Electron. **QE-24** (1988).

[10] A.A. Kaminskii: *Crystalline Lasers* (CRC Press, Boca Raton, FL), 1996.

[11] H.G. Danielmeyer and M. Blätte: *Appl. Phys.* **1**, 269 (1973).

[12] J.H. Campbell, T. Suratwala, C. Thorsness, P. Ehrmann, W. Steele, and M. McLean: *ICF Quarterly Report*, Jan–March 1999, vol. 9, p. 111. Lawrence Livermore Laboratory.

[13] R.A. Fields, M. Birnbaum, and C.L. Fincher: *Appl. Phys. Lett.* **51**, 1885 (1987).

[14] D.S. Sumida and T.Y. Fan: *Proc. Advanced Solid State Lasers* **20**, 100 (Opt. Soc. Am., Washington, DC), 1994. Also *Opt. Lett.* **20**, 2384 (1995).

[15] W.F. Krupke, *IEEE J. Select. Topics Quantum Electron.* **QE-6**, 1287 (2000).

[16] P.F. Moulton: *J. Opt. Soc. Am. B* **3**, 125 (1986). Also *Laser Focus* **14**, 83 (1983).

[17] R. Rao, G. Vaillancourt, H.S. Kwok, and C.P. Khattak: *OSA Proc. on Tunable Solid State Lasers*, Vol. 5, North Falmouth, MA, 1989, p. 39.

Exercises

1. Ruby is a three-level laser. This means that it can absorb its own laser emission. When designing a ruby laser you must be careful to minimize the unpumped parts of the gain medium to minimize this type of loss. Using the data given in the text, calculate the peak absorption coefficient of unpumped ruby at its room temperature lasing frequency. (See (1.36) and (1.37) and the data in Section 2.2.)

2. Why were some early ruby lasers cooled to liquid nitrogen temperature (~ 77 K)?

3. The absorption spectrum of most Nd^{3+}-doped materials is made up of narrow line features. To better utilize the light produced by flashlamp pump sources some materials are co-doped with Cr^{3+} that has broad and strong visible absorption features. When the transfer of energy is efficient, as it is in gadolinium scandium gallium garnet (Nd, Cr : GSGG) such co-doping can be very useful. The energy level diagrams of a system with two co-dopants such that energy absorbed by Species A (the donor species) is transferred to Species B (the acceptor species) so that Species B can lase as indicated is shown in the figure. Energy transfer depends on the populations in both the donor and the acceptor levels involved. Write the rate equations for this process. Explain your terms and be sure you do not violate any laws of physics.

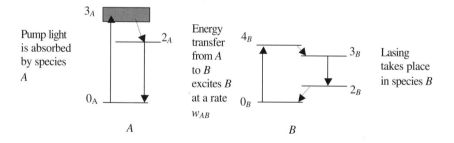

4. The R_1 laser transition of ruby has a good approximation to a Lorentzian shape with full width at half-maximum (FWHM) of 330 GHz at room temperature. The measured peak transition cross section is $\sigma_e = 2.5 \times 10^{-20}$ cm^2. If the refractive index of ruby is 1.76, what is he radiative lifetime of this transition? Since the observed lifetime is 3 ms, what is the nonradiative lifetime? What is the fluorescence quantum efficiency?

5. The Nd : YAG laser transition at $1.064\,\mu$m has a Lorentzian shape with FWHM of 195 GHz at room temperature. The upper state lifetime is $237\,\mu$s, the fluorescence quantum efficiency is 0.42, and the refractive index of Nd : YAG is 1.83. What is the peak laser transition cross section?

6. Given that the five lowest levels of a laser system have relative energies of 0, 134, 197, 311, and $848\,\text{cm}^{-1}$, what percentage of the population lies in each of these levels at room temperature (300 K)? (These levels actually define the Stark splitting of the ground state of the Nd : YAG laser.)

3

Laser Oscillator

3.1 Operation at Threshold
3.2 Gain Saturation
3.3 Circulating Power
3.4 Oscillator Performance Model
3.5 Relaxation Oscillations
3.6 Examples of Laser Oscillators
    References
    Exercises

In Chapter 1 we studied the processes which lead to optical amplification in laser materials. The regenerative laser oscillator is essentially a combination of two basic components: an optical amplifier and an optical resonator. The optical resonator, comprised of two opposing plane-parallel or curved mirrors at right angles to the axis of the active material, performs the function of a highly selective feedback element by coupling back in phase a portion of the signal emerging from the amplifying medium.

Figure 3.1 shows the basic elements of a laser oscillator. The pump lamp inverts the electron population in the laser material, leading to energy storage in the upper laser level. If this energy is released to the optical beam by stimulated emission, amplification takes place. Having been triggered by some spontaneous radiation emitted along the axis of the laser, the system starts to oscillate if the feedback is sufficiently large to compensate for the internal losses of the system. The amount of feedback is determined by the reflectivity of the mirrors. Lowering the reflectivity of the mirrors is equivalent to decreasing the feedback factor. The mirror at the output end of the laser must be partially transparent for a fraction of the radiation to "leak out" or emerge from the oscillator.

An optical structure composed of two plane-parallel mirrors is called a Fabry–Pérot resonator. In Chapter 5 we will discuss the temporal and spatial mode structures which can exist in such a resonator. For the purpose of this discussion it is sufficient to know that the role of the resonator is to maintain an electromagnetic field configuration whose losses are replenished by the amplifying

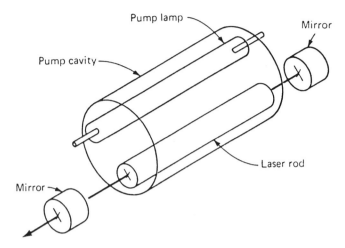

FIGURE 3.1. Major components of an optically pumped solid-state laser oscillator.

medium through induced emission. Thus, the resonator defines the spectral, directional, and spatial characteristics of the laser radiation, and the amplifying medium serves as the energy source.

In this chapter we will develop an analytical model of a laser oscillator that is based mainly on laser system parameters. A number of technical terms, which are briefly explained here, will be used. The arrangement of flashlamp and laser rod shown in Fig. 3.1 is referred to as side-pumped or transverse-pumped because the pump radiation strikes the laser rod from the side with respect to the direction of propagation of the laser radiation. The flashlamp is often water-cooled by inserting it into a flowtube or cooling jacket. Water or other suitable coolant is pumped through the annulus between the lamp envelope and the flowtube inner diameter. The pump cavity can be elliptical in cross section or closely wrapped around the lamp and laser rod. The latter design of a pump cavity is called close-coupled.

Instead of a flashlamp as shown in Fig. 3.1, laser diode arrays can be placed along the length of the laser rod. A reflective pump cavity of the type employed for flashlamps is not necessary since the radiation output from laser diodes is directional. However, in some cases focusing optics is inserted between the laser diodes and the laser crystal to shape and/or concentrate the pump beam. Since the output from laser diodes can be collimated and focused, contrary to flashlamp radiation, the pump radiation can also be introduced through the back mirror shown in Fig. 3.1. In this so-called end-pumped or longitudinal pump scheme, pump radiation and laser radiation propagate in the same direction.

The structure of the two mirrors reflecting the laser radiation is generally described as the laser resonator. However, in the literature, as well as in this book, the term cavity or laser cavity is also used. This is in contrast to the pump cavity which is the enclosure employed to contain the flashlamp radiation. The pump head is the structure and mechanical assembly that contains the laser crystal, pump source, cooling channels, and electrical connections.

3.1   Operation at Threshold

We will calculate the threshold condition of a laser oscillator composed of two mirrors having the reflectivities R_1 and R_2 and an active material of length l. We assume a gain coefficient per unit length of g in the inverted laser material. In each passage through the material the intensity gains by a factor of $\exp(gl)$. At each reflection a fraction $1 - R_1$ or $1 - R_2$ of the energy is lost. Starting at one point, the radiation will suffer two reflections before it can pass the same point in the original direction. The threshold condition is established by requiring that the photon density—after the radiation has traversed the laser material, been reflected by mirror with R_1, and returned through the material to be reflected by mirror with R_2—be equal to the initial photon density. Then on every complete two-way passage of the radiation through the laser the loss will just equal the gain. We can express the threshold condition by

$$R_1 R_2 \exp(2gl) = 1. \tag{3.1}$$

The regenerative amplifier becomes unstable when the amplification per transit exceeds the losses. In this case oscillations will build up, starting from a small disturbance. Clearly, if the round-trip gain

$$G = R_1 R_2 \exp(2gl) \tag{3.2}$$

is larger than 1, radiation of the proper frequency will build up rapidly until it becomes so large that the stimulated transitions will deplete the upper level and reduce the value of g. The condition of steady state is reached if the gain per pass exactly balances the internal and external losses. This process, called gain saturation, will be discussed in Section 3.2. In an oscillator a number of loss mechanisms are responsible for attenuating the beam; the most important ones are reflection, scattering, and absorption losses caused by the various optical components. It is convenient to lump all losses which are proportional to the length of the gain medium, such as absorption and bulk scattering, into an absorption coefficient α. The condition for oscillation is then

$$R_1 R_2 \exp(g - \alpha)2l = 1. \tag{3.3}$$

Rearranging (3.3) yields

$$2gl = -\ln R_1 R_2 + 2\alpha l. \tag{3.4}$$

Losses such as scattering at interfaces and Fresnel reflections which are not dependent on the length of the gain medium can be thought of as leakage from the rear mirror. Hence the reduced reflectivity R_2 of the rear mirror $R_2 = 1 - \delta_M$ takes into account the miscellaneous losses. In practice, δ_M does not exceed a few percent. With the approximation

$$\ln(1 - \delta_M) \approx -\delta_M, \tag{3.5}$$

one can combine the optical losses in the resonator with the losses in the crystal

$$\delta = 2\alpha l + \delta_M, \tag{3.6}$$

where δ is the two-way loss in the resonator. With the aid of (3.5), (3.6) we can express the threshold condition (3.3) in the following form:

$$2gl = \delta - \ln R_1 \approx T + \delta. \tag{3.7}$$

where $T = 1 - R_1$ is the transmission of the output mirror.

In (1.61), (1.64) the decay time τ_c of the photon flux in the resonator was introduced. This time constant, determined by photons either scattered, absorbed, or emitted, can be related to the loss term of (3.7) and the quality factor Q of the resonator. Since τ_c represents the average lifetime of the photons in the resonator, it can be expressed by the round trip time of a photon in the resonator $t_\tau = 2L/c$ divided by the fractional loss ϵ per round trip

$$\tau_c = t_\tau/\epsilon = \frac{2L}{c(T + \delta)}, \tag{3.8}$$

where $\epsilon = T + \delta$ is the loss term in (3.7) and L is the length of the resonator. Resonators are characterized by the quality factor Q, which is defined as 2π times the ratio of stored energy E_{st} to dissipated energy E_d per period T_0. The resonator Q defined in this way is $Q = 2\pi E_{st}/E_d$ where $E_d = E_{st}(1 - \exp(-T_0/\tau_c))$ from which we obtain

$$Q = 2\pi \left[1 - \exp\left(\frac{-T_0}{\tau_c}\right) \right]^{-1} \approx \frac{2\pi\tau_c}{T_0} = 2\pi\nu_0\tau_c, \tag{3.9}$$

where $\omega_0 = 2\pi\nu_0 = 2\pi/T_0$. In a typical cw Nd:YAG laser, the transmission of the output mirror is around 10% and the combined losses about 5%. If we assume a resonator length of 50 cm, we obtain a photon lifetime of $\tau_c = 22$ ns and a Q factor for the resonator of $Q = 39 \times 10^6$. Compared to electronic resonance circuits, optical resonators have very high values of Q.

As will be further discussed in Section 5.2.3 the finite lifetime of the photons also determine the minimum bandwidth of the passive resonator according to the Fourier transform $\Delta\omega = 1/\tau_c$. With this relationship the quality factor in (3.9) can also be defined as $Q = \nu_0/\Delta\nu$. The laser parameters and conditions that lead to a low threshold will be examined next.

We turn to the rate equation (1.61), which gives the photon density in the amplifying medium. It is clear from this equation that for onset of laser emission the rate of change of the photon density must be equal to or greater than zero. Thus at laser threshold for sustained oscillation the condition

$$\frac{\partial \phi}{\partial t} \geq 0 \tag{3.10}$$

must be fulfilled, which enables us to obtain from (1.61) the required inversion density at threshold,

$$n \geq \frac{1}{c \sigma \tau_c}. \tag{3.11}$$

In deriving this expression we have ignored the factor S, which denotes the small contribution from spontaneous emission to the induced emission.

According to (1.36) the gain coefficient g can be expressed by

$$g = n\sigma, \tag{3.12}$$

where n is the population inversion and σ is the stimulated emission cross section. The expression for threshold (3.11) is identical to (3.7) if we introduce τ_c from (3.8) and g from (3.12) into (3.11) and assume that the resonator mirrors are coated onto the ends of the laser rod, i.e., $L = \ell$. If we replace σ with (1.40) we may write the threshold condition (3.11) in terms of fundamental laser parameters

$$n > \frac{\tau_f 8\pi \nu_0^2}{\tau_c c^3 g(\nu_s, \nu_0)}. \tag{3.13}$$

The lineshape factor $g(\nu_s, \nu_0)$ and therefore the stimulated emission cross section σ are largest at the center of the atomic line. Thus from (3.13) we can see qualitatively how the linewidth of the laser output is related to the linewidth of the atomic system. Self-sustained oscillation that develops from noise will occur in the neighborhood of the resonant frequency because only at a narrow spectral range at the peak will the amplification be large enough to offset losses. Consequently, the output of the laser will be sharply peaked, and its linewidth will be much narrower than the atomic linewidth.

It is also obvious from this equation that an increase of the inversion by stronger pumping will increase the laser linewidth because the threshold condition can now be met for values of $g(\nu_s, \nu_0)$ farther away from the center. As we will see in Chapter 5 the linewidth of an actual laser system is related to the linewidth of the active material, the level of pump power, and the properties of the optical resonator. The threshold condition at the center of the atomic line is obtained by introducing the peak values of the amplification curve into (3.13). If $g(\nu_s, \nu_0)$ has a Lorentzian shape with full width at half-maximum of $\Delta \nu$ centered about ν_s, then $g(\nu_0) = 2/\pi \Delta \nu$ and

$$n > \frac{\tau_f 4\pi^2 \Delta \nu \nu_0^2}{\tau_c c^3}. \tag{3.14}$$

For a Gaussian lineshape, the right-hand expression of (3.14) has to be divided by $(\pi \ln 2)^{1/2}$. Again, we have assumed that the laser threshold will be reached first by a resonator mode whose resonant frequency lies closest to the center of the atomic line.

From (3.14) we can infer those factors favoring high gain and low threshold for a laser oscillator. In order to achieve a low-threshold inversion, the atomic linewidth $\Delta\nu$ of the laser material should be narrow. Furthermore, the incidental losses in the laser cavity and crystal should be minimized to increase the photon lifetime τ_c. It is to be noted that the critical inversion density for threshold depends only on a single resonator parameter, namely τ_c. A high reflectivity of the output mirror will increase τ_c and therefore decrease the laser threshold. However, this will also decrease the useful radiation coupled out from the laser. We will address the question of optimum output coupling in Section 3.4.

We will now calculate the pumping rate W_p, which is required to maintain the oscillator at threshold. For operation at or near threshold the photon density ϕ is very small and can be ignored. Setting $\phi = 0$ in the rate equation (1.58) and assuming a steady-state condition of the inversion, $\partial n / \partial t = 0$, as is the case in a conventional operation of the laser oscillator, we obtain for a three-level system

$$\frac{n}{n_{\text{tot}}} = \frac{W_p \tau_f - g_2/g_1}{W_p \tau_f + 1} \tag{3.15}$$

and for a four-level system from (1.70)

$$\frac{n_2}{n_g} = \frac{W_p \tau_f}{W_p \tau_f + 1} \approx W_p \tau_f. \tag{3.16}$$

Other factors being equal, four-level laser systems have lower pump-power thresholds than three-level systems. In a four-level system an inversion is achieved for any finite pumping rate W_p. In a three-level system we have the requirement that the pumping rate W_p exceeds a minimum or threshold value given by

$$W_{p(\text{th})} = \frac{g_2}{\tau_f g_1} \tag{3.17}$$

before any inversion at all can be obtained. Whereas for a four-level material the spontaneous lifetime has no effect on obtaining threshold inversion, in a three-level material the pump rate required to reach threshold is inversely proportional to τ_f. Thus, for three-level oscillators only materials with long fluorescence lifetimes are of interest.

The reader is reminded again that (3.15), (3.16) are valid only for a negligible photon flux ϕ. This situation occurs at operation near threshold; it will later be characterized as the regime of small-signal amplification. We will now calculate the minimum pump power which has to be absorbed in the pump bands of the crystal to maintain the threshold inversion. This will be accomplished by first calculating the fluorescence power at threshold, since near above threshold almost all the pump power supplied to the active material goes into spontaneous emission. The fluorescence power per unit volume of the laser transition in a four-level system is

$$\frac{P_f}{V} = \frac{h\nu n_{th}}{\tau_f},$$ (3.18)

where $n_2 = n_{th}$ is the inversion at threshold.

In a three-level system at threshold, $n_2 \approx n_1 \approx n_{tot}/2$ and

$$\frac{P_f}{V} \approx \frac{h\nu n_{tot}}{2\tau_f}.$$ (3.19)

In order that the critical inversion is maintained, the loss by fluorescence from the upper laser level must be supplied by the pump energy. As a result we obtain, for the absorbed pump power P_a needed to compensate for population loss of the laser level by spontaneous emission for a four-level laser

$$\frac{P_a}{V} = \frac{h\nu_p n_{th}}{\eta_Q \tau_f},$$ (3.20)

where $h\nu_p$ is the photon energy at the pump wavelength and η_Q is the quantum efficiency as defined in (1.56). In a three-level system, one requires $n_2 \approx n_1 = n_{tot}/2$ to reach transparency, therefore

$$\frac{P_a}{V} = \frac{h\nu_p n_{tot}}{2\eta_Q \tau_f}.$$ (3.21)

3.2 Gain Saturation

In the previous section we considered the conditions for laser threshold. Threshold was characterized by a steady-state population inversion, that is, $\partial n/\partial t = 0$ in the rate equations. In doing this we neglected the effect of stimulated emission by setting $\phi = 0$. This is a good assumption at threshold, where the induced transitions are small compared with the number of spontaneous processes.

As the threshold is exceeded, however, stimulated emission and photon density in the resonator build up. Far above threshold we have to consider a large photon density in the resonator. From (1.58) we can see that $\partial n/\partial t$ decreases for increasing photon density. Steady state is reached when the population inversion stabilizes at a point where the upward transitions supplied by the pump source equal the downward transitions caused by stimulated and spontaneous emission. With $\partial n/\partial t = 0$ one obtains, for the steady-state inversion population in the presence of a strong photon density ϕ,

$$n = n_{tot} \left(W_p - \frac{\gamma - 1}{\tau_f} \right) \left(\gamma c \sigma \phi + W_p + \frac{1}{\tau_f} \right)^{-1}.$$ (3.22)

The photon density ϕ is given by the sum of two beams traveling in opposite directions through the laser material.

We will now express (3.22) in terms of operating parameters. From Section 1.3 we recall that the gain coefficient $g = -\alpha$ is defined by the product of stimulated emission, cross section, and inversion population. Furthermore, we will define a gain coefficient that the system would have at a certain pump level in the absence of stimulated emission. Setting $\phi = 0$ in (3.22) we obtain the small-signal gain coefficient

$$g_0 = \sigma n_{\text{tot}}[W_p \tau_f - (\gamma - 1)](W_p \tau_f + 1)^{-1}, \tag{3.23}$$

which an active material has when pumped at a level above threshold and when lasing action is inhibited by blocking the optical beam or by removing one or both of the resonator mirrors. If feedback is restored, the photon density in the resonator will increase exponentially at the onset with g_0. As soon as the photon density becomes appreciable, the gain of the system is reduced according to

$$g = g_0 \left(1 + \frac{\gamma c \sigma \phi}{W_p + (1/\tau_f)}\right) - 1, \tag{3.24}$$

where g is the saturated gain coefficient. Equation (3.24) was obtained by introducing (3.23) into (3.22) and using $g = \sigma n$.

We can express ϕ by the power density I in the system. With $I = c\phi h\nu$ we obtain

$$g = \frac{g_0}{1 + I/I_s}, \tag{3.25}$$

where

$$I_s = \left(W_p + \frac{1}{\tau_f}\right) \frac{h\nu}{\gamma \sigma}. \tag{3.26}$$

The parameter I_s defines a flux in the active material at which the small-signal gain coefficient g_0 is reduced by one-half.

In a four-level system $W_p \ll 1/\tau_f$ and $\gamma = 1$, so (3.26) reduces to

$$I_s = \frac{h\nu}{\sigma \tau_f}. \tag{3.27}$$

For a three-level system the saturation flux is

$$I_s = \frac{h\nu[W_p + (1/\tau_{21})]}{\sigma[1 + (g_2/g_1)]}. \tag{3.28}$$

As we can see from (3.23) the small-signal gain depends only on the material parameters and the amount of pumping power delivered to the active material.

The large-signal or saturated gain depends in addition on the power density in the resonator.

Gain saturation as a function of steady-state radiation intensity must be analyzed for lasers with homogeneous and inhomogeneous line broadening. Equation (3.25) is valid only for the former case, in which the gain decreases proportionately over the entire transition line. As we have seen in Chapter 2, a ruby laser has a homogeneously broadened bandwidth, whereas in Nd : glass the interaction of the active ion with the electrostatic field of the host leads to an inhomogeneous line. However, in solid-state materials such as Nd : glass, the cross-relaxation rate is very fast. The latter is associated with any process characteristic of the laser medium that affects the transfer of excitation within the atomic spectral line so as to prevent or minimize the departure of this line from the equilibrium distribution. It has been shown that in the case of a very fast cross-relaxation within the inhomogeneous line, the saturated gain is in agreement with that of a homogeneously broadened bandwidth.

3.3   Circulating Power

In a laser resonator the power density in the active medium increases up to the point where the saturated gain equals the total losses. This power density I is obtained from the threshold conditions (3.7) and the expression for saturated gain (3.25),

$$I = I_s \left[\frac{2g_0 l}{\delta - \ln R_1} - 1 \right]. \tag{3.29}$$

The first expression in brackets is a measure of by how much the laser is pumped above threshold. From (3.29) it follows as one would expect $I = 0$ at threshold. If the gain medium is pumped above threshold, g_0 and I will increase, whereas the saturated gain coefficient g will stay constant according to (3.7). As I increases, so will the output power which is coupled out through mirror R_1.

We will now relate the laser output to the power density in the resonator. The power density I in the laser resonator comprises the power densities of two oppositely traveling waves. One can think of a circulating power density I_{circ}, which is reflected back and forth between the two resonator mirrors. Figure 3.2 illustrates a standing wave laser resonator, where $I_L(x)$ and $I_R(x)$ are the intracavity one-way intensities of the left and right traveling resonator beams as a function of the intracavity location x. The rear reflector and output coupler are located at $x = 0$ and $x = L$, respectively. The rod faces are located at $x = a$ and $x = b$. Hence, the length of the active medium is $\ell = b - a$.

The beam moving in the $(+x)$ direction in Fig. 3.2 will experience an increase according to

$$\frac{dI_L(x)}{dx} = g(x)I_L(x). \tag{3.30}$$

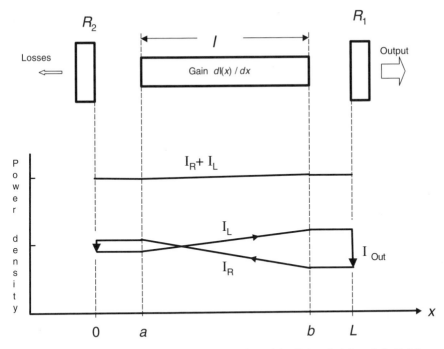

FIGURE 3.2. Circulating power traveling from left to right (I_L) and right to left (I_R) in a laser oscillator.

If the gain coefficient $g(x)$ is replaced by the expression given in (3.25) we obtain

$$\frac{dI_L(x)}{dx} = \frac{g_0 I_L(x)}{1 + [I_L(x) + I_R(x)]/I_s} \tag{3.31}$$

An identical equation, but with the sign reversed and $I_L(x)$ replaced by $I_R(x)$ is obtained for the wave traveling in the $(-x)$ direction. From (3.31) and the corresponding equation for $I_R(x)$ it follows that the product of the intensities of the two counterpropagation beams is constant, that is,

$$I_L(x) I_R(x) = \text{const.} \tag{3.32}$$

In most solid-state lasers, the internal losses are small and the reflectivity of the output mirror is high, in this case variations of $I_L(x)$ and $I_R(x)$ as a function of x are relatively small. If we treat these changes as a perturbation $I_L(x) + \Delta I$ and $I_R(x) + \Delta I$ in (3.32) we obtain, as an approximation,

$$I = I_L + I_R = \text{const,} \tag{3.33}$$

where I is the total power density in the resonator which is to a first-order independent of position along the resonator (see also Fig. 3.2).

The circulating power I_{circ} is to a good approximation equal to the average of the power densities of the counterpropagating beams

$$I_{circ} = (I_L + I_R)/2. \tag{3.34}$$

From Fig. 3.2 follows

$$I_{out} = (1 - R_1)I_L \tag{3.35a}$$

and

$$I_R = R_1 I_L. \tag{3.35b}$$

Introducing (3.35) into (3.33) yields

$$P_{out} = AI\left(\frac{1-R}{1+R}\right), \tag{3.36}$$

where R is now the reflectivity of the output mirror and A is the cross section of the gain medium. For values of R close to 1, (3.36) reduces to

$$P_{out} = AIT/2, \tag{3.37}$$

where T is the transmission of the output mirror.

3.4   Oscillator Performance Model

In this section we will develop a model for the laser oscillator. First we will discuss the various steps involved in the conversion process of electrical input to laser output. After that, we will relate these energy transfer mechanisms to parameters which are accessible to external measurements of the laser oscillator. The purpose of this section is to gain insight into the energy conversion mechanisms and therefore provide an understanding of the dependency and interrelationship of the various design parameters which may help in the optimization of the overall laser efficiency. In almost all applications of lasers, it is a major goal of the laser designer to achieve the desired output performance with the maximum system efficiency.

3.4.1   Conversion of Input to Output Energy

The flow of energy from electrical input to laser output radiation is illustrated schematically in Fig. 3.3. Also listed are the principal factors and design issues that influence the energy conversion process. There are different ways of partitioning this chain of transfer processes. This approach was chosen from an engineering point of view which divides the conversion process into steps related to individual system components. As shown in Fig. 3.3, the energy transfer from electrical input to laser output can conveniently be expressed as a four-step process.

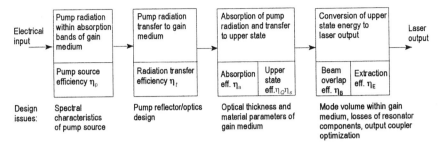

FIGURE 3.3. Energy flow in a solid-state laser system.

Conversion of Electrical Input Delivered to the Pump Source to Useful Pump Radiation

We define as useful radiation the emission from the pump source that falls into the absorption bands of the laser medium. The pump source efficiency η_P is therefore the fraction of electrical input power that is emitted as optical radiation within the absorption region of the gain medium. The output of a laser diode or a diode array represents all useful pump radiation, provided the spectral output is matched to the absorption band of gain medium. Typical values of η_p for commercially available cw and quasi-cw diode arrays are 0.3 to 0.5.

For flashlamp or cw arc lamp-pumped systems, the pump source efficiency may be defined as

$$\eta_p = P_\lambda / P_{in} = \int_{\lambda_1}^{\lambda_2} P'_\lambda \, d\lambda / P_{in}, \tag{3.38}$$

where P_λ is the spectral output power of the lamp within the absorption bands of the gain material, P_{in} is the electrical power input, and P'_λ is the radiative power per unit wavelength emitted by the lamp, and the integral is taken over the wavelength range λ_1 to λ_2, which is useful for pumping the upper laser level. The output characteristics of arc lamps and their dependency on operating parameters will be discussed in Section 6.1. The measurement of η_p for broadband sources is somewhat involved and requires either a calorimetric measurement of the power absorbed in a sample of the laser material or integration over the source emission spectrum and the absorption spectrum of the gain material. Typical values are $\eta_p = 0.04$–0.08. These numbers are typical for a 5 to 10 mm thick laser material. The magnitude of η_P is dependent on the thickness of the active material because as the thickness increases, radiation at the wings of the absorption band will start to contribute more to the pumping action.

Transfer of the Useful Pump Radiation Emitted by the Pump Source to the Gain Medium

The transfer of flashlamp pump radiation to the laser medium is accomplished by means of a completely enclosed reflective chamber or pump cavity. The radiation transfer efficiency η_t can be defined as

$$P_e = \eta_t P_\lambda, \tag{3.39}$$

where P_λ is the useful pump radiation emitted by the source and P_e is the fraction of this radiation transferred into the laser material. The factor η_t is a combination of the capture efficiency, defined by the fraction of rays leaving the source and intersecting the laser rod, and the transmission efficiency. The former is based on the geometrical shape of the pump cavity, diameter and separation of the pump source, and laser rod. The latter is a function of the reflectivity of the walls of the pump cavity, reflection losses at the rod surface and coolant jacket, absorption losses in the coolant fluid, and radiation losses through the clearance holes at the side walls of the pump cavity. For close-coupled cavities, typical values are $\eta_t = 0.3$–0.6.

In diode-pumped lasers the radiation transfer is much simpler. In so-called end-pumped lasers, the transfer system usually consists of lenses for the collection and focusing of diode radiation into the laser crystal. Furthermore, in side-pumped systems, the laser diodes are mounted in close proximity to the laser crystal without the use of any intervening optics. If we express reflection losses and spill-over losses at the optics or active medium by the parameter R, we can write

$$\eta_t = (1 - R). \tag{3.40}$$

Since the laser crystal and optical components are all antireflection coated, the radiation transfer losses are very small in these systems. Values for the radiation transfer efficiency are typically $\eta_t = 0.85$–0.98.

Absorption of Pump Radiation by the Gain Medium and Transfer of Energy to the Upper Laser Level

This energy transfer can be divided into two processes. The first is the absorption of pump radiation into the pump bands of the gain medium expressed by η_a, and the second is the transfer of energy from the pump band to the upper laser level expressed by $\eta_Q \eta_S$.

The absorption efficiency is the ratio of power P_a absorbed to power P_e entering the laser medium

$$\eta_a = P_a / P_e. \tag{3.41}$$

The quantity η_a is a function of the path length and the spectral absorption coefficient of the laser medium integrated over the emission spectrum of the pump source. If the pump radiation is totally diffuse inside the laser rod, as is the case when the lateral surface is rough ground, a good approximation can be derived by expressing (3.41) in a different form

$$\eta_a = \frac{(dP_a/dV)V}{I_w F}, \tag{3.42a}$$

where dP_a/dV is the power absorbed per volume, I_w is the power density at the cylindrical surface of the rod, and F, V are the cylindrical rod surface and volume of the rod, respectively.

The power absorbed per volume can also be expressed by $dP_a/dV = \alpha_0 I_{av}$, where α_0 is the absorption coefficient of the laser material averaged over the spectral emission range of the lamp and I_{av} is the average power density inside the rod. Radiation with an energy density W_{av} propagating at a velocity c has a power density of $I_{av} = cW_{av}$, where $c = c_0/n_0$ is the speed of light in the medium and n_0 is the refractive index. The energy density in the laser rod has a radial dependence due to absorption. The average value W_{av} can be closely approximated by taking the energy density W_s at the surface weighted by the factor $\exp(-\alpha_0 R)$, where R is the rod radius and α_0 is the absorption coefficient. From [3] it follows that this is a valid approximation over a large range of α_0. From these considerations follows

$$dP_a/dV = \alpha_0 c W_s \exp(-\alpha_0 R). \tag{3.42b}$$

Owing to grinding the lateral surface of the rod, the pump radiation upon entering the rod will be diffused. For an enclosure with diffusely reflecting walls, we obtain from blackbody radiation theory a relationship between energy W_s inside the enclosure and the intensity I_w emitted by the walls

$$I_w = cW_s/4. \tag{3.42c}$$

Introducing (3.42b, c) into (3.42a) yields our final result

$$\eta_a = 2\alpha_0 R \exp(-\alpha_0 R). \tag{3.43}$$

For diode pumped lasers, the absorption efficiency can be approximated by

$$\eta_a = 1 - \exp(-\alpha_0 l), \tag{3.44}$$

where α_0 is the absorption coefficient of the laser crystal at the wavelength emitted by the laser diode and l is the path length in the crystal. Detailed data on η_a for both flashlamp and laser diode radiation can be found in Chapter 6.

The upper state efficiency may be defined as the ratio of the power emitted at the laser transition to the power absorbed into the pump bands. This efficiency is the product of two contributing factors, the quantum efficiency η_Q, which is defined as the number of photons contributing to laser emission divided by the number of pump photons, and η_S, the quantum defect efficiency. The latter is sometimes referred to as the Stokes factor, which represents the ratio of the photon energy emitted at the laser transition $h\nu_L$ to the energy of a pump photon $h\nu_p$, that is,

$$\eta_S = \left(\frac{h\nu_L}{h\nu_p}\right) = \frac{\lambda_P}{\lambda_L}, \tag{3.45}$$

where λ_P and λ_L is the wavelength of the pump transition and the laser wavelength, respectively. For Nd : YAG emitting at 1064 nm that is pumped by a laser diode array at 808 nm, we obtain $\eta_S = 0.76$, and $\eta_Q = 0.90$.

In a flashlamp-pumped system, the value of η_S is an average value derived from considering the whole absorption spectrum of the laser.

Conversion of the Upper State Energy to Laser Output

The efficiency of this process can be divided into the fractional spatial overlap of the resonator modes with the pumped region of the laser medium and the fraction of the energy stored in the upper laser level which can be extracted as output.

The beam overlap efficiency η_B is defined as the resonator mode volume divided by the pumped volume of the active material. A value of η_B less than 1 indicates that part of the inversion will decay by spontaneous, rather than stimulated, emission. In an oscillator η_B expresses the mode-matching efficiency, that is, the spatial overlap between the resonator modes and the gain distribution. In an amplifier η_B is a measure of the spatial overlap between the input beam and the gain distribution in the laser material.

This subject usually does not receive a lot of attention in laser literature, but a poor overlap of the gain region of the laser with the laser-beam profile is often the main reason that a particular laser performs below expectations. For example, the generally disappointing performance of slab lasers can often be traced to an insufficient utilization of the rectangular gain cross-section by the laser beam. Likewise, the low overall efficiency of lasers with a TEM_{00} mode output is the result of a mode volume that occupies only a small fraction of the gain region of the laser rod. On the other hand, so-called end-pumped lasers, where the output from a laser diode pump is focused into the gain medium, achieve near perfect overlap.

Instead of comparing the pump with the mode volume, it is often sufficient to compare the cross-sections, if we assume that the radial distribution of the resonator mode does not change appreciably along the length of the active medium. For example, in a 50 cm long resonator, with two curved mirrors of 5 m radius at each end, the TEM_{00} mode changes only by 5% over the length of the resonator (Section 5.1.3).

The overlap between the resonator mode intensity distribution $I(r)$ and the gain distribution $g(r)$ is given by an overlap integral

$$\eta_B = \int g(r)I(r)2\pi r\,dr \Big/ \int\int I^2(r)2\pi r\,dr. \tag{3.46}$$

For two practical cases (3.46) can be easily evaluated. The first case is a uniformly pumped gain medium operated in a highly multimode resonator. If we assume a top hat distribtution for both the gain region and resonator beam (i.e., independent of r), then η_B is simply the ratio of the overlapping cross sections, $\eta_B = w_m^2/w_p^2$, where πw_m^2 and πw_p^2 is the cross section of the resonator mode and pump region, respectively. The efficiency η_B can at best equal unity.

Another simple case is found in end-pumped lasers. In this case a Gaussian pump beam and a Gaussian (TEM$_{00}$) resonator mode propagate coaxially inside the active medium. Introducing in (3.46) a normalized Gaussian resonator mode

$$I(r) = (2/\pi w_m^2) \exp[-2(r/w_m)^2], \qquad (3.47)$$

and a similar expression for the pump beam yields

$$\eta_B = \frac{2w_m^2}{w_p^2 + w_m^2} \quad \text{for} \quad w_p > w_m, \quad \text{and} \quad \eta_B = 1 \quad \text{for} \quad w_p \leq w_m, \qquad (3.48)$$

where w_p and w_m are the spot sizes for the pump beam and resonator mode profiles. For $w_m = w_p$, the TEM$_{00}$ mode and pump beam occupy the same volume and $\eta_B = 1$.

Values for η_B can range from as low as 0.1, for example, for a laser rod of 5 mm diameter operated inside a large radius mirror resonator containing a 1.5 mm aperture for fundamental mode control, to 0.95 for an end-pumped laser operating at TEM$_{00}$. Innovative resonator designs, employing unstable resonators, internal lenses, variable reflectivity mirrors, and so forth, can achieve TEM$_{00}$ mode operation typically at $\eta_B = 0.3$–0.5. Multimode lasers typically achieve $\eta_B = 0.8$–0.9. If a laser oscillator is followed by an amplifier, a telescope is usually inserted between these two stages in order to match the oscillator beam to the diameter of the amplifier and thereby optimize η_B.

The circulating power in an optical resonator is diminished by internal losses described by the round trip loss δ (Section 3.2) and by radiation coupled out of the resonator. For reasons that will become apparent in the next section, we will define an extraction efficiency η_E which describes the fraction of total available upper state energy or power which appears at the output of the laser

$$\eta_E = P_{out}/P_{avail}. \qquad (3.49)$$

Expressions for η_E will be given in Section 3.4.2 and in Chapters 4 and 8 for the cw oscillator, laser amplifier, and Q-switch oscillator, respectively.

An indication of the reduction of available output power due to losses in resonator can be obtained from the coupling efficiency

$$\eta_c = T/(T + \delta). \qquad (3.50)$$

As will be explained in the next section, the slope of the output versus input curve of a laser is directly proportional to this factor, whereas the overall system efficiency of a laser is directly proportional to η_E.

The conversion processes described so far are equally applicable to cw and pulsed lasers, provided that the power terms are replaced with energy and integration over the pulse length is carried out where appropriate. The energy flow depicted in Fig. 3.3 can also be extended to laser amplifiers and Q-switched systems because the discussion of pump source efficiency, radiation transfer, and so forth, is equally applicable to these systems. Even the definition for the extraction efficiency remains the same, however, the analytical expressions for η_E are different as will be discussed in Chapter 4 for the laser amplifier.

If the laser oscillator is Q-switched, additional loss mechanisms come into play which are associated with energy storage at the upper laser level and with the transient behavior of the system during the switching process. The Q-switch process will be described in detail in Chapter 8. However, for completeness of the discussion on energy transfer mechanisms in a laser oscillator, we will briefly discuss the loss mechanisms associated with Q-switch operations.

In a Q-switched laser, a large upper state population is created in the laser medium, and stimulated emission is prevented during the pump cycle by the introduction of a high loss in the resonator. At the end of the pump pulse, the loss is removed (the resonator is switched to a high Q) and the stored energy in the gain medium is converted to optical radiation in the resonator from which it is coupled out by the output mirror.

There are losses prior to opening of the Q-switch, such as fluorescence losses and Amplified Spontaneous Emission (ASE) losses that will depopulate the upper state stored energy. Also, not all of the stored energy available at the time of Q-switching is converted to optical radiation. For a Q-switched laser, the extraction efficiency η_E in Fig. 3.3 can be expressed as

$$\eta_E = \eta_{St}\eta_{ASE}\eta_{EQ}, \tag{3.51}$$

where η_{St} and η_{ASE} account for the fluorescence and ASE losses prior to the opening of the Q-switch and η_{EQ} is the extraction efficiency of the Q-switch process.

Assuming a square pump pulse of duration t_p, the maximum upper state population reached at the end of the pump cycle is given by

$$n_2(t_p) = n_g W_p \tau_f \left[1 - \exp(-t_p/\tau_f)\right]. \tag{3.52}$$

Since the total number of ions raised to the upper level during the pump pulse is $n_g W_p t_p$, the fraction available at the time of Q-switching ($t = t_p$) is

$$\eta_{St} = \frac{[1 - \exp(-t_p/\tau_f)]}{t_p/\tau_f}. \tag{3.53}$$

The storage efficiency η_{St} is therefore the ratio of the energy stored in the upper laser level at the time of Q-switching to the total energy deposited in the upper laser level. From the expression for η_{St} it follows that for a pump pulse equal to the fluorescence lifetime ($t_p = \tau_f$) the storage efficiency is 0.63. Clearly, a short pump pulse increases the overall efficiency of a Q-switched laser. However, a shorter pump pulse puts an extra burden on the pump source because the pump has to operate at a higher peak power to deliver the same energy to the laser medium.

If the inversion reaches a critical value, the gain can be so high such that spontaneous emission after amplification across the gain medium may be large enough to deplete the laser inversion. Furthermore, reflections from internal surfaces can increase the path length or allow multiple passes inside the gain section which will make it easier for this unwanted radiation to build up. In high-gain oscillators, multistage lasers, or in laser systems having large-gain regions, ASE coupled with parasitic oscillations present the limiting factor for energy storage.

We can define η_{ASE} as the fractional loss of the stored energy density to ASE and parasitic oscillations

$$\eta_{ASE} = 1 - E_{ASE}/E_{St}. \tag{3.54}$$

Minimizing reflections internal to the laser medium by AR coatings and providing a highly scattering, absorbing, or low-reflection (index matching) surface of the laser rod, coupled with good isolation between amplifier stages, will minimize ASE and parasitic losses. The occurrence of ASE can often be recognized in a laser oscillator or amplifier as a saturation in the laser output as the lamp input is increased (see also Section 4.4.1).

The fraction of the stored energy E_{St} available at the time of Q-switching to energy E_{EX} extracted by the Q-switch, can be expressed as the Q-switch extraction efficiency

$$\eta_{EQ} = E_{EX}/E_{St}. \tag{3.55}$$

The fraction of initial inversion remaining in the gain medium after emission of a Q-switched pulse is a function of the initial threshold and final population inversion densities. These parameters are related via a transcendental equation, as shown in Chapter 8.

3.4.2 Laser Output

In this subsection, we will describe the basic relationships between externally measurable quantities, such as laser output, threshold, and slope efficiency, and internal systems and materials parameters.

After the pump source in a laser oscillator is turned on, the radiation flux in the resonator that builds up from noise will increase rapidly. As a result of the increasing flux, the gain coefficient decreases according to (3.25) and finally stabilizes at a value determined by (3.7). A fraction of the intracavity power is coupled out of the resonator and appears as useful laser output according to (3.36). If we combine (3.29), (3.36), the laser output takes the form

$$P_{out} = A \left(\frac{1-R}{1+R} \right) I_S \left(\frac{2g_0 l}{\delta - \ln R} - 1 \right). \tag{3.56}$$

In this equation, I_S is a materials parameter, A and l are the cross-section and length of the laser rod, respectively, and R is the reflectivity of the output coupler. These quantities are usually known, whereas the unsaturated gain coefficient g_0 and the resonator losses δ are not known. We will now relate g_0 to system parameters and describe methods for the measurement of g_0 and the losses δ in an oscillator.

The Four-Level System

The population inversion in a four-level system as a function of pump rate is given by (3.16). Making the assumption that $W_P \tau_f \ll 1$ and multiplying both sides of this equation by the stimulated emission cross section yields

$$g_0 = \sigma n_0 W_P \tau_f. \tag{3.57}$$

Now we recall from Chapter 1 that $W_P n_g$ gives the number of atoms transferred from the ground level to the upper laser level per unit time and volume, i.e., pump power density divided by photon energy. This can be expressed as

$$W_P n_g = \eta_Q \eta_S \eta_B P_a / h \nu_L V, \tag{3.58}$$

where P_a is the total absorbed pump power and V is the volume of the gain medium. Not all of the atoms transferred from the ground level to the upper laser level contribute to gain, this is accounted for by η_Q. The Stokes factor η_S enters in (3.58) because the photon energy is expressed in emitted laser photons rather than pump photons, and the overlap efficiency η_B defined in (3.46) accounts for the fact that not all of the population inversion interacts with the photon density of the resonator modes.

If we introduce (3.58) into (3.57), we can express the small signal gain coefficient in terms of absorbed pump power

$$g_0 = \sigma \tau_f \eta_Q \eta_S \eta_B P_{ab} / h \nu_L V. \tag{3.59}$$

The absorbed pump power in the laser material is related to the electrical input to the pump source by

$$P_{ab} = \eta_P \eta_t \eta_a P_{in}. \tag{3.60}$$

With (3.59), (3.60) we can establish a simple relationship between the small signal, single pass gain and lamp input power

$$g_0 = \sigma \tau_f \eta P_{in} / h \nu_L V, \tag{3.61}$$

where, for convenience, we have combined all the efficiency factors into

$$\eta = \eta_p \eta_t \eta_a \eta_Q \eta_S \eta_B. \tag{3.62}$$

If we replace the small signal coefficient g_0 in (3.56) with the system and materials parameters given in (3.61) we obtain our final result

$$P_{out} = \sigma_S (P_{in} - P_{th}), \tag{3.63}$$

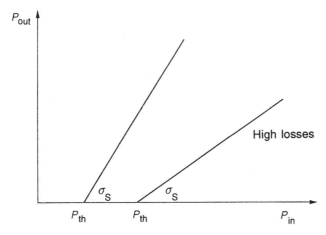

FIGURE 3.4. Laser output versus pump input characterized by a threshold input power P_{th} and a slope efficiency σ_S.

where σ_S is the slope efficiency of the output versus input curve, as shown in Fig. 3.4,

$$\sigma_S = \left(\frac{-\ln R}{\delta - \ln R}\right)\eta \approx \frac{T}{T + \delta}\eta. \tag{3.64}$$

The electrical input at threshold is

$$P_{th} = \left(\frac{\delta - \ln R}{2}\right)\frac{Ah\nu_L}{\eta\sigma\tau_f} \approx \left(\frac{T + \delta}{2}\right)\frac{Ah\nu_L}{\eta\sigma\tau_f}. \tag{3.65}$$

In deriving (3.64), (3.65) we have made use of the approximation $2(1 - R)/(1 + R) = -\ln R$. Also for reflectivities of the output coupler of $R = 0.9$ or higher, the term $(-\ln R)$ can be replaced in most cases by the mirror transmission T with sufficient accuracy. In the laser literature, (3.65) is sometimes expressed in terms of pump photon energy, in this case $\eta_Q\eta_S$ have to be excluded from η since $h\nu_L = \eta_Q\eta_S h\nu_p$.

From (3.65) it follows that a laser material with a large product of the stimulated emission cross section and fluorescence lifetime ($\sigma\tau_f$) will have a low laser threshold. The slope efficiency σ_S is simply the product of all the efficiency factors discussed in Section 3.4.1. The input power P_{th} required to achieve laser threshold is inversely proportional to the same efficiency factors. Therefore, a decrease of any of the η terms will decrease the slope efficiency and increase the threshold, as shown in Fig. 3.4. In the expressions for σ_S and P_{th} we have left T and δ in explicit form, because these parameters are subject to optimization in a laser resonator. As expected, higher optical losses δ, caused by reflection, scattering, or absorption, increase the threshold input power and decrease the slope efficiency.

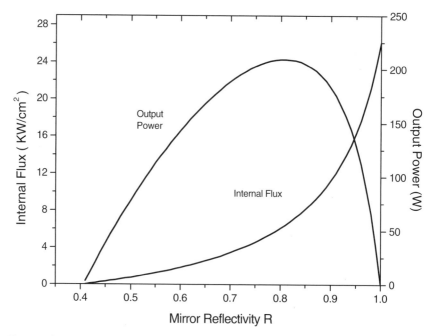

FIGURE 3.5. Laser output power and total flux inside the resonator as a function of mirror reflectivity. Parameters: $I_5 = 2.9$ kW/cm^2, $2g_0\ell = 1$, $\delta = 0.1$, $A = 0.4$ cm^2.

From (3.29) and (3.36) we can calculate the power density inside the resonator and the laser output as a function of output coupling. The general shape of I and P_{out} as a function of T is shown in Fig. 3.5. The power density I is maximum if all the radiation is contained in the resonator, that is, $R = 1$. At $2g_0\ell = \delta - \ln R$ the flux inside the resonator is zero. This occurs if the reflectivity of the output mirror is so low, namely $R_{\text{th}} = \exp(\delta - 2g_0\ell)$, that the laser just reaches threshold at the given input. The laser output P_{out} is zero for $R = 1$ and R_{th}. As shown in Fig. 3.5, the output reaches a maximum for a specific value of R.

Differentiation of (3.56) determines the output coupling R_{opt} which maximizes P_{out}, i.e.,

$$-\ln R_{\text{opt}} = (\sqrt{2g_0\ell/\delta} - 1)\delta. \tag{3.66}$$

From this expression it can be seen that low-gain systems require a high reflectivity and vice versa. For cw-pumped Nd : YAG lasers the gain coefficient is typically on the order of $g_0 = (0.05 - 0.1)$ cm^{-1} and the optimum output coupling ranges from 0.80 to 0.98. The lower number is typical for multi-hundred watt Nd : YAG lasers pumped with up to 10 kW of input power. The very high reflectivity is for very small systems. Pulsed-pumped systems operate at much higher pump powers and, correspondingly, have a higher gain coefficient. For example, a typical industrial Nd : YAG laser for cutting sheet metal has a pulse length of 100 μs and an output energy of 2 J per pulse. Assuming a 2% system efficiency, the pump pulse has a peak power of 1 MW. Gain coefficients for pulsed systems are on the order of $(0.3–0.5)\,\mathrm{cm}^{-1}$, and the output mirror reflectivity ranges from 0.6 to 0.8. Pulsed systems that are also Q-switched have the highest gain coefficients, typically 1.5 to $2.5\,\mathrm{cm}^{-1}$, and the output coupler reflectivity is between 0.4 and 0.6.

Introducing the expression for R_{opt} into (3.56) gives the laser output at the optimum output coupling

$$P_{\mathrm{opt}} = g_0 l I_S A (1 - \sqrt{\delta/2g_0 l})^2. \tag{3.67}$$

If we use the definition for g_0 and I_S as given before, we can readily see that $g_0 I_S = n_2 h\nu/\tau_f$, which represents the total excited state power per unit volume. (A similar expression will be derived in Chapter 4 for the available energy in a laser amplifier.) Therefore, the maximum available power from the oscillator is

$$P_{\mathrm{avail}} = g_0 l I_S A. \tag{3.68}$$

The optimum power output can be expressed as

$$P_{\mathrm{opt}} = \eta_E P_{\mathrm{avail}}, \tag{3.69}$$

where

$$\eta_E = \left(1 - \sqrt{\delta/2g_0 l}\right)^2 \tag{3.70}$$

is the extraction efficiency already mentioned in Section 3.4.1. The behavior of η_E as a function of the loss-to-gain ratio $\delta/2g_0 l$ is depicted in Fig. 3.6. The detrimental effect of even a very small internal loss on the extraction efficiency is quite apparent. For example, in order to extract at least 50% of the power available in

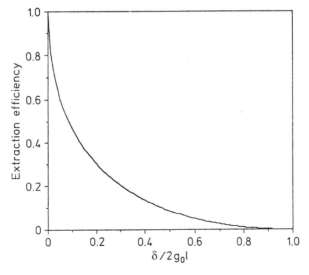

FIGURE 3.6. Extraction efficiency η_E as a function of loss-to-gain ratios $\delta/2g_0 l$.

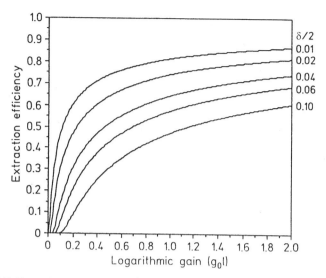

FIGURE 3.7. Extraction efficiency for the optimized resonator as a function of the single-pass logarithmic gain. Parameter is the one-way resonator loss.

the laser material, the internal loss has to be less than 10% of the unsaturated gain. Achievement of a high extraction efficiency is particularly difficult in cw systems because the gain is relatively small, and unavoidable resonator losses can represent a significant fraction of the gain. In Fig. 3.7 the extraction efficiency for the optimized resonator is plotted for different values of δ and $g_0 l$.

The overall system efficiency of a solid-state laser is directly proportional to the extraction efficiency

$$\eta_{sys} = P_{out}/P_{in} = \eta_E \eta. \tag{3.71}$$

If one compares (3.71) with the expression for the slope efficiency of a laser given by (3.65), then the first term, namely the coupling efficiency, is replaced by the extraction efficiency η_E.

The sensitivity of the laser output to values of T that are either above or below T_{opt} is illustrated in Fig. 3.8. For resonators which are either over- or undercoupled, the reduction of power compared to that available at T_{opt} depends on how far the system is operated above threshold. For oscillators far above threshold, the curve has a broad maximum and excursions of $\pm 20\%$ from T_{opt} do not reduce the output by more than a few percent.

As we have seen, the resonator losses and the gain in the laser material play an important part in the optimization process of a laser system. Following a method first proposed by Findlay and Clay [4], the resonator losses can be determined by using output mirrors with different reflectivities and determining threshold power for lasing for each mirror. According to (3.7) we can write

$$-\ln R = 2g_0 \ell - \delta. \tag{3.72}$$

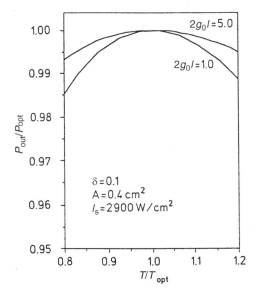

FIGURE 3.8. Sensitivity of laser output for nonoptimal output coupling.

Extrapolation of the straight-line plot of $-\ln R$ versus P_{th}, at $P_{th} = 0$, yields the round-trip resonator loss δ as shown in Fig. 3.9. With δ determined, the small signal gain g_0 as a function of input power can be plotted. From the slope of the straight line we can also calculate the efficiency factor η. With (3.61) and (3.27)

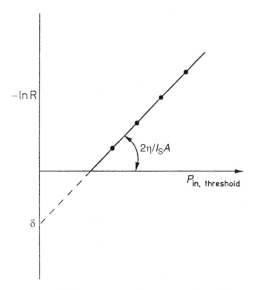

FIGURE 3.9. Measurement of the resonator losses as well as the product of all the efficiency factors involved in the energy transfer mechanism of a laser.

we obtain $2g_0 l = 2\eta P_{in}/I_S A$, and (3.72) can be written

$$-\ln R = (2\eta/I_S A)P_{in} - \delta. \tag{3.73}$$

The Three-Level System

The population inversion in a three-level system as a function of the pump rate is given by (3.15). If we multiply both sides of this equation by the stimulated emission cross section σ_{21}, we obtain

$$g_0 = \alpha_0 \frac{W_P \tau_{21} - 1}{W_P \tau_{21} + 1}, \tag{3.74}$$

where $\alpha_0 = \sigma_{21} n_{tot}$ is the absorption coefficient of the material when all atoms exist in the ground state. In the absence of pumping, (3.74) simply becomes $g_0 = -\alpha_0$. In order to simplify our analysis, we assumed $g_2 = g_1$.

We assume now that the pump rate W_P is a linear function of lamp input P_{in},

$$W_P \tau_f = (\eta/A I_S)P_{in}. \tag{3.75}$$

Introducing (3.75) into (3.74) yields

$$g_0 = \alpha_0 \frac{(\eta/A I_S)P_{in} - 1}{(\eta/A I_S)P_{in} + 1}. \tag{3.76}$$

In a four-level system, the small signal gain is directly proportional to the pump rate and therefore to the input power of the pump source. From (3.61) we obtain

$$g_0 = (\eta/A I_S)(P_{in}/\ell). \tag{3.77}$$

Equation (3.76) and (3.77) are plotted in Fig. 3.10. Since both lasers are assumed to be pulsed, the input power P_{in} has been replaced by the flashlamp input energy $E_{in} = P_{in} t_p$, where $t_p = 1$ ms is the pump pulse width. The other parameters are $(\eta/I_S A) = 10^{-6}$ W^{-1}, $\alpha_0 = 0.2$ cm^{-1}, and $\ell = 15$ cm.

3.5   Relaxation Oscillations

So far in this chapter we have considered only the steady-state behavior of the laser oscillator. Let us now consider some aspects of transient or dynamic behavior. Relaxation oscillations are by far the most predominant mechanisms causing fluctuations in the output of a solid-state laser. Instead of being a smooth pulse, the output of a pumped laser is comprised of characteristic spikes. In cw-pumped solid-state lasers the relaxation oscillations, rather than causing spiking of the output, manifest themselves as damped, sinusoidal oscillations with a well-defined decay time.

In many solid-state lasers the output is a highly irregular function of time. The output consists of individual bursts with random amplitude, duration, and sep-

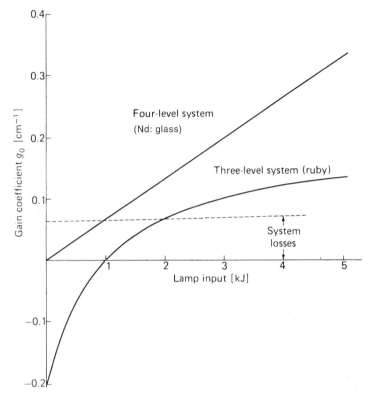

FIGURE 3.10. Gain versus lamp input for a four-level and a three-level system.

aration. These lasers typically exhibit what is termed "spiking" in their output. We will explain the phenomena of the spike formation with the aid of Fig. 3.11. When the laser pump source is first turned on, there are a negligible number of photons in the cavity at the appropriate frequency. The pump radiation causes a linear buildup of excited ions and the population is inverted.

Although under steady-state oscillation conditions N_2 can never exceed $N_{2,th}$, under transient conditions the pump can raise N_2 above the threshold level because no laser oscillation has yet been built up and no radiation yet exists in the cavity to pull N_2 back down by means of stimulated emission.

The laser oscillation does not begin to build up, in fact, until after N_2 passes $N_{2,th}$, so that the net round-trip gain in the laser exceeds unity. Then, however, because N_2 is considerably in excess of $N_{2,th}$, the oscillation level will actually build up very rapidly to a value of the photon flux ϕ substantially in excess of the steady-state value for the particular pumping level.

But, when $\phi(t)$ becomes very large, the rate of depletion of the upper-level ions due to stimulated emission becomes correspondingly large, in fact, considerably larger than the pumping rate W_p. As a result, the upper-level population $N_2(t)$ passes through a maximum and begins to decrease rapidly, driven downward by

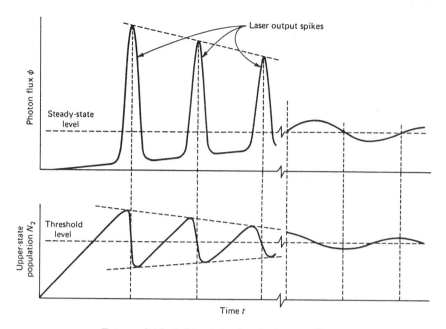

FIGURE 3.11. Spiking behavior of a laser oscillator.

the large radiation density. The population $N_2(t)$ is driven back below the threshold level $N_{2,\text{th}}$; the net gain in the laser cavity becomes less than unity, and so the existing oscillation in the laser cavity begins to die out.

To complete the cycle of this relaxation process, once the radiation level has decreased below the proper steady-state level, the stimulated emission rate again becomes small. At this point the pumping process can begin to build the population level N_2 back up toward and through the threshold value again. This causes the generation of another burst of laser action, and the system can again go through a repeat performance of the same or a very similar cycle.

Turning now to the rate equation, we can interpret these curves as follows: At the beginning of the pump pulse we can assume that the induced emission is negligible because of the low-photon density. During this time we may neglect the term containing ϕ in (1.53) and write

$$\frac{dn}{dt} = W_p n_{\text{tot}}. \tag{3.78}$$

The population inversion therefore increases linearly with time before the development of a large spiking pulse. As the photon density builds up, the stimulated emission terms become important and for the short duration of one pulse the effect of the pumping can be neglected. Therefore, during the actual spiking pulse the rate equations can be written by neglecting both the pumping rate for the excess population and the cavity loss rate in (1.58), (1.61):

$$\frac{dn}{dt} = -\gamma c \sigma n \phi, \qquad \frac{d\phi}{dt} = +c \sigma n \phi. \tag{3.79}$$

The photon density thus grows with time and the population inversion decreases with time. The photon density reaches a peak when the decreasing inversion reaches the threshold value n_{th}. The inversion reaches a minimum at $\gamma n c \sigma \phi \approx W_p n_{tot}$. The cycle repeats itself, forming another spike. The inversion fluctuates in a zigzag fashion around the threshold value n_{th}. As time passes, the peaks become smaller and the curve becomes damped sinusoidal.

Solutions of the laser rate equations predict a train of regular and damped spikes at the output of the laser. Most lasers, however, show completely irregular, undamped spikes. This discrepancy between theory and experiment is due to the fact that the spiking behavior dies out very slowly in most solid-state lasers and therefore persists over the complete pump cycle. Furthermore, mechanical and thermal shocks and disturbances present in real lasers act to continually reexcite the spiking behavior and keep it from damping out. Hence many lasers, especially the ruby laser, spike continuously without ever damping down to the steady state. Depending on the system parameters such as mode structure, resonator design, pump level, and so forth, the spiking may be highly irregular in appearance or it may be regular.

In cw-pumped lasers, such as Nd : YAG, the relaxation oscillations are much weaker and usually consist of damped sinusoidal oscillations around the steady-state value. These oscillations may be treated as perturbations of the steady-state population inversion and photon density given in the rate equations (1.58), (1.61). Compared to the fluorescence time τ_f, the relaxation oscillations have a much shorter period; therefore the term including τ_f in the rate equations can be ignored (i.e., $\tau_f \to \infty$).

We now introduce a small perturbation Δn into the steady-state value of the population inversion n; similarly, a perturbation $\Delta \phi$ is introduced into the steady state of the photon density ϕ. Thus we may write

$$n' = n + \Delta n \qquad \text{and} \qquad \phi' = \phi + \Delta \phi. \tag{3.80}$$

We now proceed to eliminate the population inversion n from (1.61). This is done by first differentiating the equation and then substituting $\partial n / \partial t$ from (1.58). The differential equation is then linearized by introducing n and ϕ from (3.80). Neglecting products of $(\Delta n \Delta \phi)$, we finally obtain

$$\frac{d^2(\Delta \phi)}{dt^2} + c \sigma \phi \frac{d(\Delta \phi)}{dt} + (\sigma c)^2 \phi n (\Delta \phi) = 0. \tag{3.81}$$

The solution of this equation gives the time variation of the photon density

$$\Delta \phi \approx \exp\left[\left(\frac{-\sigma c \phi}{2}\right)\right] t \sin[\sigma c (\phi n)^{1/2} t]. \tag{3.82}$$

The frequency $\omega_s = \sigma c(\phi n)^{1/2}$ and the decay time constant $\tau_R = 2/\sigma c\phi$ of this oscillation can be expressed in terms of laser parameters by noting that $I = c\phi h\nu$ and $n = 1/c\sigma\tau_c$. The latter expression follows from (1.61) for the steady-state condition, i.e., $\partial\phi/\partial t = 0$, and ignoring the initial noise level S. With the introduction of the intracavity power density I and the photon decay time τ_c we obtain

$$\omega_s = \sqrt{\frac{\sigma I}{\tau_c h\nu}} \qquad \text{and} \qquad \tau_R = \frac{2h\nu}{I\sigma}. \qquad (3.83)$$

These expressions can be further simplified for the case of a four-level system by introducing the saturation power density I_s, leading to

$$\omega_s = \sqrt{\frac{I}{I_s \tau_f \tau_c}} \qquad \text{and} \qquad \tau_R = 2\tau_f \left(\frac{I_s}{I}\right). \qquad (3.84)$$

Note, that the greater the power density I and therefore the output power from the laser, the higher the oscillation frequency. The decay time τ_R will decrease for higher output power.

From these equations it follows that the damping time is proportional to the spontaneous lifetime. This is the reason that relaxation oscillations are observed mainly in solid-state lasers where the upper-state lifetime is relatively long. Oscilloscope traces of the relaxation oscillations of a cw-pumped Nd : YAG laser are presented in Fig. 3.12.

Like most solid-state lasers, Nd : YAG exhibits relaxation oscillation. Figure 3.12 exhibits oscilloscope traces of the relaxation oscillations of a small cw-pumped Nd : YAG laser. The oscillation is a damped sine wave with only a small content of harmonics. Figure 3.12(b) displays the spectrum of the relaxation oscillations, as obtained by a spectrum analyzer. From (3.84) it follows that the resonant frequency is proportional to $(P_{out}^{1/2})$ of the laser. Figure 3.12(b) illustrates this dependence. With (3.84) we can calculate the center frequency ν_S and the time constant τ_R of the relaxation oscillations. With the laser operated at 1 W output, a mirror transmission of $T = 0.05$, and beam diameter of 0.12 cm, one obtains $I = 4.0 \times 10^3$ W/cm^2; $l = 35$ cm is the length of the cavity and $\delta = 0.03$ are the combined cavity losses. With these values, and $I_S = 2900$ W/cm^2 and $\tau_f = 230\,\mu$s, it follows that $\tau_R = 333\,\mu$s and $\nu_S = 72$ kHz. The measured center frequency of the relaxation oscillation is, according to Fig. 3.12(b), about 68 kHz.

3.6   Examples of Laser Oscillators

The Nd : YAG laser is by far the most widely used and most versatile solid-state laser. Pumping can be accomplished by means of flashlamps, cw arc lamps, or laser diodes, and the laser can be operated cw or pulsed thereby achieving pulse repetition rates from single shot to several hundred megahertz. In this section

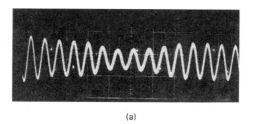

(a)

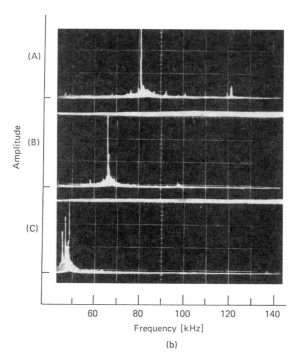

(b)

FIGURE 3.12. Relaxation oscillation of a cw-pumped Nd : YAG laser. (a) Oscilloscope trace showing the temporal behavior of a relaxation oscillation: time scale, $20\,\mu s$/div. (b) Frequency spectrum of relaxation oscillations at different output power levels: (A) 1.3 W, (B) 1.0 W, (C) 0.25 W.

we will relate the performance characteristics of several of these systems to the oscillator model described in this chapter.

3.6.1 Lamp-Pumped cw Nd : YAG Laser

A type of laser which is widely used for materials processing consists of an Nd : YAG crystal 7.5 cm or 10 cm long and with a diameter of 6.2 mm pumped by two krypton filled arc lamps. Electrical input to the lamps can be up to 12 kW. The optical resonator is typically comprised of two dielectrically coated mirrors

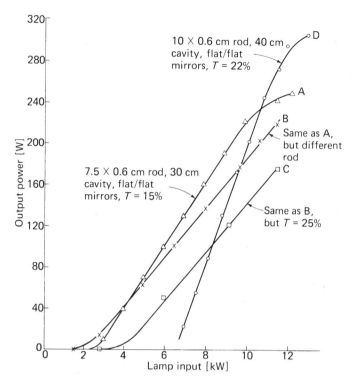

FIGURE 3.13. Continuous output versus lamp input of a powerful Nd : YAG laser.

with a separation of 30 to 40 cm. Curvature of the mirrors and transmission of the output coupler is selected for optimum performance, that is, output power and beam quality. In Fig. 3.13 the output versus input is plotted for a number of different laser crystals and mirror combinations. Curve A shows the output curve of a system that has a slope efficiency of $\sigma_S = 3.1\%$ and an extrapolated threshold of $P_{th} = 2.5$ kW. The nonlinear portion of the curve close to threshold is due to the focusing action of the elliptical reflection. At first, only the center of the rod lases. The nonlinear behavior of the output curve at the high input end is due to thermal lensing within the laser rod. This lensing effect, caused by the thermal gradients in the laser crystal, increases the resonator losses (see Chapter 7).

In Fig. 3.14 the lamp input power required to achieve laser threshold was measured for different mirror reflectivities for the laser system mentioned above. If one plots $\ln(1/R)$ versus P_{in} one obtains a linear function according to (3.72). From this measurement follows a combined loss of $\delta = 0.075$ in the resonator. With this value and the slope efficiency measured for curve A in Fig. 3.13 one can calculate the efficiency factor $\eta = 4.6\%$. The same value can also be obtained from the slope of the curve in Fig. 3.14. Best agreement is obtained if one assumes a saturation power density of $I_S = 2.2$ kW/cm^2. This implies an effective emission cross section somewhere between the two values listed in Table 2.2. From

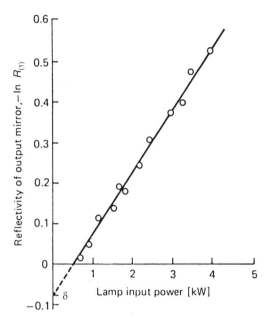

FIGURE 3.14. Threshold power input as a function of mirror reflectivity.

the data of Fig. 3.14 one can plot the small-signal, single-pass gain and the gain coefficient as a function of lamp input power as shown in Fig. 3.15.

It is instructive to calculate the inversion density in the Nd : YAG crystal that has to be sustained to achieve the maximum small-signal gain coefficient of $g_0 = 0.11\,\text{cm}^{-1}$. Assuming a neodymium concentration of $1.38 \times 10^{20}\,\text{cm}^{-3}$ and a stimulated emission cross section of $2.8 \times 10^{-19}\,\text{cm}^2$, it follows from (3.12) that at maximum pump input only 0.28% of the total neodymium atoms are inverted. The small percentage of atoms at the upper laser level is due to the four-level nature of the transition and the large cross section of Nd : YAG.

For a mirror transmission of $T = 0.15$ the small-signal gain coefficient at threshold is about $g = 0.02\,\text{cm}^{-1}$, which results in an upper-state population density of $n_t = 10^{16}\,\text{cm}^{-3}$. From this value and using (3.18) we can calculate the total fluorescence output of the laser at threshold. With $V = 2.3\,\text{cm}^3$, $\tau_f = 230\,\mu\text{s}$, $h\nu = 1.86 \times 10^{-19}$ W s, one obtains $P_f = 130$ W.

If δ and η have been determined for a particular laser crystal and resonator combination, the laser output at different input levels and mirror reflectivities can be calculated from (3.63)–(3.65). An example of such a calculation is shown in Fig. 3.16. The mirror reflectivity which gives the highest output power for the different input powers is located along the dashed curve. This curve has been obtained from (3.66) with (3.61), (3.27). Figure 3.17 shows the intracavity power density as a function of laser output power for a fixed lamp input power of 12 kW. The parameter is the reflectivity of the front mirror. This curve is obtained from Fig. 3.16 (uppermost curve) and using (3.36). As we can see from this figure, the

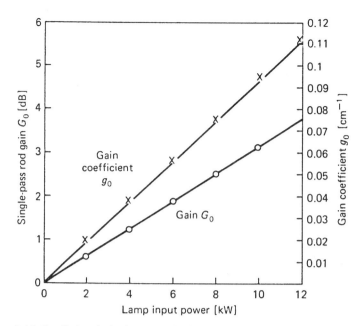

FIGURE 3.15. Small-signal, single-pass rod gain and gain coefficient as a function of lamp input.

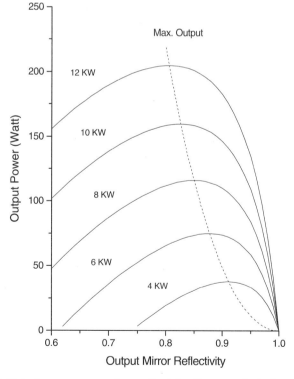

FIGURE 3.16. Output power versus mirror reflectivity. Parameter is the lamp input power. Parameters: $\eta = 0.031$, $\delta = 0.075$, $I_S = 2.2 \, \text{KW/cm}^2$.

110

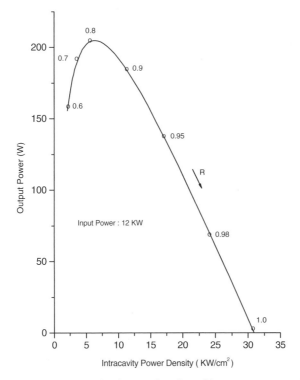

FIGURE 3.17. Intracavity power density as a function of laser output power. Parameter is the reflectivity of the output coupler.

circulating power density in the resonator increased for the higher reflectivities despite the reduction of output power.

Figure 3.18 shows a photograph of a typical, commercially available, cw-pumped Nd : YAG laser. The laser head contains a single arc lamp. The mechanical shutter is used to stop laser oscillations for short periods of time without having to turn off the arc lamp. The purpose of the mode selector and Q-switch will be discussed in Chapters 5 and 8. Bellows are employed between each optical element in the laser head in order to seal out dust and dirt particles from the optical surfaces. The laser head cover is sealed with a gasket in order to further reduce environmental contamination.

3.6.2   Diode Side-Pumped Nd : YAG Laser

The system illustrated here produces an energy per pulse of about 0.5 J at a repetition rate of 40 Hz. Critical design issues for this laser include heat removal from the diode arrays and laser rod and the overlapping of the pump and resonator mode volumes. In side-pumped configurations, laser-diode arrays are not required to be coherent, and pump power can be easily scaled with multiple arrays around the circumference of the rod or along its axis. Instead of one diode array

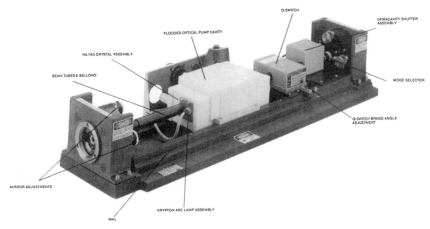

FIGURE 3.18. Photograph of a commercial cw-pumped Nd : YAG laser (Quantronix, Series 100).

pumping the laser crystal, this particular laser employs 16 diode arrays located symmetrically around the rod. As shown in Fig. 3.19 the diode pumps are arranged in four rings, each consisting of four arrays. Since each array is 1 cm long, the total pumped length of the 6.6 cm × 0.63 cm Nd : YAG crystal is 4 cm. This arrangement permits the incorporation of large water-cooled heat sinks required for heat dissipation, and it also provides for a very symmetrical pump profile. An eight-fold symmetry is produced by rotating adjacent rings of diodes by 45°. A photograph of the extremely compact design is also shown in Fig. 3.19. The symmetrical arrangement of the pump sources around the rod produces a very uniform pump distribution, as illustrated in Fig. 3.20. The intensity profile displays the fluorescence output of the rod taken with a CCD camera. In Fig. 3.21, the output versus optical pump input is plotted for long pulse multimode and TEM_{00}

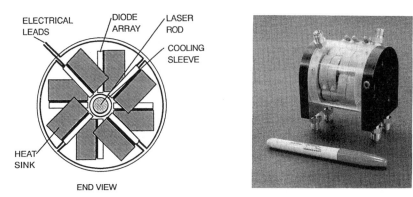

FIGURE 3.19. Cross section (left) and photograph (right) of diode-pumped Nd : YAG laser head.

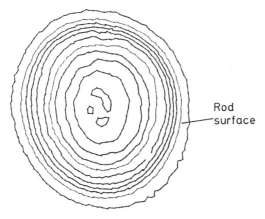

FIGURE 3.20. Pump distribution of a 16-diode array side-pumped Nd : YAG crystal (each line represents a 10% change in intensity).

mode operation. Also shown is the output for Q-switch TEM_{00} mode operation. The resonator configuration for the long pulse, multimode operation is depicted in Fig. 3.22. The TEM_{00} mode performance was achieved with a variable reflectivity mirror and a concave–convex resonator structure which will be described in Chapter 5.

The multimode laser output can be expressed by

$$E_{out} = 0.5(E_{opt} - 180 \, \text{mJ}),$$

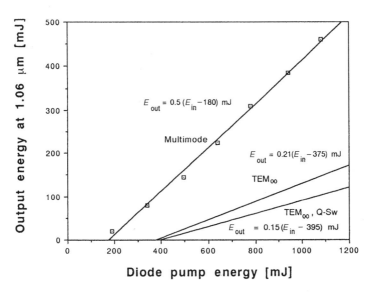

FIGURE 3.21. Output versus input energy for diode-pumped Nd : YAG oscillator.

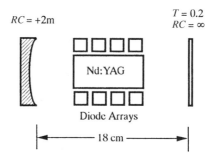

FIGURE 3.22. Resonator configuration for multimode operation.

where E_{opt} is the optical pump energy from the 16 diode arrays. The electrical input energy E_{in} required to achieve E_{opt} is

$$E_{opt} = 0.5(E_{in} - 640\,mJ).$$

Combining these two output–input curves relates the laser output with the electrical input energy

$$E_{out} = 0.25(E_{in} - 1000\,mJ).$$

The slope efficiency of the laser is 25%, and the overall electrical efficiency at the maximum output of 460 mJ per pulse is 16%. The optimum output coupling was experimentally determined. Figure 3.23 shows a plot of the laser output for different values of the reflectivity. The different efficiency factors of the system

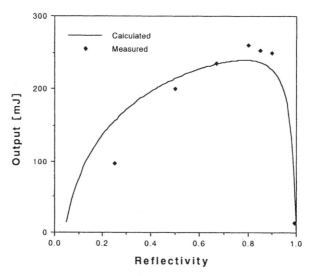

FIGURE 3.23. Oscillator output vs. mirror reflectivity (calculated curve based on $T_{opt} = 0.2$, $g_0 l = 1.4$, $\delta = 0.03$)

TABLE 3.1. Energy transfer efficiencies of a large
diode-pumped Nd : YAG oscillator.

Transfer process		Multimode	Single mode
Diode slope efficiency	η_P	0.50	0.50
Transfer efficiency	η_t	0.95	0.95
Absorption efficiency	η_a	0.90	0.90
Stokes shift	η_S	0.76	0.76
Quantum efficiency	η_Q	0.95	0.95
Coupling efficiency	η_C	0.90	0.90
Beam overlap efficiency	η_B	0.90	0.38
Electrical slope efficiency	σ_S	0.25	0.10

are listed in Table 3.1. The slope efficiency of the diode array was measured with
a power meter. The absorbed pump radiation in the 6.3 mm diameter crystal was
calculated from a computer code.

The coupling efficiency η_C follows from $T_{opt} = 0.20$ and the measured round-
trip loss of $\delta = 2.2\%$. The gain/mode overlap efficiency η_B was estimated by
comparing the beam profile with the pump distribution in the laser rod. The final
value was adjusted for the product of the η terms to agree with the measured slope
efficiency of 25%. For TEM_{00} mode operation, the major difference is a substantially reduced value for η_B as a result of a smaller beam. All other parameters
remain unchanged.

3.6.3   End-Pumped Systems

In this so-called end-pumped configuration, the radiation from a single laser diode
or diode bar is focused to a small spot on the end of a laser rod. With a suitable
choice of focusing optics, the diode pump radiation can be adjusted to coincide
with the diameter of the TEM_{00} resonator mode. The end pumping configuration
thus allows the maximum use of the energy from the laser diodes.

Using this longitudinal pumping scheme, the fraction of the active laser volume
excited by the diode laser can be matched quite well to the TEM_{00} lasing volume.
A solid-state laser pumped in this manner operates naturally in the fundamental
spatial mode without intracavity apertures.

Many excellent systems with extremely compact packaging and high efficiency that utilize the mode matching pump profile are commercially available.
By way of illustrating the concept, we will consider the configuration illustrated
in Fig. 3.24 which was originally proposed by Sipes [5]. The output from a
diode array with 200 mW output at 808 nm is collimated and focused into a
1 cm long × 0.5 cm diameter 1% Nd : YAG sample. The resonator configuration
is plano-concave, with the pumped end of the Nd : YAG rod being coated for
high reflection at 1.06 μm, and with an output coupler having a 5 cm radius of
curvature and a reflectivity at 1.06 μm of 95%.

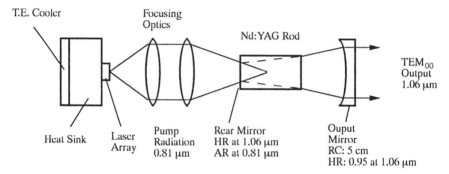

T.E. Cooler

Focusing
Optics

Nd:YAG Rod

TEM$_{00}$
Output
1.06 μm

Heat Sink

Laser
Array

Pump
Radiation
0.81 μm

Rear Mirror
HR at 1.06 μm
AR at 0.81 μm

Ouput
Mirror
RC: 5 cm
HR: 0.95 at 1.06 μm

FIGURE 3.24. End-pumped laser oscillator.

Figure 3.25 displays electrical input power versus 1.06 μm output power for the configuration illustrated in Fig. 3.24. We see that for approximately 1 W of electrical input power, 80 mW of Nd : YAG output is measured.

The electrical slope efficiency of the laser is 13%, and the overall efficiency at 80 mW output is 8%. Also shown in Fig. 3.25 is the performance of the diode array that has a slope efficiency of 34% and an overall efficiency of 22% at 220 mW output. The energy transfer steps of the laser are listed in Table 3.2. The slope efficiency η_P of the diode laser output is a measured quantity. The transfer efficiency η_T includes the collection of diode radiation by the lens system and reflection losses at the surfaces. The pump radiation is completely absorbed in the 1-cm long crystal, i.e., $\eta_a = 1$. The coupling efficiency η_C follows from the measured

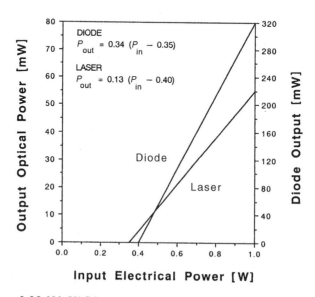

FIGURE 3.25. Nd : YAG laser and diode output versus electrical input power.

TABLE 3.2. Energy transfer efficiencies of
an end-pumped Nd : YAG oscillator.

Transfer process		
Diode slope efficiency	η_p'	0.34
Transfer efficiency	η_t	0.90
Absorption efficiency	η_a	1.00
Stokes shift	η_S	0.76
Quantum efficiency	η_Q	0.95
Coupling efficiency	η_C	0.71
Beam overlap efficiency	η_B	0.85
Electrical slope efficiency	σ_S	0.13

1% one-way loss and an output coupling of 5%. As stated by the authors, the large ellipticity of the diode array beams made it difficult to focus the entire pump beam into the laser resonator mode. The value of η_B is therefore lower than can be achieved with an optimized system. The measured slope efficiency of the laser is the product of the η-terms listed in Table 3.2.

This example of an end-pumped laser demonstrates the high overall performance at TEM$_{00}$ mode which can be achieved with this type of system. The attractive features of end-pumped lasers are a very compact design combined with high beam quality and efficiency, as a result of the good overlap between the pumped region and the TEM$_{00}$ laser mode.

Figure 3.26 displays a photograph of a typical end-pumped laser. The optical components shown separately are the laser diode array, two cylindrical and two

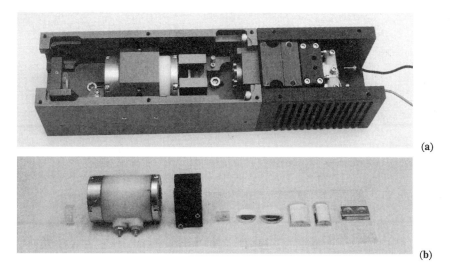

(a)

(b)

FIGURE 3.26. Photograph of (a) assembled end-pumped laser and (b) disassembled optical components. [Fibertek, Inc.].

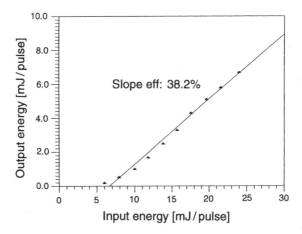

FIGURE 3.27. Laser output as a function of pump energy at the laser crystal.

spherical lenses, an Nd:YAG crystal, polarizer, Q-switch, and output coupler. The end mirror of the resonator is coated onto the laser crystal. The diode array comprises three quasi-cw 1 cm bars. The three-bar stack had a maximum output energy of 27 mJ in a $200\,\mu s$ pulse. At that pulse energy, the laser output was 6.7 mJ in the TEM_{00} mode and 8.4 mJ multimode (Fig. 3.27). The pulse repetition was 50 pps. The resonator loss was determined by measuring the pump threshold as a function of the reflectivities of the output coupler, as described in Section 3.4.2. The measurement yielded a value of 2.4% for the resonator round-trip loss. The maximum output was obtained for an output coupler reflectivity of 95%. The optical design of this laser is depicted in Figs. 6.39.

Summary

Basically a laser oscillator consists of a medium with optical gain inside a resonator formed by two mirrors. The two mirrors perform the function of optical feedback by reflecting the optical radiation back and forth through the gain medium.

The oscillator starts to lase after having been triggered by some spontaneous radiation emitted along the axis of the resonator. This radiation is amplified and a fraction leaks out of the output mirror. Most of the radiation is reflected and amplified again as it passes through the active material. To continue oscillating, the gain per double pass must exceed the losses. As the intensity builds up inside the resonator, the gain is reduced because of gain saturation. A stable operating point is reached when the round trip gain and the losses are equal. Dynamic interactions between the gain of the active medium and the photon flux in the resonator causes damped oscillations around the steady state value.

In order to achieve maximum output from an oscillator under given operating conditions the reflectivity of the output mirror has to be optimized. Increasing the output coupling ($T = 1 - R$) is equivalent to decreasing the feedback factor and increasing the loss in the resonator. The limiting condition is that feedback must be sufficiently large to compensate for the internal and output coupling losses in the system. The transmission of the output mirror will have an optimum value that maximizes the output power from the oscillator. This value is between zero and a transmission at which the losses exceed the gain and the oscillator will cease to lase.

Threshold and laser output as a function of materials and systems parameters are discussed in this chapter. The conversion of electrical input energy to laser output proceeds through a number of steps that can be related to the pump source, pump radiation transfer system, laser medium, and optical resonator. The efficiency of the energy conversion or transfer process associated with each step contributes to the overall efficiency of the laser. The energy flow within a laser oscillator is analyzed in detail, and an illustration of the practical realization of different oscillators concludes the chapter.

References

[1] W.W. Rigrod: *J. Appl. Phys.* **36**, 2487 (1965); also *IEEE J. Quantum Electron.* **QE-14**, 377 (1978).

[2] A.E. Siegman: *Lasers* (University Science Books, Mill Valley, CA), 1986.

[3] W. Koechner, *Solid-State Laser Engineering*, 5th ed. (Springer-Verlag, Berlin), 1999, Fig. 6.82, p. 389.

[4] D. Findlay, R.A. Clay: *Phys. Lett.* **20**, 277 (1966).

[5] D.L. Sipes: *Appl. Phys. Lett.* **47**, 74–76 (1985).

Exercises

1. Prove that when the gain length is equal to the resonator length, the condition giving the required inversion density at lasing threshold, $n = 1/c\sigma\tau_c$, is identical to the threshold condition $2gl = \delta - \ln(R_1) \approx T + \delta$. Indicate any equation you use from the text and define all the quantities that you use.

2. For the system described in Fig. 3.2 show that the quantity $I_{rms} = \sqrt{I_R I_L}$ is constant throughout the resonator.

3. Derive equations (3.64) and (3.65). (*Hint:* Use the approximations given in the text.)

4. The beam overlap efficiency is a very important parameter for evaluating many lasers. Derive equations (3.47) and (3.48). Do this again assuming the beam and gain distributions to be uniform.

5. Derive the expression for the output mirror reflectivity (the commensurate transmission, e.g., $T = 1 - R$, is called the output coupling) that results in

optimum output from a laser (e.g., equation (3.66)). Then insert your result into equation (3.56) and find the output obtained when using this reflectivity (e.g., equation (3.67)).

6. Derive the expressions in equation (3.84). Describe the relaxation oscillations of a laser pumped just a tiny bit above oscillation threshold when it is perturbed a small amount for a brief instant and then returned to its original condition. Sketch the output as a function of time and explain your result.

7. Describe the relaxation oscillations of a laser pumped with a 100 ns duration pump pulse that achieves an intensity twice the saturation power density. The resonator mirrors are 100% and 75% reflecting, and the cavity length is 30 cm. The length of the gain medium is 1 cm, and the nonproductive losses are equivalent to having distributed losses throughout the resonator with a loss coefficient of $0.05 \, \text{cm}^{-1}$. The fluorescent lifetime of the gain medium is 8 ns. Sketch the output as a function of time and explain your result.

8. Starting from the rate equation, for the photon density in a resonator having an average photon lifetime of τ_c and the definition of cavity Q, derive equation (3.9).

4

Laser Amplifier

4.1 Pulse Amplification

4.2 Nd : YAG Amplifiers

4.3 Nd : Glass Amplifiers

4.4 Depopulation Losses

4.5 Self-Focusing

    References

    Exercises

The use of lasers as pulse amplifiers is of great interest in the design of high-energy, high-brightness radiation sources. In the pulse amplifiers described in this chapter the input Q-switched or mode-locked pulse is considerably shorter than the fluorescent lifetime of the active medium. Hence the effect of spontaneous emission and pumping rate on the population inversion during the amplification process can be neglected. Furthermore, energy is extracted from the amplifier, which was stored in the amplifying medium prior to the arrival of the pulse.

Because of the energy storage, pulse amplifiers have high-gain and high-energy extraction. In contrast, the small gain and energy storage of a cw amplifier requires many passes through the medium to achieve useful amplification. This can best be achieved in an oscillator. Therefore, in order to increase the output from a cw oscillator, several laser pump modules are operated in series between the resonant mirrors.

Although cw amplifiers are seldom employed in the traditional bulk solid-state lasers considered in this text, the situation is different for rare-earth-doped fiber lasers. In a fiber laser the pump and signal beam are confined by the fiber core along the entire length of the fiber, which is typically several meters. The long gain path combined with the high pump beam density achievable in the fiber core make high-gain amplification possible. Fiber-optic cw amplifiers represent a rapidly expanding branch of optics and the erbium-doped fiber amplifier (EDFA) is a critical component for telecommunication applications.

FIGURE 4.1. Schematic diagram of a laser oscillator–amplifier configuration.

The generation of high-energy pulses is based on the combination of a master oscillator and a power amplifier. For the purpose of illustrating the amplifier concept and principles we assume a straightforward system, as shown in Fig. 4.1. In this scheme an amplifier is driven by an oscillator which generates an initial laser pulse of moderate power and energy. In the power amplifier with a large volume of active material, stored energy is extracted during the passage of the oscillator pulse.

In an oscillator–amplifier system, pulse width, beam divergence, and spectral width are primarily determined by the oscillator, whereas pulse energy and power are determined by the amplifier. Operating an oscillator at relatively low-energy levels reduces beam divergence and spectral width. Therefore, from an oscillator–amplifier combination one can obtain either a higher energy than is achievable from an oscillator alone, or the same energy in a beam that has a smaller beam divergence and narrower linewidth. Generally speaking, the purpose of adding an amplifier to a laser oscillator is to increase the brightness $B_r[\mathrm{W\,cm}^{-2}\,\mathrm{sr}^{-1}]$ of the output beam

$$B_r = \frac{P_{\text{out}}}{A\Omega},\tag{4.1}$$

where P_{out} is the power of the output beam emitted from the area A and Ω is the solid-angle divergence of the beam. Multiple-stage amplifier systems can be built if higher amplifications are required.

In the design of laser amplifiers the following aspects must be considered:

• Gain and energy extraction.
• Wavefront distortions.
• Feedback which may lead to parasitic oscillations.

In the following section we will address these design considerations.

4.1 Pulse Amplification

Of primary interest in the design of amplifiers is the gain which can be achieved and the energy which can be extracted from the amplifier. The rod length in an amplifier is determined primarily by the desired gain, while the rod diameter, set by damage threshold considerations, is dependent on the output energy.

To a first approximation we can assume the growth of input energy to be exponential, for the amount of stimulated emission is proportional to the exciting photon flux. It will be seen, however, that exponential amplification will occur only at low photon-flux levels. A deviation from the exponential gain regime in an amplifier occurs when an optical pulse traveling in the inverted medium becomes strong enough to change the population of the laser levels appreciably. The optical amplifier will exhibit saturation effects as a result of depletion of the inversion density by the driving signal. Taking an extreme case, we can see that if a high-intensity light pulse is incident on an amplifier, the stimulated emission can completely deplete the stored energy as it progresses. Then the gain can be expected to be linear with the length of the active medium rather than exponential. The events during the amplifier action are assumed to be fast compared with the pumping rate W_p and the spontaneous emission time τ_f. Therefore $t_p \ll \tau_f$, W_p^{-1}, t_p being the width of the pulse which passes through the amplifying medium.

Thus the amplification process is based on the energy stored in the upper laser level prior to the arrival of the input signal. As the input pulse passes through the amplifier, the ions are stimulated to release the stored energy. The amplification process can be described by the rate equations (1.58), (1.61). If we ignore the effect of fluorescence and pumping during the pulse duration, we obtain for the population inversion

$$\frac{\partial n}{\partial t} = -\gamma n c \sigma \phi. \tag{4.2}$$

The growth of a pulse traversing a medium with an inverted population is described by the nonlinear, time-dependent photon-transport equation, which accounts for the effect of the radiation on the active medium and vice versa,

$$\frac{\partial \phi}{\partial t} = c n \sigma \phi - \frac{\partial \phi}{\partial x} c. \tag{4.3}$$

The rate at which the photon density changes in a small volume of material is equal to the net difference between the generation of photons by the stimulated emission process and the flux of photons which flows out from that region. The latter process is described by the second term on the right of (4.3). This term that characterizes a traveling-wave process is absent in (1.61).

Consider the one-dimensional case of a beam of monochromatic radiation incident on the front surface of an amplifier rod of length L. The point at which the beam enters the gain medium is designated the reference point, $x = 0$. The two differential equations (4.2), (4.3) must be solved for the inverted electron population n and the photon flux ϕ. Frantz and Nodvik [1] solved these nonlinear equations for various types of input pulse shapes.

If we take, for the input to the amplifier, a square pulse of duration t_p and initial photon density ϕ_0, the solution for the photon density is

$$\frac{\phi(x,t)}{\phi_0} = \left\{ 1 - [1 - \exp(-\sigma n x)] \exp\left[-\gamma \sigma \phi_0 c \left(t - \frac{x}{c} \right) \right] \right\}^{-1}, \tag{4.4}$$

where n is the inverted population density, assumed to be uniform throughout the laser material at $t = 0$. The energy gain for a light beam passing through a laser amplifier of length $x = l$ is given by

$$G = \frac{1}{\phi_0 t_p} \int_{-\infty}^{+\infty} \phi(l, t) \, dt. \tag{4.5}$$

After introducing (4.4) into (4.5) and integrating, we obtain

$$G = \frac{1}{c\gamma\sigma\phi_0 t_p} \ln\{1 + [\exp(\gamma\sigma\phi_0 t_p c) - 1]e^{n\sigma l}\}. \tag{4.6}$$

We shall cast this equation in a different form such that it contains directly measurable laser parameters. The input energy per unit area can be expressed as

$$E_i = c\phi_0 t_p h\nu. \tag{4.7}$$

A saturation fluence E_s can be defined by

$$E_s = \frac{h\nu}{\gamma\sigma} = \frac{J_{st}}{\gamma g_0}, \tag{4.8}$$

where $J_{st} = h\nu n$ is the stored energy per volume and $g_0 = n\sigma$ is the small-signal gain coefficient.

In a four-level system $\gamma = 1$ and the total stored energy per unit volume in the amplifier is

$$J_{st} = g_0 E_s. \tag{4.9}$$

The extraction efficiency η_E is the energy extracted from the amplifier divided by the stored energy in the upper-laser level at the time of pulse arrival. With this definition, we can write

$$\eta_E = \frac{E_{out} - E_{in}}{g_0 \ell E_s}. \tag{4.10}$$

In this expression E_{out}, E_{in} is the amplifier signal output and input fluence, respectively. In a four-level system, all the stored energy can theoretically be extracted by a signal. In a three-level system $\gamma = 1 + g_2/g_1$, and only a fraction of the stored energy will be released because as the upper laser level is depleted, the lower-level density is building up.

Introducing (4.7), (4.8) into (4.6), one obtains

$$G = \frac{E_s}{E_{in}} \ln\left\{1 + \left[\exp\left(\frac{E_{in}}{E_s}\right) - 1\right]G_0\right\}. \tag{4.11}$$

This expression represents a unique relationship between the gain G, the input pulse energy density E_{in}, the saturation parameter E_s, and the small-signal, single-pass gain $G_0 = \exp(g_0 l)$.

Equation (4.11), which is valid for rectangular input pulses, encompasses the regime from small-signal gain to complete saturation of the amplifier. The equation can be simplified for these extreme cases. Consider a low-input signal E_{in} such that $E_{in}/E_s \ll 1$ and, furthermore, $G_0 E_{in}/E_s \ll 1$; then (4.11) can be approximated to

$$G \approx G_0 \equiv \exp(g_0 l). \tag{4.12}$$

In this case, the "low-level gain" is exponential with amplifier length and no saturation effects occur. This, of course, holds only for rod lengths up to a value where the output energy density $G_0 E_{in}$ is small compared to E_s.

For high-level energy densities such that $E_{in}/E_s \gg 1$, (4.11) becomes

$$G \simeq 1 + \left(\frac{E_s}{E_{in}} \right) g_0 l. \tag{4.13}$$

Thus, the energy gain is linear with the length of the gain medium, implying that every excited state contributes its stimulated emission to the beam. Such a condition obviously represents the most efficient conversion of stored energy to beam energy, and for this reason amplifier designs that operate in saturation are used wherever practical, with the major limitation being the laser damage threshold.

We will now recast (4.11) into a form that makes it convenient to model the energy output and extraction efficiency for single- and multiple-amplifier stages operated either in a single- or double-pass configuration. With the notation indicated in Fig. 4.2, E_{in} is now the input to the amplifier and E_{out} is the output fluence which are related by

$$E_{out} = E_S \ln \left\{ 1 + \left[\exp\left(\frac{E_{in}}{E_S} \right) - 1 \right] \exp(g_0 l) \right\}. \tag{4.14}$$

The extraction efficiency is, according to (4.10),

$$\eta_E = (E_{out} - E_{in})/g_0 l E_S. \tag{4.15}$$

In a laser system which has multiple stages, these equations can be applied successively, whereby the output of one stage becomes the input for the next stage.

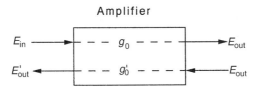

FIGURE 4.2. Notation for the calculation of energy output fluence and extraction efficiency for one- and two-pass single or multiple amplifier stages.

As already mentioned, efficient energy extraction from an amplifier requires that the input-fluence is comparable to the saturation-fluence of the laser transition. For this reason, amplifiers are often operated in a double-pass configuration; a mirror at the output returns the radiation a second time through the amplifier. A $\lambda/4$ waveplate is usually inserted between the amplifier and the mirror; this causes a 90° rotation of the polarization of the return beam. A polarizer in front of the amplifier separates the input from the output signal. In some situations, as shall be discussed in Section 10.4, the simple reflective mirror may be replaced by a phase conjugate mirror, in which case optical distortions in the amplifier chain will be reduced.

The output fluence E'_{out} from a two-pass amplifier can be calculated as follows:

$$E'_{\text{out}} = E_S \ln\left\{1 + \left[\exp(E_{\text{out}}/E_S) - 1\right]\exp(g'_0 l)\right\}. \tag{4.16}$$

The input for the return pass is now E_{out}, which is obtained from (4.14) as the output of the first pass. The gain for the return pass is now lower because energy has been extracted from the gain medium on the first pass

$$g'_0 = (1 - \eta_E)g_0. \tag{4.17}$$

The extraction efficiency of the double-pass amplifier is

$$\eta'_E = (E'_{\text{out}} - E_{\text{in}})/g_0 l E_S. \tag{4.18}$$

The extraction efficiency calculated from (4.14)–(4.18) for one- and two-pass amplifiers, for different values of $g_0 l$ and normalized input fluences, are plotted in Fig. 4.3. The results show the increase in extraction efficiency with higher input energies, and the considerable improvement one can achieve with double-pass amplifiers. Equations (4.14)–(4.18) can be readily applied to multistage systems by writing a simple computer program which sequentially applies these equations to the different amplifier stages. In Section 4.2, we will illustrate the results of such a modeling effort for a four-stage double-pass amplifier chain.

It should be noted that the above equations assume a uniform gain coefficient and beam intensity profile. In most systems, both quantities will have a radially dependent profile. In this case, an effective gain coefficient can be calculated according to

$$g_{\text{eff}} = \int g_0(r) I_B(r) 2\pi r \, dr \Big/ \int I_B(r) 2\pi r \, dr, \tag{4.19}$$

where $g_0(r)$ is the radial gain distribution and $I_B(r)$ is the radial intensity profile of the beam. In laser amplifier technology the small-signal gain coefficient $g_0 = n\sigma$ is sometimes expressed as

$$g_0 = \beta J_{\text{st}}, \tag{4.20}$$

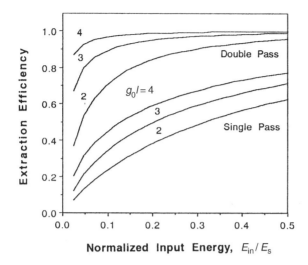

FIGURE 4.3. Extraction efficiency for a one- and two-pass amplifier as a function of input intensity E_{in} and small signal logarithmic gain $g_0 l$. Input is normalized to saturation fluence E_S.

where J_{st} is the previously discussed stored energy per unit volume and

$$\beta = \frac{\sigma}{h\nu} \tag{4.21}$$

is a parameter relating the gain to the stored energy.

4.2   Nd : YAG Amplifiers

Amplified spontaneous emission (ASE) and parasitic oscillations due to the high gain of Nd : YAG effectively limit the energy storage density and therefore the useful energy that can be extracted from a given crystal. The small-signal gain coefficient g_0 and stored energy are related according to $g_0 = \beta J_{st}$. With the materials parameters listed in Table 2.2 we obtain for Nd : YAG, $\beta = 4.73\,cm^2/J$, a value 30 times higher than that for Nd : glass. If we want to extract 500 mJ from an Nd : YAG rod 6.3 mm in diameter and 7.5 cm long, the minimum stored energy density has to be $J_{st} = 0.21\,J/cm^3$. The small-signal single-pass gain in the rod will be $G_0 = \exp(\beta J_{st} l) = 1720$. An Nd : glass rod of the same dimensions would have a gain of 1.3.

As a result of the high gain in Nd : YAG, only small inversion levels can be achieved. Once the gain reaches a certain level, amplification of spontaneous emission effectively depletes the upper level. Furthermore, small reflections from the end of the rod or other components in the optical path can lead to oscillations. These loss mechanisms, which will be discussed in more detail in Section 4.4, lead

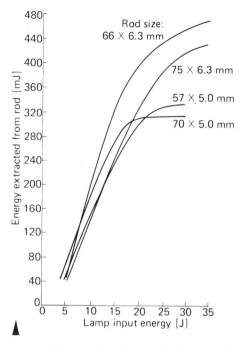

FIGURE 4.4. Energy extraction from Nd : YAG amplifiers with different rod sizes [2].

to a leveling off of the output energy versus the pump input energy curve in an Nd : YAG amplifier. Figure 4.4 shows plots of energy extracted from an Nd : YAG amplifier versus lamp input energy. As can be seen from these curves, the maximum energy that can be extracted from the different rods reaches a saturation level.

The data show that a long rod will provide a long path and therefore high gain for the spontaneous emission to build up, whereas in a relatively short rod of large diameter more total energy can be stored for the same total gain. Increasing the temperature of an Nd : YAG rod will reduce its gain, and therefore more energy can be stored. For example, the extracted energy from an amplifier was increased from 770 to 926 mJ/cm^2 by raising the temperature from 26° to 96° C.

The relative performances of a laser rod when used as a normal-mode oscillator, as a Q-switched oscillator, and as a single-pass amplifier are displayed in Fig. 4.5. A 0.63 by 6.6 cm Nd : YAG laser rod in a silver-plated, single-ellipse, single-lamp pump cavity was used in all modes of operation. Normal-mode performance was achieved with two plane-parallel dielectric-coated mirrors. The pump pulse had a duration of approximately 100 μs at the half-power points. The Q-switched performance was obtained with a rotating prism switch. The single-pass amplifier performance represents the energy extracted from the rod with a 300-mJ input from an oscillator. These data show that all modes of operation are approximately

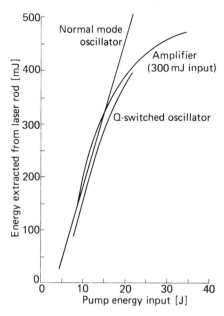

FIGURE 4.5. Energy extraction from Nd : YAG operated as a normal mode and Q-switched oscillator, and amplifier of Q-switched pulses [2].

equivalent until the 320 mJ output level is reached. Above this level depopulation losses decrease the extractable energy.

If the Nd : YAG crystal is pumped with an input of 14 J, a total of 300 mJ can be extracted from the amplifier (top curve in Fig. 4.4). Since the signal input is 300 mJ, the amplifier has a saturated gain of $G = 2$ and a total output of 600 mJ. In order to extract 300 mJ from this rod, at least 0.15 J/cm^3 must be stored in the upper level. This corresponds to a small-signal gain of $G_0 = 108$. Using (4.13) we can compare the measured saturated gain with the theoretical gain. For an amplifier operated in the saturation regime $E_{in} \gg E_s$, the theoretical gain is $G = 1 + (E_s/E_{in})g_0 l$, where the saturation energy for Nd : YAG is $E_s = 1/\beta = 0.2$ J/cm^2. With $E_{in} = 1$ J/cm^2 based on a cross-section area of 0.3 cm^2, $g_0 = 0.71$ cm^{-1}, and $l = 6.6$ cm for this amplifier, we obtain $G = 2$, in agreement with the measurement.

As was mentioned at the beginning of this chapter, the fraction of stored energy that can be extracted from a signal depends on the energy density of the incoming signal. In this particular case the amplifier is completely saturated since $E_{in} = 5E_s$. According to our previous discussion, this should result in a high extraction efficiency. If we introduce $E_0 = 2$ J/cm^2, $E_{in} = 1$ J/cm^2, and g_0, ℓ, and E_S from above into (4.10) we obtain an extraction efficiency of $\eta_E \approx 1$. Figure 4.6 shows the measured extraction efficiency of this amplifier as a function of signal input.

The transition region in which the amplifier output as a function of signal input changes from an exponential (small-signal) relationship to a linear (saturated)

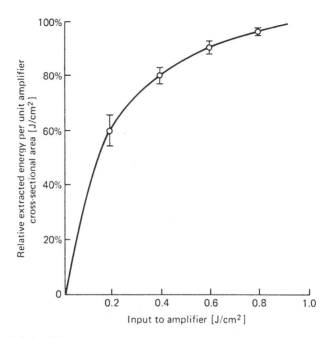

FIGURE 4.6. Amplifier energy extraction as a function of oscillator energy density [2].

relationship occurs for input energy densities of approximately $0.2 \, \text{J/cm}^2$, in accordance with the theoretical value. The pulse width for the data presented in Fig. 4.6 ranged from 15 to 22 ns. The curve in this figure shows that for efficient energy extraction an Nd:YAG amplifier should be operated with an input signal of around $1 \, \text{J/cm}^2$.

In our next example, we will describe a modern, multistage Nd:YAG master-oscillator power-amplifier (MOPA) design, as depicted in Fig. 4.7. The laser produces an output in the TEM_{00} mode of 750 mJ at $1.064 \, \mu\text{m}$ at a repetition rate of 40 Hz. A harmonic generator converts this output to 532 nm with 65% conversion efficiency.

The system features an oscillator and four amplifiers in a double-pass configuration. The linearly polarized output from the oscillator is expanded by a telescope to match the amplifier rods. A Faraday rotator and a $\lambda/2$ wave plate act as a one-way valve for the radiation, thereby isolating the oscillator from laser radiation and amplified spontaneous emission reflected back by the amplifier chain. The output from the oscillator passes through the four amplifiers and, after reflection by a mirror, the radiation passes through the amplifiers a second time. A quarter-wave plate introduces a 90° rotation of the polarized beam after reflection by the rear mirror; this allows the radiation to be coupled out by the polarizer located at the input of the amplifier chain. After a slight expansion, the output beam is passed through a KTP crystal for second-harmonic generation. Located between the two pairs of amplifiers is a 90° rotator which serves the purpose of minimizing

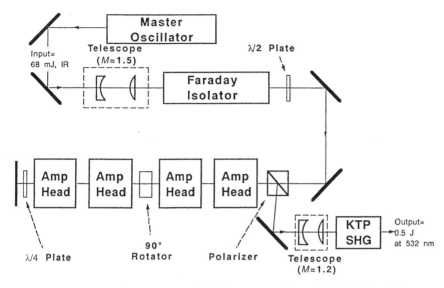

FIGURE 4.7. Optical schematic of a high-power multistage Nd : YAG laser [3].

thermally induced birefringence losses, as will be explained in Section 7.1.1. The changes in the polarization of the beam as it travels back and forth through the amplifier chain is illustrated in Fig. 4.8.

Each amplifier contains 16 linear diode arrays for side pumping of the Nd : YAG crystal. The optical pump energy for each amplifier is 900 mJ at 808 nm, or 4.5 kW at the pump pulse width of 200 μs. In each amplifier, the arrays are arranged in an eightfold symmetrical pattern around the 7.6 mm diameter and 6.5 cm long laser rod in order to produce a uniform excitation. The active length of the rod pumped by the arrays is 4 cm. The small-signal, single-pass gain of one amplifier as a function of pump energy is plotted in Fig. 4.9.

For low input signals, the gain increases exponentially with g_0l according to (4.12). The logarithmic gain g_0l is proportional to the pump energy, as derived

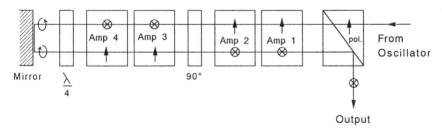

FIGURE 4.8. Two-pass amplifier chain with polarization output coupling and birefringence compensation.

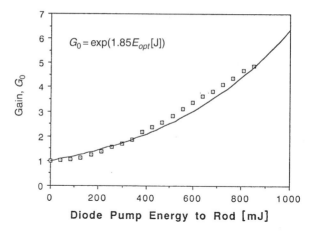

FIGURE 4.9. Small signal, single pass gain as a function of optical pump energy.

in (3.61),

$$g_0 l = \eta' E_p, \qquad (4.22)$$

where E_p is the optical pump energy from the diode arrays and the η terms for an amplifier operated in the energy storage mode are

$$\eta' = \eta_t \eta_a \eta_S \eta_Q \eta_B \eta_{St} \eta_{ASE}/A E_S. \qquad (4.23)$$

For this particular amplifier design, the numerical values are $\eta_t = 0.88$ for the transfer efficiency, $\eta_a = 0.85$ for the absorption of pump radiation in the 7.6 mm diameter Nd : YAG crystal, $\eta_S = 0.76$ and $\eta_Q = 0.90$ for the Stokes shift and quantum efficiency, $\eta_B = 0.62$ for the overlap between the beam and gain region of the rod, and $\eta_{ASE} = 0.90$ and $\eta_{St} = 0.68$ for the storage efficiency. The latter is calculated from (3.53) for a pump pulse length of $t_p = 200\,\mu s$ and a fluorescence lifetime of $\tau_f = 230\,\mu s$ for Nd : YAG. With $A = 0.25\,\mathrm{cm}^2$ and $E_S = 0.44\,\mathrm{J/cm}^2$, one obtains $\eta' = 1.85$. The curve in Fig. 4.9 is based on this value of η', i.e.,

$$E_{out}/E_{in} = \exp[1.85 E_{opt}(J)]. \qquad (4.24)$$

The energy output as a function of the signal input of one amplifier stage single pass and two amplifiers in a double-pass configuration is plotted in Fig. 4.10. The amplifiers are operated at a fixed pump energy of 900 mJ each. Also plotted in this figure is the gain for the two-amplifier double-pass configuration. The amplifiers are highly saturated as can be seen from the nonlinear shape of the E_{out} versus E_{in} curve and the drop in gain at the higher input levels.

The increase in energy as the signal pulse travels forward and backward through the amplifier chain is plotted in Fig. 4.11. Shown are the measured data points and a curve representing the values calculated from (4.14)–(4.18) with $E_{in} = 50\,\mathrm{mJ}$,

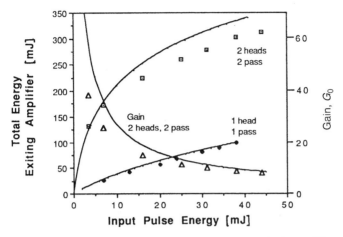

FIGURE 4.10. Energy output as a function of signal input for a single-amplifier stage and a two-amplifier, double pass configuration. Also shown is the gain for the latter configuration.

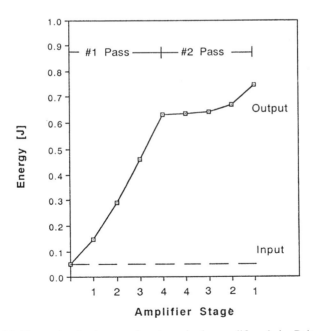

FIGURE 4.11. Energy levels at successive stages in the amplifier chain. Points are measured values and the curve is obtained from a computer model.

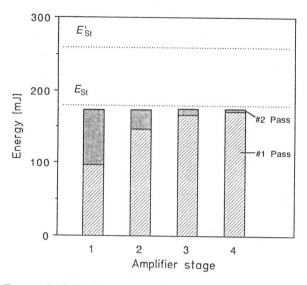

FIGURE 4.12. Total energy extraction from each amplifier stage

$A = 0.25\,\text{cm}^2$, $G_0 = 4.8$, and $E_S = 0.44\,\text{J/cm}^2$. Energy extraction from each stage for both passes is depicted in Fig. 4.12. The data indicate that the amplifiers are totally saturated and all the stored energy within the beam is extracted. As the input beam travels the first time through the amplifiers, successively more energy is extracted from each stage since the ratio of E_{in}/E_S increases. On the return pass, very little energy is removed from the last two stages because stored energy has already been depleted.

The logarithmic gain for the four stages having a 16 cm active length is $g_0 l = 6.0$. The double-pass configuration increases the extraction efficiency from 0.8 to 1.0. It should be noted that the small-signal gain obtained from Fig. 4.9 is measured over the cross section of the beam. The gain and beam profile are both centrally peaked in this design. Both the beam profile and the gain profile have a Gaussian shape with spot sizes of 2.85 and 4.25 mm, respectively. In order to avoid diffraction effects, provide for adequate beam alignment tolerance, and accommodate a slightly expanding beam, the beam cross section at the $1/e^2$ intensity points is $A = 0.25\,\text{cm}^2$, whereas the rod cross section is $0.45\,\text{cm}^2$.

Although the extraction is complete within the beam, stored energy is left at the outer regions of the rod. The beam fill factor takes into account this fact. Since both gain and beam profiles are radially dependent, the beam fill factor cannot be calculated from the ratio of the areas, as would be the case with a uniform gain and beam profile. The beam overlap efficiency can be calculated provided the gain and beam profiles are known. These distributions were obtained from images taken with a CCD camera which recorded the profiles of the fluorescence and laser beam output. With the radial spot sizes of the Gaussian approximations

TABLE 4.1. Energy conversion efficiency of the two-pass multistage Nd : YAG amplifier.

Laser diode array electrical efficiency η_P		0.35
Conversion of optical pump energy to upper state energy		
transfer efficiency	$\eta_t = 0.88$	
absorption efficiency	$\eta_a = 0.85$	0.51
Stokes efficiency	$\eta_S = 0.75$	
quantum efficiency	$\eta_Q = 0.90$	
Conversion of upper state energy to laser output		
beam overlap efficiency	$\eta_B = 0.62$	
storage efficiency	$\eta_{St} = 0.68$	0.38
fractional loss	$\eta_{ASE} = 0.90$	
extraction efficiency	$\eta_{EQ} = 1.00$	
Amplifier efficiency η_{sys}		0.068

given above, one obtains from (3.47) a value of $\eta_B = 0.62$ for the beam overlap efficiency.

According to (4.9) the total stored energy at the upper laser level is $E_{st} = g_0 l E_S A$. With $g_0 l = 1.55$, $E_S = 0.44 \, \text{J/cm}^2$, and $A = 0.25 \, \text{cm}^2$, one obtains $E_{St} = 170 \, \text{mJ}$ of stored energy within the volume addressed by the beam. The stored energy in the full cross section of the rod is $E'_{St} = E_{St}/\eta_B = 260 \, \text{mJ}$. The values for E_{St} and E'_{St} are indicated in Fig. 4.12.

The electrical system efficiency η_{sys} of the amplifier chain is the product of the laser diode efficiency η_p, the conversion efficiency of optical pump power to the upper laser level at the time of energy extraction, and the extraction efficiency of the stored energy into laser output, i.e.,

$$\eta_{sys} = \frac{E_{out} - E_{in}}{E_{EL}} = \eta_p \eta_T \eta_a \eta_S \eta_Q \eta_B \eta_E, \tag{4.25}$$

where

$$\eta_E = \eta_{St} \eta_{ASE} \eta_{EQ}. \tag{4.26}$$

Table 4.1 lists the individual efficiencies of the amplifier system.

4.3   Nd : Glass Amplifiers

An enormous database regarding the design of Nd : Glass amplifiers exists since these systems have become the lasers of choice for inertial confinement fusion research. Motivated by requirements to drive inertial confinement fusion targets at ever higher powers and energies, very large Nd : glass laser systems have been designed, built, and operated at a number of laboratories throughout the world over the past 30 years.

The architectural design of these master oscillator-pulse amplifier (MOPA) systems is based on an oscillator where a single pulse is produced and shaped, then amplified and multiplexed to feed a number of amplifier chains. The preamplifier stages consist of flashlamp-pumped Nd : glass rods, while at the higher power levels amplifier stages employ Nd : glass slabs pumped by flash lamps. The rectangular slabs are mounted at Brewster's angle to minimize Fresnel losses at the surfaces. The clear apertures of the power amplifiers increase stepwise down the chain to avoid optical damage as the beam energy grows. Located between the amplifier stages are spatial filters and Faraday isolators.

Spatial filters are important elements in a high-peak-power laser system and are required to serve three purposes: removal of small-scale spatial irregularities from the beam before they grow exponentially to significant power levels; reduction of the self-induced phase front distortion in the spatial envelope of the beam; and expansion of the beam to match the beam profile to amplifiers of different apertures. Laser oscillation in the chain is prevented by the appropriate placement along the chain of Faraday rotators and polarizer plates. These devices prevent radiation from traveling upstream in the amplifier chain.

The system shown in Fig. 4.13 is an example of such a design. Shown is one 10 kJ beam line of the NOVA glass laser system built at Lawrence Livermore Laboratory. The complete system has 10 such identical beam lines producing a total output energy in excess of 100 kJ in a 2.5 ns pulse. The National Ignition Facility currently under development at Livermore National Laboratory will have 192 beamlines with about 10 kJ output from each aperture in a pulse of 20 ns duration.

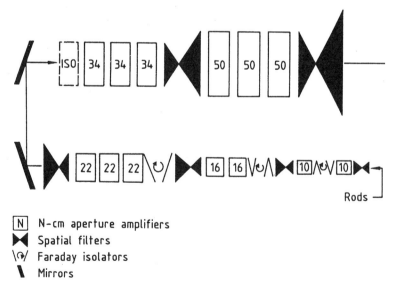

FIGURE 4.13. Component layout of one beam line of the NOVA system (Lawrence Livermore National Laboratory).

The design of these systems is governed by the goal of simultaneously achieving high stored energy and efficient extraction. Both are essential to minimize the number and cost of amplifiers.

High-energy storage implies high gain, which can lead to amplified spontaneous emission owing to the large gain path across the slab dimension. Actually, the maximum energy storage that can be achieved in a large-aperture amplifier is determined by the onset of parasitic lasing, which is due to the formation of a laser resonator by Fresnel reflections at the interfaces of the gain medium. In general, many complex resonator configurations with multiple reflected beams can lase simultaneously. However, it has been shown that the lowest-loss path lies in a plane across the diameter of the disk. With D the diameter of the disk, parasitics will start if

$$\exp(g_0 D) = 1/R, \tag{4.27}$$

where g_0 is the small signal gain coefficient and R is the reflectivity of the edge of the disk.

In Nd : glass disks without special edge cladding the magnitude of the Fresnel reflections is on the order of 5%. The onset of parasitic lasing becomes a problem when the transfer gain approaches 20 or $g_0 D = 3$. Cladding the edge of the disks with index-matched absorbing glass results in a dramatic reduction of parasitic losses. A design parameter and upper limit for state-of-the-art edge cladded large-slab amplifiers is a product of $g_0 D = 4.2$. This value allows for a 1.5% reflection from the disk edge.

Efficient energy extraction requires operation of the amplifier at very high input fluence, as was discussed in Section 4.1. Extraction is particularly efficient at a signal fluence higher than the saturation fluence of Nd : glass. The upper limit is surface and bulk damage in the glass due to optical inhomogeneities, inclusions, and bubbles and laser-induced damage caused by the nonlinear refractive index.

Actually, the principal limit on the performance of a high-power glass laser, such as employed in fusion research, is the nonlinear refractive index. In the presence of an intense light wave, with a power density of gigawatts per centimeter squared or more, the refractive index of a transparent dielectric is increased. This intensity-induced index change is small, on the order of 5 ppm at 10^{10} W/cm^2, but it has profound effects. Changes in the refractive index can be described to first order by the equation

$$n_0(E) = n_0 + \Delta n_0 \tag{4.28}$$

with

$$\Delta n_0 = n_2 \langle E^2 \rangle = \gamma I, \tag{4.29}$$

where n_0 and n_2 is the linear and nonlinear refractive index, respectively; γ is the nonlinear refractive coefficient, E is the electric field strength, and I is the beam intensity. In the literature the nonlinear optical term is expressed by either n_2 or γ, the conversion is

$$\gamma\,[\text{cm}^2/\text{W}] = 4.19 \times 10^{-3} n_2/n_0 \,[\text{esu}].\tag{4.30}$$

From (4.29) it follows that a laser beam propagating in a transparent medium induces an increase in the index of refraction by an amount proportional to the beam intensity.

This nonlinearity gives rise to so-called small-scale ripple growth, in which case the amplitude of small perturbations grows exponentially with propagation distance. In this process, a small region of higher intensity in the beam raises the index locally, which tends to focus light toward it and raises the intensity even more. It has been shown that ripples grow from their original amplitude by a factor of

$$G = \exp(B),\tag{4.31}$$

where the B factor (or breakup integral) is the cumulative nonlinear phase retardation over the optical path length

$$B = \frac{2\pi}{\lambda_0} \int \frac{\Delta n_0}{n_0}\, dl.\tag{4.32}$$

Experience has shown that B needs to be less than one wavelength of delay. For example, for the complete beam line of the NOVA system illustrated in Fig. 4.13 the value is $B = 4$. The control of nonlinear ripple growth leading to beam breakup is an essential part of the design and operation of high-power solid-state lasers.

Figure 4.14 illustrates the progressive growth of beam modulation as the power is increased in high-power glass lasers. The exponential growth of ripples in a high-peak-power system, if left uncontrolled, will quickly lead to component damage and beam degradation.

High-power solid-state laser systems must not only be carefully designed and built to minimize beam irregularities caused by dirt particles, material imperfections, diffraction from apertures, and other "noise" sources; some provision must

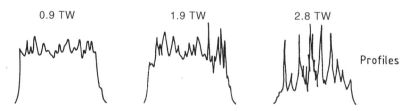

FIGURE 4.14. Beam profiles of a large Nd : glass laser amplifier showing progressive modulation growth (small scale beam breakup) as the power is increased [4].

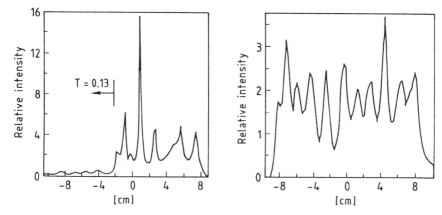

FIGURE 4.15. Relative intensity of one beam line of the SHIVA system at 1.5 TW, before and after installation of the output spatial filter [5].

also be made for removing the remaining small-scale structure, before it becomes large enough to cause severe degradation of the laser's focusing properties.

As was already mentioned, spatial filters accomplish the task of removal of unwanted ripples, imaging, and beam magnification. A spatial filter is a low-pass filter, where pin-hole diameter limits high spatial frequencies that can be passed. The spatial filter is designed to remove the most damaging ripples while passing most of the energy. Therefore ripples, rather than growing exponentially, are reset at each spatial filter. Figure 4.15 shows the relative intensity of one Shiva beam line before and after installation of a spatial filter at the output. As one can see from these densitometer scans, the intensity of the high-frequency spike was reduced by a factor of 4 by the low-pass filter.

A few examples of typical rod and disk amplifiers will illustrate the characteristics of these devices. The front end of each beam line of a high-power Nd:glass laser consists of rod amplifiers. The performance of such an amplifier is illustrated in Fig. 4.16. The Nd:glass rod is pumped by flashlamps with an arc length of 30 cm. At the maximum pump-energy of 22.5 kJ a small signal gain of 59 was measured with a probe beam. From this measurement we calculate a small signal gain of $g_0 = 0.136 \, \text{cm}^{-1}$ from (4.12). Since the saturation fluence from this Nd:glass $E_S = 4.8 \, \text{J/cm}^2$, we obtain from (4.9) a stored energy density of $J_{st} = 0.66 \, \text{J/cm}^3$. The pump efficiency of this amplifier, defined as the ratio of stored energy to pump energy, is $\eta_{amp} = 0.5\%$.

We will illustrate the energy extraction and operating characteristics of disk amplifiers by comparing the measured data of two different amplifier stages of the NOVA beam line illustrated in Fig. 4.13 [7].

The three 22 cm slab amplifiers together have a small signal gain of $G_0 = 9.1$, from which follows a small signal gain coefficient of $g_0 = 0.18 \, \text{cm}^{-1}$ for the combined optical path length of 12 cm. With this gain coefficient, parasitic oscillations can be suppressed since the product $g_0 D$ is below 4. The average input

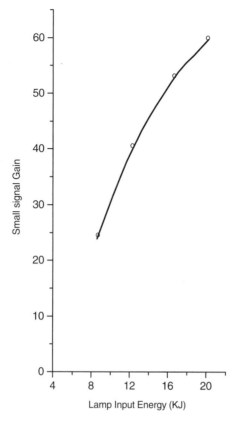

FIGURE 4.16. Small signal gain of an Nd : glass amplifier. Clear aperture 2.5 cm, rod length 38 cm, Hoya laser glass LHG-7 [6].

fluence to this stage is 0.27 J/cm^2 and the output fluence is 1.96 J/cm^2 across the 233 cm^2 of clear aperture area. This total extracted energy is 393 J from this stage, whereas the energy stored in the upper lower level is 5060 J. The energy extraction efficiency from this amplifier stage is about 8%. This efficiency is the product of the beam fill factor and the actual extraction of stored energy within the beam area. To avoid diffraction effects, the laser beam is smaller than the clear aperture of the amplifier. In this particular case the beam fill factor is 0.69. The input fluence, to this amplifier stage, is considerably below the saturation fluence, therefore the energy extraction efficiency across the beam area is relatively low, i.e., 12%. With an emission cross section of $\sigma = 3.8 \times 10^{-20}$ cm^2 for the phosphate glass employed in this laser we obtain a saturation fluence of $E_S = 5.0$ J/cm^2 from (4.8).

High extraction efficiency requires a fluence level at or above the saturation level. At the amplifier stage with a 50 cm aperture the fluence levels are considerably higher compared to the smaller amplifiers. For the combined three amplifiers, shown in Fig. 4.13, the input fluence is $2.6 \, J/cm^2$ and the output fluence is $9.70 \, J/cm^2$ over a clear aperture area of $1330 \, cm^2$. Owing to these high fluence levels, the energy extraction efficiency has risen to 0.54. The beam fill factor is 0.8, which means that about 43% of the total stored energy is extracted from the amplifier. The fluence at the output of this amplifier stage is about twice the saturation fluence, and is about as high as one can safely operate these systems before the potential for optical damage becomes too high. The fluence level at the output of this amplifier stage translates to a peak power of close to $4 \, GW/cm^2$ for a 2.5 ns pulse. The total output at this pulse length is 12.9 kJ. The combined stored energy in all three amplifiers is 23 kJ, which results in a small signal gain of $G_0 = 7.5$. Consistent with the larger aperture the small gain coefficient has been reduced to $g_0 = 0.1 \, cm^{-1}$ to avoid parasitics.

The measured output fluence levels listed above agree with the calculated values from (4.11) to within a few percent if one introduces the small signal gain, input fluence, and saturation fluence for each amplifier into this equation.

4.4 Depopulation Losses

One of the main considerations in the design of an amplifier or amplifier chain is its stability under the expected operating conditions. In a laser operating as an amplifier or as a Q-switched oscillator the active material acts as an energy storage device, and to a large extent its utility is determined by the amount of population inversion that can be achieved. As the inversion is increased in the active material, a number of different mechanisms begin to depopulate the upper level.

Depopulation can be caused by amplified stimulated emission (ASE). A process that will be enhanced by radiation from the pump source falling within the wavelength of the laser transition, or by an increase of the pathlength in the gain medium either by internal or external reflections.

Also, if sufficient feedback exists in an amplifier because of reflections from optical surfaces, the amplifier will start to lase prior to the arrival of the signal pulse. If the reflections are caused by surfaces which are normal to the beam direction, the amplifier will simply become an oscillator and prelasing will occur prior to Q-switching or the arrival of a signal pulse. Reflections that include the cylindrical surfaces of a rod will lead to parasitic modes, propagating at oblique angles with respect to the optical axis of the laser.

4.4.1 Amplified Spontaneous Emission

The level of population inversion that can be achieved in an amplifier is usually limited by depopulation losses which are caused by amplified spontaneous emission. The favorable condition for strong ASE is a high gain combined with a long path length in the active material. In these situations, spontaneous decay occurring near one end of a laser rod can be amplified to a significant level before leaving the active material at the other end. A threshold for ASE does not exist; how-

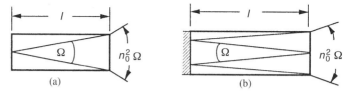

FIGURE 4.17. Directionality and maximum pathlength for ASE in a laser rod (a) without a reflector and (b) with a reflector on one end.

ever, the power emitted as fluorescence increases rapidly with gain. Therefore, as the pump power increases, ASE becomes the dominant decay mechanism for the laser amplifier. At that point, an intense emission within a solid angle Ω around the axis of the active material is observed from each end of the rod

$$\Omega = A/l^2, \qquad (4.33)$$

where l and A are the rod length and cross-sectional area, respectively. As a result of refraction at the end faces, the geometrical aperture angle of the rod is increased by n_0^2, as shown in Fig. 4.17. No mirrors are required for ASE to occur; however, reflections from a mirror or internal reflections from the cylindrical surfaces of the rod will increase the pathlength for amplified spontaneous emission, which will lead to a further increase in intensity. This aspect will be treated in Section 4.4.2.

In high-gain, multistage amplifier systems ASE may become large enough to deplete the upper state inversion. ASE is particularly important in large Nd : glass systems employed in fusion experiments. An analytical expression for the fluorescence flux I_{ASE} from a laser rod as a function of small signal gain, which has been found very useful in estimating the severity of ASE, is given in (4.34) with the approximation $G_0 > 1$ [8]

$$\frac{I_{ASE}}{I_S} = \frac{\Omega}{4} \frac{G_0}{(\ln G_0)^{1/2}}, \qquad (4.34)$$

where I_S is the saturation flux and G_0 is the small signal gain of the active medium. We will apply this formula to a multistage Nd : YAG amplifier and compare the calculated with the measured values of ASE.

In Fig. 4.18, the measured ASE from a four-stage, double-pass Nd : YAG amplifier chain is plotted as a function of diode-pump input. At an input of about 500 mJ into each amplifier ASE starts to become noticeable, and quickly increases in intensity to reach 75 mJ at an input of 900 mJ per amplifier. The detrimental effect of ASE can be seen in Fig. 4.19, which shows the output from the amplifier chain as a function of pump input. As the amplified spontaneous emission increases, the slope of output versus input decreases and the difference can be accounted for by the loss due to ASE.

From (4.34) in conjunction with Fig. 4.9 which shows the gain versus pump input for one amplifier head, we can calculate the ASE from the amplifier chain.

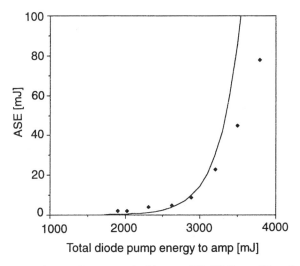

FIGURE 4.18. ASE from a four-stage double pass Nd : YAG amplifier chain. Dots are measured values and the solid line represents calculated values from (4.34).

We have to recall that $\Omega = A/4l^2$ since the pathlength for ASE is doubled in a double-pass system, and $G_0 = (G_0')^8$, where G_0' is the small signal gain in one amplifier. With $A = 0.45\,\mathrm{cm}^2$, $l = 50\,\mathrm{cm}$, $I_S = 2.9 \times 10^3\,\mathrm{W/cm^2}$, and $t_P = 200\,\mu s$, one obtains the curve in Fig. 4.18, which closely matches the observed ASE output.

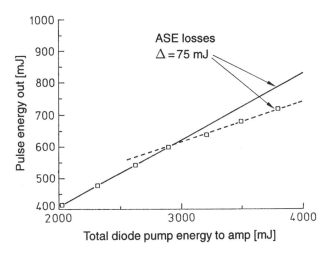

FIGURE 4.19. Signal output versus pump input from a multistage Nd : YAG amplifier chain. (Onset of ASE shown in Fig. 4.18 accounts for drop in output energy.)

4.4.2   Prelasing and Parasitic Modes

Prelasing and parasitic modes are consequences of the interaction of a high-gain material with reflecting surfaces.

Prelasing

Laser action, occurring during the pump phase in an amplifier, results from the residual feedback of the various interfaces in the optical path. The condition for stable operation can be derived by considering the gain in an amplifier rod of length ℓ pumped to an inversion level characterized by a gain coefficient g per unit length. Let the end faces of the rod exhibit residual reflectivities of R_1 and R_2. Then spontaneous emission emanating from any given location in the rod, traveling normal to the rod end faces, will exhibit a loop gain after one round trip in the rod of approximately $R_1 R_2 \exp(2g\ell)$. If this loop gain is greater than unity, oscillations will build up until the usable stored energy is depleted. The requirement for an amplifier is thus

$$R_1 R_2 \exp(2g\ell) < 1, \tag{4.35}$$

which sets an upper limit on rod length.

Parasitic Oscillations

Internal reflections at the boundaries of the active medium can drastically enhance the onset of ASE, particularly if these reflections lead to a closed path, that is, a ray that is reflected upon itself. In this case, we have a strong feedback mechanism and as soon as the gain in the laser medium exceeds the reflection losses parasitic oscillations will set in. For example, in rods with polished walls there can exist internal modes completely contained by total internal reflection. These are particularly troublesome.

One type of internal mode, the whisper mode, propagates circumferentially and has no longitudinal component. This mode can penetrate into the rod as far as r_0/n_0, where n_0 is the index of refraction and r_0 is the rod radius. Another type of internal mode, the light-pipe mode, propagates down the rod in a zigzag fashion. The gain limitations due to internal radial modes are considerably improved in immersed rods particularly in the case of an index matching fluid, or rods with rough surfaces, grooves, or antireflection coatings. Longitudinal modes can be suppressed if the faces of the rod are either wedged or antireflection coated. Parasitic modes in laser disks have already been described in Section 4.3.

4.5   Self-Focusing

Self-focusing is a consequence of the dependence of the refractive index in dielectric materials on the light intensity.

As we discussed in Section 4.3, an intense laser beam propagating in a transparent medium induces an increase in the index of refraction by an amount proportional to the beam intensity. For optical glass a typical value of the nonlinear refractive coefficient is $\gamma = 4 \times 10^{-7}$ cm^2/GW. Thus at a power density of 2.5 GW/cm^2 the fractional index change is 1 ppm. Although a change of this magnitude seems small, it can dramatically affect beam quality and laser performance.

Since laser beams tend to be more intense at the center than at the edge, the beam is slowed at the center with respect to the edge and consequently converges. If the path through the medium is sufficiently long, the beam will be focused to a small filament, and the medium will usually break down via avalanche ionization.

A variety of physical mechanisms result in intensity-dependent contributions to the refractive index, such as electronic polarization, electrostriction, and thermal effects. The latter two mechanisms are responsible for stimulated Brillouin and Rayleigh scattering which form the basis for optical phase conjugation treated in Chapter 10. For short-pulse operation in solids, electronic polarization, also called the optical Kerr effect, which is caused by the nonlinear distortion of the electron orbits around the average position of the nuclei, usually dominates. The optical Kerr effect is utilized in Kerr lens mode-locking discussed in Chapter 9. Here we treat the negative aspects of this physical process as far as the design of lasers is concerned.

Self-focusing of laser radiation occurs when the focusing effect due to the intensity-dependent refractive index exceeds the spreading of the beam by diffraction. For a given value of n_2, two parameters characterize the tendency of a medium to exhibit self-focusing. The first, the so-called critical power of self-focusing, is the power level that will lead to self-focusing that just compensates for diffraction spreading. The second is the focusing length that represents the distance at which an initially collimated beam will be brought to a catastrophic self-focus within the medium for power levels in excess of the critical power.

A first-order description of the phenomenon can be obtained by considering a circular beam of constant intensity entering a medium having an index nonlinearity Δn proportional to power density. We define the critical power P_c to be that for which the angle for total internal reflection at the boundary equals the far-field diffraction angle. The critical angle for total internal reflection is

$$\cos \Theta_c = \frac{1}{1 + (\Delta n_0 / n_0)}. \tag{4.36}$$

If we expand the cosine for small angles and the right-hand side of (4.36) for small $\Delta n / n$, we obtain

$$\Theta_c^2 = 2 \Delta n_0 / n_0. \tag{4.37}$$

The beam expands due to diffraction with a half-cone angle according to

$$\Theta_D \approx 1.22 \frac{\lambda_0}{n_0 D}, \tag{4.38}$$

where D is the beam diameter and λ_0 is the wavelength in vacuum. For $\Theta_c = \Theta_D$, and with

$$P = I(\pi/4)D^2 \qquad (4.39)$$

one obtains from (4.29), (4.37), (4.38) the critical power

$$P_{cr} = \frac{\pi(1.22)^2\lambda_0^2}{8n_0\gamma}. \qquad (4.40)$$

For Gaussian beams the equation for P_{cr} has the same functional form as (4.40) and differs by just a numerical factor.

With the assumption that the shape of the amplitude profile is unchanged under self-focusing, one finds the following relation between power and self-focusing length: For a Gaussian profile whose spot size $w(z)$ changes by a scale factor $f(z) = w(z)/w_0$, the width of the beam varies with z according to [9]

$$f^2(z) = 1 - \left(\frac{P}{P_{cr}} - 1\right)\left(\frac{\lambda z}{\pi w_0^2}\right)^2, \qquad (4.41)$$

where w_0 is the beam waist radius at the entrance of the nonlinear material. For $P \ll P_{cr}$ this expression is identical to (5.5) and describes the divergence of the beam determined by diffraction. For a beam with $P = P_{cr}$ diffraction spreading is compensated for by self-focusing and $f(z) = 1$ for all z; this is referred to as self-trapping of the beam. For larger powers $P > P_{cr}$ self-focusing overcomes diffraction and the beam is focused at a distance

$$z_f = \frac{\pi w_0^2}{\lambda}\left(\frac{P}{P_{cr}} - 1\right)^{-1/2} \approx w_0(2n_0\gamma I_0)^{-1/2}. \qquad (4.42)$$

As can be seen, the self-focusing length is inversely dependent on the square root of the total laser power. Equation (4.42) may be used even when the constant phase surface passing through $z = 0$ is curved, corresponding to a converging or diverging incident beam. One need only replace z_f by

$$z_f(R) = (1/z_f - 1/R)^{-1}, \qquad (4.43)$$

where R is the radius of curvature of the incident phase surface ($R < 0$ for converging beams) and $z_f(R)$ is the new position of the self-focus.

Of course, (4.42) is only valid when the length of the medium is greater than z_f so that catastrophic self-focusing can occur. If the medium is shorter, it acts as a nonlinear lens. If the medium is so short that the beam profile does not change significantly, the equivalent focal length of the nonlinear lens is

$$f = \frac{w^2}{4\gamma I_0 \ell}, \qquad (4.44)$$

where I_0 is the peak intensity and ℓ is the length of the medium. This equation is the same as (9.12) which will be used in the section on Kerr lens mode-locking.

Summary

The generation of high-power laser pulses is based on the combination of a master oscillator and power amplifier. The master oscillator generates an initial pulse of moderate power and energy with the desired pulse duration, beam divergence, and spectral width. In a single or multistage amplifier with a large volume of active material the pulse energy is increased to the desired level. In almost all cases amplifiers are employed to increase the energy of Q-switched or mode-locked pulses. Therefore the pulse width is short compared to the lifetime of the upper laser level, and the small signal gain is determined by the stored energy in the amplifier prior to the arrival of the pulse. For optimal operation the timing between the oscillator and amplifier must be such that there is a maximum population inversion in the amplifier when the pulse from the oscillator enters it.

The fraction of stored energy that can be extracted from the amplifier depends on the ratio of pulse energy to saturation energy at each point in the amplifier. At input energies well below the saturation energy, the pulse energy grows exponentially with length in the active material, at energy levels above the saturation density the energy grows linearly. In the exponential regime, the signal gain is high but the energy extraction is low; the reverse is true for operation of the amplifier in saturation. Typically the growth of energy of a pulse entering an amplifier or amplifier chain changes from an exponential increase at the input side to a linear energy increase with length at the output end. Pulse amplification in the regime of amplifier saturation is determined by eq. (4.11).

In the few cases where an amplifier is employed to amplify continuous radiation, such as in multikilowatt Nd : YAG lasers, or where the pulse duration is long compared to the upper state lifetime, the small signal gain is a function of the pump power, and the gain saturation depends on the ratio of power density to saturation power density at each point as given by eq. (3.25).

The design of an amplifier chain is mainly governed by two considerations: elimination of any feedback path that may lead to oscillations and avoidance of optical damage. Self-oscillation is eliminated by dividing the amplifier chain into a number of stages that are separated by optical isolators. Damage of the amplifier medium due to high power radiation limits the energy and power for a given cross section. Therefore the cross section of the active medium is increased in successive stages of the amplifier chain.

References

[1] L.M. Frantz and J.S. Nodvik: *J. Appl. Phys.* **34**, 2346 (1963).

[2] T.G. Crow and T.J. Snyder: Techniques for achieving high-power Q-switched operation in YAG : Nd. *Final Tech. Report* AFAL-TR-70-69, Air Force, WPAFB (1970); see also *Laser J.* **18** (1970).

[3] W. Koechner, R. Burnham, J. Kasinski, P. Bournes, D. DiBiase, K. Le, L. Marshall, and A. Hays: High-power diode-pumped solid-state lasers for optical space communications, in *Proc. Int'l. Conf. Optical Space Com.* (Munich, 10–14 June 1991) (edited by J. Franz), *SPIE Proc.* **1522**, 169 (1991).

[4] W.W. Simmons, J.T. Hunt, and W.E. Warren: *IEEE J. Quantum Electron.* **17**, 1727 (1981).

[5] D.R. Speck, E.S. Bliss, J.A. Glaze, J.W. Herris, F.W. Holloway, J.T. Hunt, B.C. Johnson, D.J. Kuizenga, R.G. Ozarski, H.G. Patton, P.R. Rupert, G.J. Suski, C.D. Swift, and C.E. Thompson: *IEEE J. Quantum Electron.* **17**, 1599 (1981).

[6] C. Yamanaka, Y. Kato, Y. Izawa, K. Yoshida, T. Yamanaka, T. Sasaki, M. Nakatsuka, T. Mochizuki, J. Kuroda, and S. Nakai: *IEEE J. Quantum Electron.* **17**, 1639 (1981).

[7] C.Bibeau, D.R. Speck, R.B. Ehrlich, C.W. Laumann, D.T. Kyrazis, M.A. Henesian, J.K. Lawson, M.D. Perry, P.J. Wegner, and T.L. Weiland: *Appl. Opt.* **31**, 5799 (1992).

[8] G.J. Linford, E.R. Peressini, W.R. Sooy, and M.L. Spaeth: *Appl. Opt.* **13**, 379 (1974).

[9] S.A. Akhmanov, R.V. Khokhlov, and A.P. Sukhorukov: Self-focusing, self-defocusing and self-modulation of laser beams, in *Laser Handbook* (edited by E.T. Arecchi, E.O. Schulz-DuBois) (North-Holland, Amsterdam), 1972, p. 1151.

Exercises

1. Suppose the square input pulse used to derive (4.11) and its limiting cases, (4.12) and (4.13), is distributed about the amplifier axis located at $r = 0$ with a Gaussian dependence on the radial distance from the axis, r. That is, the input energy density depends on the radius according to

$$E(r) = E_0 \exp\left(-\frac{2r^2}{w_0^2}\right),$$

where E_0 is the value of energy density on the beam axis, e.g., where $r = 0$, and w_0 is the radius at which the electric field in the beam has fallen to $1/e$ of its value on axis. When $E_0 = E_{se}$ and allowing $E(r) = 0$ for $r > 2w_0$, describe the distortion in words to explain your reasoning and with sketches to show what you mean. (*Hint:* Consider the gain seen by input light in various rings going out from the center.)

2. Suppose a square pulse is input into an optical amplifier. What temporal shape would you expect it to have when it exited the amplifier in both the limiting cases in (4.12) and (4.13)? (*Hint:* Consider the effect of amplification on the population density and the sequential amplification of successive time intervals in the pulse.)

3. Find the saturation intensity of Q-246 Nd : glass from the data in Table 2.4. If it is to be used as an amplifier with a small signal gain of a factor of 20, find the input intensity such that the saturated gain of the amplifier is 10. What is the resulting ouput intensity?

5

Optical Resonator

5.1 Transverse Modes
5.2 Longitudinal Modes
5.3 Unstable Resonators
   References
   Exercises

The radiation emitted by most lasers contains several discrete optical frequencies, separated from each other by frequency differences, which can be associated with different modes of the optical resonator. Each mode is defined by the variation of the electromagnetic field perpendicular and along the axis of the resonator. The symbols TEM_{mnq} or TEM_{plq} are used to describe the transverse and longitudinal mode structure of a wave inside the resonator for cylindrical and rectangular coordinates, respectively. The capital letters stand for "transverse electromagnetic waves," and the first two indices identify a particular transverse mode, whereas q describes a longitudinal mode. Although a resonator mode consists of a transverse and axial field distribution, it is useful to consider these two components separately because they are responsible for different aspects of laser performance.

The spectral characteristics of a laser, such as linewidth and coherence length, are primarily determined by the longitudinal modes; whereas beam divergence, beam diameter, and energy distribution are governed by the transverse mode structure. In general, lasers are multimode oscillators unless specific efforts are made to limit the number of oscillating modes. The reason for this lies in the fact that a very large number of longitudinal resonator modes fall within the bandwidth exhibited by the laser transition, and a large number of transverse resonator modes can occupy the cross section of the active material.

5.1 Transverse Modes

The theory of modes in optical resonators has been treated in [1]–[3]; comprehensive reviews of the subject can also be found in [4], [5].

5.1.1   Intensity Distribution

In an optical resonator, electromagnetic fields can exist whose distribution of amplitudes and phases reproduce themselves upon repeated reflections between the mirrors. These particular field configurations comprise the transverse electromagnetic modes of a passive resonator.

Transverse modes are defined by the designation TEM_{mn} for Cartesian coordinates. The integers m and n represent the number of nodes of zeros of intensity transverse to the beam axis in the vertical and horizontal directions. In cylindrical coordinates the modes are labeled TEM_{pl} and are characterized by the number of radial nodes p and angular nodes l. The higher the values of m, n, p, and l, the higher the mode order. The lowest-order mode is the TEM_{00} mode, which has a Gaussian intensity profile with its maximum on the beam axis. For modes with subscripts of 1 or more, intensity maxima occur that are off-axis in a symmetrical pattern. To determine the location and amplitudes of the peaks and nodes of the oscillation modes, it is necessary to employ higher-order equations which involve either Hermite (H) or Laguerre (L) polynomials. The Hermite polynomials are used when working with rectangular coordinates, while Laguerre polynomials are more convenient when working with cylindrical coordinates.

In cylindrical coordinates, the radial intensity distribution of allowable circularly symmetric TEM_{pl} modes is given by the expression

$$I_{pl}(r, \phi, z) = I_0 \varrho^l [L_p^l \varrho]^2 (\cos^2 l\phi) \exp(-\varrho) \tag{5.1}$$

with $\varrho = 2r^2(z)/w^2(z)$, where z is the propagation direction of the beam and r, ϕ are the polar coordinates in a plane transverse to the beam direction. The radial intensity distributions are normalized to the spot size of a Gaussian profile; that is, $w(z)$ is the spot size of the Gaussian beam, defined as the radius at which the intensity of the TEM_{00} mode is $1/e^2$ of its peak value on the axis. L_p is the generalized Laguerre polynomial of order p and index l.

The intensity distribution given in (5.1) is the product of a radial part and an angular part. For modes with $l = 0$ (i.e., TEM_{p0}), the angular dependence drops out and the mode pattern contains p dark concentric rings, each ring corresponding to a zero of $L_p^0(\varrho)$. The radial intensity distribution decays because of the factor $\exp(-\varrho)$. The center of a pl mode will be bright if $l = 0$, but dark otherwise because of the factor ϱ^l. These modes, besides having p zeros in the radial direction, also have $2l$ nodes in azimuth.

The only change in a (pl) mode distribution comes through the dependence of the spot size $w(z)$ on the axial position z. However, the modes preserve the general shape of their electric field distributions for all values of z. As w increases with z, the transverse dimensions increase so that the sizes of the mode patterns stay in constant ratio to each other.

From (5.1) it is possible to determine any beam mode profile. Figure 5.1(a) depicts various cylindrical transverse intensity patterns as they would appear in the output beam of a laser. Note that the area occupied by a mode increases with the mode number. A mode designation accompanied by an asterisk indicates a

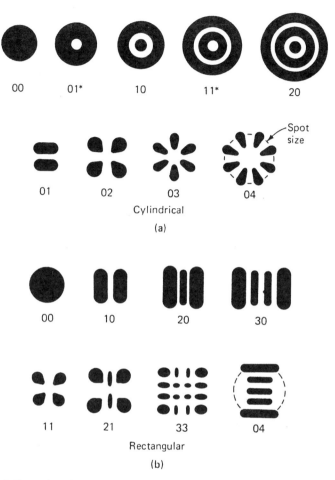

FIGURE 5.1. Examples of (a) cylindrical and (b) rectangular transverse mode patterns. For cylindrical modes, the first subscript indicates the number of dark rings, whereas the second subscript indicates the number of dark bars across the pattern. For rectangular patterns, the two subscripts give the number of dark bars in the x- and y-directions [6].

mode which is a linear superposition of two like modes, one rotated $90°$ about the axis relative to the other. For example, the TEM mode designated 01^* is made up of two TEM_{01} modes.

The intensity distribution of the modes shown in Fig. 5.1(a) can be calculated if we introduce the appropriate Laguerre polynomials into (5.1), that is,

$$L_0^l(\varrho) = 1, \qquad L_1^0(\varrho) = 1 - \varrho; \qquad L_2^0(\varrho) = 1 - 2\varrho + \tfrac{1}{2}\varrho^2.$$

A plot of the intensity distributions of the lowest-order mode and the next two higher-order transverse modes, that is, TEM_{00}, TEM_{01^*}, and TEM_{10}, is illustrated in Fig. 5.2.

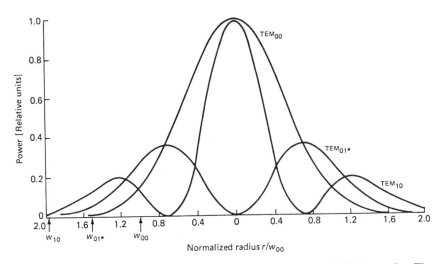

FIGURE 5.2. Radial intensity distribution for TEM$_{00}$, TEM$_{01*}$, and TEM$_{10}$ modes. The radii are normalized to the beam radius w_{00} of the fundamental mode [7].

In rectangular coordinates the intensity distributions of an (m, n) mode is given by

$$I_{mn}(x, y, z) = I_0 \left[H_m \left(\frac{x(2)^{1/2}}{w(z)} \right) \exp \left(\frac{-x^2}{w^2(z)} \right) \right]^2$$

$$\times \left[H_n \left(\frac{y(2)^{1/2}}{w(z)} \right) \exp \left(\frac{-y^2}{w^2(z)} \right) \right]^2. \tag{5.2}$$

As before, $w(z)$ is the spot size at which the transverse intensity decreases to $1/e^2$ of the peak intensity of the lowest-order mode. The function $H_m(s)$ is the mth-order Hermite polynomial, for example, $H_0(s) = 1, H_1(s) = 2s$, and $H_2(s) = 4s^2 - 2$. At a given axial position z, the intensity distribution consists of the product of a function of x alone and a function of y alone. The intensity patterns of rectangular transverse modes are sketched in Fig. 5.1(b). The m, n values of a single spatial mode can be determined by counting the number of dark bars crossing the pattern in the x- and y-directions. Note that the fundamental mode ($m = n = 0$) in this geometry is identical with the fundamental mode in cylindrical geometry.

The transverse modes shown in Fig. 5.1 can exist as linearly polarized beams, as illustrated by Fig. 5.3. By combining two orthogonally polarized modes of the same order, it is possible to synthesize other polarization configurations; this is shown in Fig. 5.4 for the TEM$_{01}$ mode.

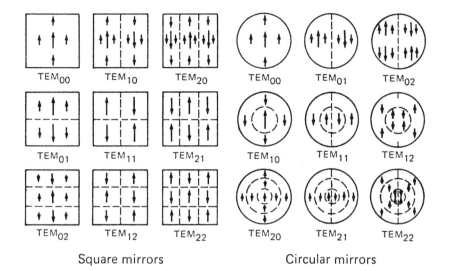

FIGURE 5.3. Linearly polarized resonator mode configurations for square and circular mirrors [1].

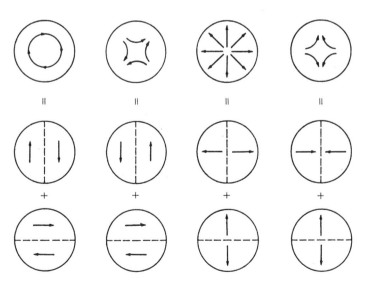

FIGURE 5.4. Synthesis of different polarization configurations from the linearly polarized TEM$_{01}$ mode.

5.1.2 Characteristics of a Gaussian Beam

A light beam emitted from a laser with a Gaussian intensity profile is called the "fundamental mode" or TEM_{00} mode. Because of its importance it is discussed here in greater detail. The decrease of the field amplitude with distance r from the axis in a Gaussian beam is described by

$$E(r) = E_0 \exp\left(\frac{-r^2}{w^2}\right).$$ (5.3)

Thus, the distribution of power density is

$$I(r) = I_0 \exp\left(\frac{-2r^2}{w^2}\right).$$ (5.4)

The quantity w is the radial distance at which the field amplitude drops to $1/e$ of its value on the axis, and the power density is decreased to $1/e^2$ of its axial value. The parameter w is often called the beam radius or "spot size," and $2w$, the beam diameter. The fraction of the total power of a Gaussian beam, which is contained in a radial aperture of $r = w$, $r = 1.5w$, and $r = 2w$, is 86.5%, 98.9%, and 99.9%. If a Gaussian beam is passed through a radial aperture of $3w$, then only $10^{-6}\%$ of the beam power is lost due to the obstruction. For our subsequent discussion, an "infinite aperture" will mean a radial aperture in excess of three spot sizes.

If we consider now a propagating Gaussian beam, we note that although the intensity distribution is Gaussian in every beam cross section, the width of the intensity profile changes along the axis. The Gaussian beam contracts to a minimum diameter $2w_0$ at the beam waist where the phase front is planar. If one measures z from this waist, the expansion laws for the beam assume a simple form. The spot size a distance z from the beam waist expands as a hyperbola, which has the form

$$w(z) = w_0\left[1 + \left(\frac{\lambda z}{\pi w_0^2}\right)^2\right]^{1/2}.$$ (5.5)

Its asymptote is inclined at an angle $\Theta/2$ with the axis, as shown in Fig. 5.5, and defines the far-field divergence angle of the emerging beam. The full divergence angle for the fundamental mode is given by

$$\Theta = \lim_{z \to \infty} \frac{2w(z)}{z} = \frac{2\lambda}{\pi w_0}.$$ (5.6)

From these considerations it follows that, at large distances, the spot size increases linearly with z and the beam diverges at a constant cone angle Θ. A most interesting point here is that, the smaller the spot size w_0 at the beam waist, the greater the divergence.

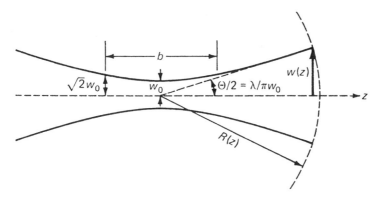

FIGURE 5.5. Contour of a Gaussian beam.

At sufficiently large distances from the beam waist, the wave has a spherical wavefront appearing to emanate from a point on the beam axis at the waist. If $R(z)$ is the radius of curvature of the wavefront that intersects the axis at z, then

$$R(z) = z \left[1 + \left(\frac{\pi w_0^2}{\lambda z} \right)^2 \right]. \tag{5.7}$$

It is important to note that in a Gaussian beam the wavefront has the same phase across its entire surface.

Sometimes the properties of a TEM$_{00}$ mode beam are described by specifying a confocal parameter

$$b = \frac{2\pi w_0^2}{\lambda}, \tag{5.8}$$

where b is the distance between the points at each side of the beam waist for which $w(z) = (2)^{1/2} w_0$ (Fig. 5.5).

In dealing with Gaussian beams the relationship between peak power density and total power is used in several places in this book. A Gaussian beam with a peak power density I_0 at the center and a beam waist radius w_0 is characterized by

$$I(r) = I_0 \exp(-2r^2/w_0^2). \tag{5.9}$$

The total power P in the beam is obtained by integrating over the cross section

$$P = \int I(r) 2r\pi \, dr, \tag{5.10}$$

which leads to

$$P = \pi w_0^2 I_0 / 2. \tag{5.11}$$

For a Gaussian beam, the total power is half the peak intensity times the area defined by the beam radius w_0.

5.1.3 Resonator Configurations

The most commonly used laser resonators are composed of two spherical or flat mirrors facing each other. We will first consider the generation of the lowest-order mode by such a resonant structure. Once the parameters of the TEM_{00} mode are known, all higher-order modes simply scale from it in a known manner. Diffraction effects due to the finite size of the mirrors will be neglected in this section.

The Gaussian beam indicated in Fig. 5.6 has a wavefront curvature of R_1 at a distance t_1 from the beam waist. If we put a mirror at L_1, whose radius of curvature equals that of the wavefront, then the mode shape has not been altered. To proceed further, we can go along the z-axis to another point L_2 where the TEM_{00} has a radius of curvature R_2, and place there a mirror whose radius of curvature R_2 equals that of the spherical wavefront at L_2. Again the mode shape remains unaltered.

Therefore, to make a resonator, we simply insert two reflectors that match two of the spherical surfaces defined by (5.7). Alternatively, given two mirrors separated by a distance L, if the position of the plane $z = 0$ and the value of the parameter w_0 can be adjusted so that the mirror curvatures coincide with the wavefront surfaces, we will have found the resonator mode.

We will now list formulas, derived by Kogelnik and Li [4], which relate the mode parameters w_1, w_2, w_0, L_1, and L_2 to the resonator parameters R_1, R_2, and L. As illustrated in Fig. 5.6, w_1 and w_2 are the spot radii at mirrors M_1 and M_2, respectively; L_1 and L_2 are the distances of the beam waist described by w_0 from mirrors M_1 and M_2, respectively; and R_1 and R_2 are the curvatures of mirrors M_1 and M_2 that are separated by a distance L. Labeling conventions are that concave curvatures are positive.

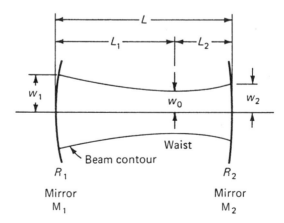

FIGURE 5.6. Mode parameters of interest for a resonator with mirrors of unequal curvature.

The beam radii at the mirrors are given by

$$w_1^4 = \left(\frac{\lambda R_1}{\pi}\right)^2 \frac{R_2 - L}{R_1 - L}\left(\frac{L}{R_1 + R_2 - L}\right),$$

$$w_2^4 = \left(\frac{\lambda R_2}{\pi}\right)^2 \frac{R_1 - L}{R_2 - L}\left(\frac{L}{R_1 + R_2 - L}\right). \qquad (5.12)$$

The radius of the beam waist, which is formed either inside or outside the resonator, is given by

$$w_0^4 = \left(\frac{\lambda}{\pi}\right)^2 \frac{L(R_1 - L)(R_2 - L)(R_1 + R_2 - L)}{(R_1 + R_2 - 2L)^2}. \qquad (5.13)$$

The distances L_1 and L_2 between the waist and the mirrors, measured positive (Fig. 5.6), are

$$L_1 = \frac{L(R_2 - L)}{R_1 + R_2 - 2L}, \qquad L_2 = \frac{L(R_1 - L)}{R_1 + R_2 - 2L}. \qquad (5.14)$$

These equations treat the most general case of a resonator. There are many optical resonator configurations for which (5.12)–(5.14) are greatly simplified. Figure 5.7 shows some of the most commonly used geometries.

Mirrors of Equal Curvature

With $R_1 = R_2 = R$ we obtain, from (5.12),

$$w_{1,2}^2 = \frac{\lambda R}{\pi}\left(\frac{L}{2R - L}\right)^{1/2}. \qquad (5.15)$$

The beam waist that occurs at the center of the resonator $L_1 = L_2 = R/2$ is

$$w_0^2 = \frac{\lambda}{2\pi}[L(2R - L)]^{1/2}. \qquad (5.16)$$

If we further assume that the mirror radii are large compared to the resonator length $R \gg L$, the above formula simplifies to

$$w_{1,2}^2 = w_0^2 = \left(\frac{\lambda}{\pi}\right)\left(\frac{RL}{2}\right)^{1/2}. \qquad (5.17)$$

As follows from (5.17) in a resonator comprised of large-radius mirrors, the beam diameter changes very little as a function of distance.

A resonator comprised of mirrors having a radius of curvature on the order of 2 to 10 m, that is, several times longer than the length of the resonator, is one of the most commonly employed configurations. Such a large-radius mirror resonator has a reasonable alignment stability and a good utilization of the active medium.

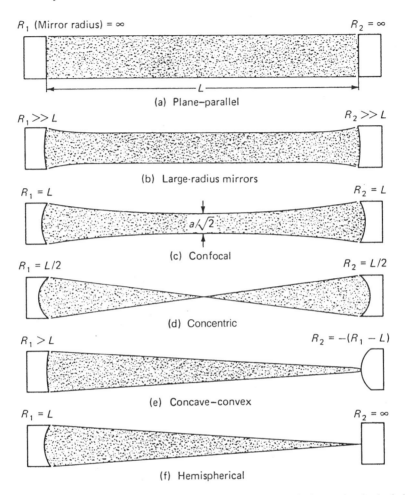

R_1 (Mirror radius) $= \infty$ $R_2 = \infty$

(a) Plane–parallel

$R_1 \gg L$ $R_2 \gg L$

(b) Large-radius mirrors

$R_1 = L$ $R_2 = L$

$a/\sqrt{2}$

(c) Confocal

$R_1 = L/2$ $R_2 = L/2$

(d) Concentric

$R_1 > L$ $R_2 = -(R_1 - L)$

(e) Concave–convex

$R_1 = L$ $R_2 = \infty$

(f) Hemispherical

FIGURE 5.7. Common resonator configurations (intracavity radiation pattern is shaded).

A special case of a symmetrical configuration is the concentric resonator that consists of two mirrors separated by twice their radius, that is, $R = L/2$. The corresponding beam consists of a mode whose dimensions are fairly large at each mirror and which focus down to a diffraction-limited point at the center of the resonator. A concentric resonator is rather sensitive to misalignment, and the small spot can lead to optical damage.

Another very important special case of a resonator with mirrors of equal curvature is the confocal resonator. For this resonator the mirror separation equals the curvature of the identical mirrors, that is, $R = L$. From (5.15), (5.16) we obtain the simplified relation

$$w_{1,2} = \left(\frac{\lambda R}{\pi}\right)^{1/2} \quad \text{and} \quad w_0 = \frac{w_{1,2}}{(2)^{1/2}}. \tag{5.18}$$

The confocal configuration gives the smallest possible mode dimension for a resonator of given length. For this reason, confocal resonators are not often employed since they do not make efficient use of the active material.

Plano-Concave Resonator

For a resonator with one flat mirror ($R_1 = \infty$) and one curved mirror we obtain

$$w_1^2 = w_0^2 = \left(\frac{\lambda}{\pi}\right)[L(R_2 - L)]^{1/2} \quad \text{and} \quad w_2^2 = \left(\frac{\lambda}{\pi}\right)R_2\left(\frac{L}{R_2 - L}\right)^{1/2}.$$

$$(5.19)$$

The beam waist w_0 occurs at the flat mirror (i.e., $L_1 = 0$ and $L_2 = L$). A special case of this resonator configuration is the hemispherical resonator. The hemispherical resonator consists of one spherical mirror and one flat mirror placed approximately at the center of curvature of the sphere. The resultant mode has a relatively large diameter at the spherical mirror and focuses to a diffraction-limited point at the plane mirror. In practice, one makes the mirror separation L slightly less than R_2 so that a value of w_1 is obtained that gives reasonably small diffraction losses.

In solid-state lasers, the small spot size can lead to optical damage at the mirror. A near hemispherical resonator has the best alignment stability of any configuration; therefore it is often employed in low-power lasers such as HeNe lasers.

Concave–Convex Resonator

The pertinent beam parameters for concave–convex resonators can be calculated if we introduce a negative radius $(-R_2)$ for the convex mirror into (5.12)–(5.14). A small-radius convex mirror in conjunction with a large-radius concave or plane mirror is a very common resonator in high-average-power solid-state lasers. As follows from the discussion in the next section, as a passive resonator such a configuration is unstable. However, in a resonator that contains a laser crystal, this configuration can be stable since the diverging properties of the concave mirror are counteracted by the focusing action of the laser rod. Since the concave mirror partially compensates for thermal lensing, a large mode volume can be achieved as will be shown in an example later in this chapter.

Plane–Parallel Resonator

The plane–parallel or flat–flat resonator, which can be considered a special case of the large-radius mirror configuration ($R_1 = R_2 = \infty$), is extremely sensitive to perturbation. However, in an active resonator, that is, a resonator containing a laser crystal, this configuration can be quite useful. Heat extraction leads to thermal lensing in the active medium, this internal lens has the effect of transforming the plane–parallel resonator to a curved mirror configuration as explained later in this section. Therefore the thermally induced lens in the laser material brings the flat–flat resonator into geometric stability.

5.1.4   Stability of Laser Resonators

For certain combinations of R_1, R_2, and L, the equations summarized in the previous subsection give nonphysical solutions (that is, imaginary spot sizes). This is the region where low-loss modes do not exist in the resonator.

Light rays that bounce back and forth between the spherical mirrors of a laser resonator experience a periodic focusing action. The effect on the rays is the same as in a periodic sequence of lenses. Rays passing through a stable sequence of lenses are periodically refocused. For unstable systems the rays become more and more dispersed the further they pass through the sequence. In an optical resonator operated in the stable region, the waves propagate between reflectors without spreading appreciably. This fact can be expressed by a stability criterion [8]

$$0 < \left(1 - \frac{L}{R_1}\right)\left(1 - \frac{L}{R_2}\right) < 1. \tag{5.20}$$

To show graphically which type of resonator is stable and which is unstable, it is useful to plot a stability diagram in which each particular resonator geometry is represented by a point. This is shown in Fig. 5.8, where the parameters

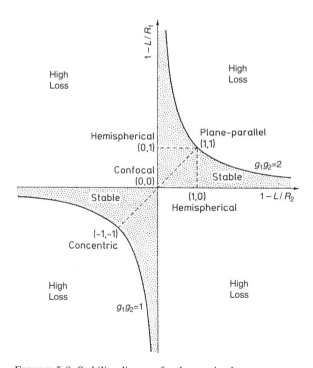

FIGURE 5.8. Stability diagram for the passive laser resonator.

$$g_1 = 1 - \frac{L}{R_1}, \qquad g_2 = 1 - \frac{L}{R_2}, \tag{5.21}$$

are drawn as the coordinate axes.

All cavity configurations are unstable unless they correspond to points located in the area enclosed by a branch of the hyperbola $g_1 g_2 = 1$ and the coordinate axes. The origin of the diagram represents the confocal system.

The resonators located along the dashed line, oriented at 45° with respect to the coordinate axis, are symmetric configurations, that is, they have mirrors with the same radius of curvature.

The diagram is divided into positive and negative branches defining quadrants for which $g_1 g_2$ is either positive or negative. The reason for this classification becomes clear when we discuss unstable resonators.

5.1.5 Higher Order Modes

The fundamental TEM$_{00}$ mode has the smallest beam radius and divergence in the resonator. The beam radius of each mode increases with increasing mode number. Owing to the complicated beam patterns it is not possible to define the beam size by a $1/e$ drop in amplitude as is the case for the TEM$_{00}$ mode. However, an idea of the relationship of beam divergence and mode structure can be obtained if one defines the spot size by a circle containing 90% of the energy for cylindrical modes, and for rectangular modes by a rectangle having dimensions of twice the standard deviation.

With these definitions we obtain, for cylindrical modes [9],

$$w_{pl} = (2p + \ell + 1)^{1/2} w_0 \tag{5.22}$$

and, for rectangular modes [10],

$$w_m = (2m + 1)^{1/2} w_0, \qquad w_n = (2n + 1)^{1/2} w_0, \tag{5.23}$$

where p, ℓ, m, n have been defined in Section 5.1.1 and w_0 is the TEM$_{00}$ mode radius. The beam divergence of each higher-order mode also increases according to the scaling law expressed by (5.22), (5.23). The increase of beam diameter and divergence of a multimode beam can be expressed by

$$\Theta = M\theta_0 \tag{5.24a}$$

and

$$D = MD_0, \tag{5.24b}$$

where the multimode beam divergence Θ and the beam diameter D are related to the fundamental mode beam parameters θ_0 and D_0 by by a factor of M.

It is not sufficient to characterize a laser beam only by its divergence because with a telescope it can always be reduced. The beam property that cannot be corrected by an optical system is the brightness, i.e., the beam intensity per unit solid angle. The brightness theorem states that the product of beam diameter and far-field angle is constant

$$\theta D = M^2 \theta_0 D_0, \tag{5.25a}$$

where M^2 is a dimensionless beam-quality figure of merit and θD is typically expressed as the beam-parameter product (mm-mrad). A laser operating in the TEM$_{00}$ mode is characterized by $M^2 = 1$ and from (5.6) we obtain

$$\theta_0 D_0 = 4\lambda/\pi. \tag{5.25b}$$

The value of M^2 expresses the degree by which the actual beam is above the diffraction limit of an ideal Gaussian beam.

For an Nd : YAG laser emitting at $1.064\,\mu$m this product is $\theta_0 D_0 = 1.35$ mm-mr. An Nd : YAG laser with a low-order mode output such as, for example, TEM$_{20}$ shown in Fig. 5.1 has a beam quality factor of $M^2 = 5$ or, in other words, the beam is five times diffraction limited. The beam-parameter product is about 6.8 mm-mrad.

Actually the output from a multimode laser rarely consists of a single higher-order mode, typically the output is comprised of the incoherent superposition of several modes. Multimode beams comprised of the superposition of modes with beam patterns, as shown in Fig. 5.1, have the property that the beam radius will retain a fixed ratio with respect to the Gaussian beam radius $w(z)$ over all distances. The multimode beam will therefore propagate with distance in the same form as described by (5.5) for a Gaussian beam

$$w'(z) = w_0' \sqrt{1 + \left(\frac{z M^2 \lambda}{\pi w_0'}\right)^2}, \tag{5.26}$$

where $w'(z)$ is the beam radius at z and w_0' is the radius at the waist of the multimode beam.

Because the envelope of a multimode and TEM$_{00}$ beam change in the same ratio over distance, calculations of the propagation of a multimode beam through a resonator can first be performed for a Gaussian beam and then multiplied by a factor $w_{0'}/w_0$ in order to obtain the multimode beam diameter at each point.

5.1.6 Diffraction Losses

In any real laser resonator some part of the laser beam will be lost either by spillover at the mirrors or by limiting apertures, such as the lateral boundaries of the active material. These losses will depend on the diameter of the laser beam in the plane of the aperture and the aperture radius. If we take a finite aperture of radius a within the resonator into account, the diffraction losses depend on four parameters, R_1, R_2, L, and a, which describe the resonator; and on three parameters λ, m, and n, characterizing the particular mode present in the resonator. Fortunately, the losses depend only on certain combinations of these parameters. These combinations are the so-called Fresnel number,

$$N = \frac{a^2}{\lambda L}, \tag{5.27}$$

and the quantities g_1 and g_2, which were defined in (5.21). The parameter N can be thought of as the ratio of the acceptance angle (a/L) of one mirror as viewed from the center of the opposing mirror to the diffraction angle (λ/a) of the beam. Therefore, when N is small, especially if $N < 1$, the loss factor will be high because only a portion of the beam will be intercepted by the mirrors.

Conversely when N is large, diffraction losses will be low. The fractional energy loss per transit because of diffraction effects for the lowest-order mode (TEM$_{00}$) is shown in Fig. 5.9 for resonators with equally curved mirrors and apertures located in front of the mirrors ($g_1 = g_2 = g$, $a_1 = a_2 = a$). The plane-parallel and concentric resonator ($|g| = 1$) have the highest losses for a

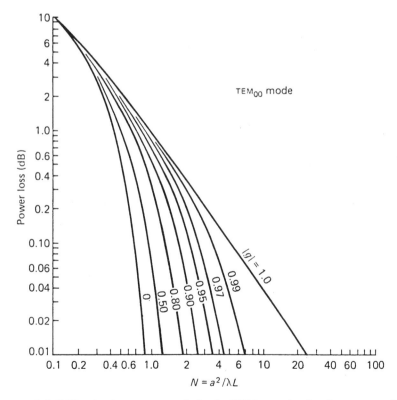

FIGURE 5.9. Diffraction losses per transit for the TEM$_{00}$ mode of various symmetrical resonators [8].

given aperture according to Fig. 5.9. This is not surprising because both resonator configurations have mode sizes which approach infinity at the limit. On the other hand, the confocal resonator ($g = 0$) has the smallest mode dimension for a given resonator length as discussed before. Therefore a given aperture will also cause the lowest diffraction losses. It can also be seen from Fig. 5.9 that the diffraction losses are very sensitive to changes in mirror curvature, and losses decrease very rapidly for all resonator configurations as N increases.

Mode selection or discrimination of higher order modes, by choosing the appropriate Fresnel number, that is, intracavity aperture, is illustrated in Fig. 5.10. In this figure, the diffraction losses in a confocal resonator for a number of low-order modes are plotted versus the Fresnel number. For $N = 1$, only the TEM_{00} and the TEM_{01} modes have a power loss per transit of less than 1% per pass. All other modes have losses above 10%. A laser with this resonator would emit on only these two modes if the gain per pass were less than 10%. Single-mode emission would require a slightly smaller aperture to reduce the value of N just under 1. Going in the other direction, if the aperture is increased by about a factor of 1.4 in diameter to yield $F = 2$, then 10 modes from TEM_{00} to TEM_{05} have less than 1% loss per transit. This resonator would clearly have a multimode output.

5.1.7 Active Resonator

So far we have discussed the modes in a passive resonator consisting of a pair of mirrors. Introducing an active element into the resonator, such as a laser crystal, in addition to altering the optical length of the cavity, will perturb the mode configuration. In solid-state lasers the governing mechanisms that distort the mode structure in the resonator are the thermal effects of the laser rod. As will be discussed in more detail in Chapter 7, optical pumping leads to a radial temperature gradient in the laser rod. As a result, in cw and high average power systems, the rod is acting like a distributed positive thick lens with an effective focal length f, which is inversely proportional to the pump power.

The theory necessary to analyze resonators that contain optical elements other than the end mirrors has been developed by Kogelnik [12]. We will apply this theory to the case of a resonator containing an internal thin lens. To a first approximation, this lens can be thought of as representing the thermal lensing introduced by the laser rod.

Beam properties of resonators containing internal optical elements are described in terms of an equivalent resonator composed of only two mirrors. The pertinent parameters of a resonator equivalent to one with an internal thin lens are

$$g_1 = 1 - \frac{L_2}{f} - \frac{L_0}{R_1}; \qquad g_2 = 1 - \frac{L_1}{f} - \frac{L_0}{R_2}, \qquad (5.28)$$

where $L_0 = L_1 + L_2 - (L_1 L_2 / f)$ and f is the focal length of the internal lens; L_1 and L_2 are the spacings between mirrors M_1, M_2 and the lens, as shown in Fig. 5.11(a).

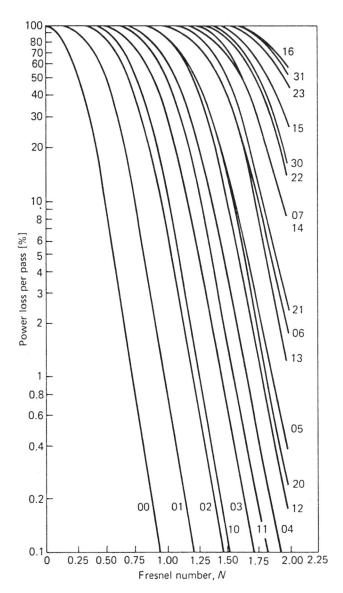

FIGURE 5.10. Diffraction loss for various resonator modes as a function of the Fresnel number [11].

In any resonator, the TEM_{00} mode spot size at one mirror can be expressed as a function of the resonator parameters

$$w_1^2 = \frac{\lambda L}{\pi} \left(\frac{g_2}{g_1(1 - g_1 g_2)} \right)^{1/2}. \tag{5.29}$$

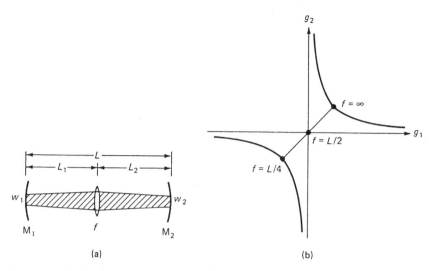

FIGURE 5.11. (a) Geometry and (b) stability diagram of a resonator containing a thin positive lens.

The ratio of the spot sizes at the two mirrors is

$$\frac{w_1^2}{w_2^2} = \frac{g_2}{g_1}. \tag{5.30}$$

The stability condition (5.20) remains unchanged.

As an example we will consider a resonator with flat mirrors ($R_1 = R_2 = \infty$) and a thin lens in the center ($L_1 = L_2 = L/2$). From (5.28) and (5.29) we obtain

$$g = g_1 = g_2 = 1 - \frac{L}{2f}, \qquad w_1^2 = w_2^2 = \left(\frac{\lambda L}{\pi}\right)(1 - g^2)^{-1/2}. \tag{5.31}$$

For $f = \infty$ the resonator configuration is plane-parallel; for $f = L/2$ we obtain the equivalent of a confocal resonator; and for $f = L/4$ the resonator corresponds to a concentric configuration.

The mode size in the resonator will grow to infinity as the mirror separation approaches four times the focal length of the laser rod. Figure 5.11(b) shows the location of a plane-parallel resonator with an internal lens of variable focal length in the stability diagram.

5.1.8 Mode-Selecting Techniques

If the transverse dimension of the gain region of a laser material is larger than the TEM$_{00}$ mode dimension, an oscillator will lase at several modes.

Many applications of solid-state lasers require operation of the laser at the TEM_{00} mode since this mode produces the smallest beam divergence, the highest power density, and, hence, the highest brightness.

For side-pumped lasers output at the fundamental mode usually requires the insertion of an aperture in the resonator to prevent oscillation of higher order modes. The pumped volume of the active material is therefore considerably larger than the TEM_{00} mode volume. The poor utilization of the stored energy in the active medium leads to a poor overall efficiency. A large amount of research has been devoted to the design of optical resonator configurations, which maximize energy extraction from solid-state lasers at the TEM_{00} mode. One finds that a resonator designed for TEM_{00} mode operation will represent a compromise between the conflicting goals of large mode radius, insensitivity to perturbation, good mode discrimination, and compact resonator length. One obvious solution to increase the TEM_{00} mode volume is to make the resonator as long as physical constraints permit, since for a given Fresnel number, the mode cross-sectional area increases proportional with length. Another approach is to utilize resonators, such as the concentric and hemispherical configuration shown in Fig. 5.7. These resonators, due to their focusing action, support large mode size differences along the axis.

For example, in a hemispherical cavity the spot size in the limit can theoretically become zero at the flat mirror and grow to infinity for $L = R$. Location of the laser rod close to the curved mirror permits utilization of a large active volume. An example of this type of resonator is indicated in Fig. 5.12. Mode selection in this resonator, which was employed in a cw Nd : YAG laser, is achieved by axially moving the laser rod until it becomes the limiting aperture for TEM_{00} operation.

Resonators with strong internal focusing suffer from several disadvantages which make them unattractive for most applications. In particular, since it is necessary to operate quite close to the edge of the optically stable region, the configurations are extremely sensitive to mechanical and optical perturbations.

When designing a resonator for large mode size it is very important to consider sensitivity to optical and mechanical perturbations such as the pump-power-dependent thermal lensing in the laser medium and misalignment of the resonator mirrors. We note from (5.28)–(5.30) that for a resonator to have low sensitivity to axial perturbations, for example, caused by thermal lensing, it is necessary that

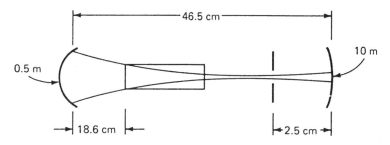

FIGURE 5.12. Focusing resonator geometry [13].

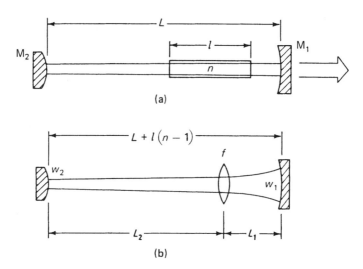

FIGURE 5.13. Convex–concave resonator containing (a) a laser rod and (b) an equivalent passive resonator [14].

$dw_1/df = 0$. This condition is met for a resonator configuration, which satisfies the condition [14]

$$g_1 g_2 = 0.5. \tag{5.32}$$

Equation (5.32) represents a hyperbola located in the middle of the shaded areas of Fig. 5.8. Note from (5.29) that large spot sizes w_1 are obtained for resonators with large g_2 values. From (5.21) it follows that in order for $g_2 > 1$, the radius of curvature of mirror R_2 has to become negative, which indicates a convex mirror according to our labeling conventions. Optical resonators with one convex and one concave mirror in close proximity to the laser medium represent a good compromise for achieving a large mode volume, low sensitivity to perturbation, and a compact design.

As an example we will calculate the parameters for a concave–convex resonator that has been used for a high-repetition-rate Nd : YAG laser (Fig. 5.13(a)). The laser rod, with a diameter of 5 mm, was measured to have a focal length of 6 m for a particular input. The optical length of the cavity was restricted to 0.8 m. The optimum value of w_1 has been found empirically to be equal to one-half the laser rod radius. With the rod as close as possible to the output mirror, the following design parameters are obtained for the equivalent resonator (Fig. 5.13(b)):

$$w_1 = 1.25\,\text{mm}, \qquad f = 6\,\text{m}, \qquad L_1 = 0.1\,\text{m}, \qquad L_2 = 0.7\,\text{m}.$$

Introducing the stability criterion $g_1 g_2 = 0.5$ into (5.29) yields an expression for the mode size as a function of g_1,

$$w_1 = \left(\frac{\lambda L}{\pi g_1}\right)^{1/2}. \tag{5.33}$$

Introducing the values for L and w_1 into this equation, one obtains $g_1 = 0.16$ and, from (5.32), $g_2 = 3.12$. From (5.28) it follows that $R_1 = 1.1\,\text{m}$, $R_2 = -0.36\,\text{m}$; and from (5.30) we obtain $w_2 = 0.28\,\text{mm}$. The mirror M_1 and the rod surface can be combined by grinding a curvature $R'_1 = nR_1$ onto the end of the rod.

As we have seen in side-pumped lasers, expansion of the TEM_{00} mode in combination with an intracavity aperture and operation of the resonator in a region that is less sensitive to perturbation are usually the options available for fundamental mode operation.

In the end-pumped lasers discussed in Sections 3.6.3, 6.2.3, and 7.3 a most elegant solution to the problem of mode selection can be implemented, namely spatial matching between the pump source beam and the resonator TEM_{00} mode. Because the beam characteristics of diode lasers allow for tight focusing of their output radiation into the active material, a near perfect overlap between the pump or gain region and the TEM_{00} mode volume can be achieved. Typically, the pump beam and resonator axis are oriented collinear within the active material. This way highly efficient TEM_{00} mode operation can be achieved.

The equations provided in this chapter for the design of laser resonators are usually sufficient to characterize the most common configurations. Complicated resonator structures containing many optical elements can be evaluated with the aid of ABCD ray transfer matrices or by means of the Jones calculus if the state of polarization is critical.

The ABCD law based on geometrical optics allows us to trace paraxial rays through a complicated sequence of lenslike elements. Each element is characterized by its ABCD matrix, and the transfer function of the whole system is obtained by multiplying all the matrices [4].

If a resonator contains a number of optical elements that change the polarization of the beam, such as waveplates, porro prisms, corner cube reflectors, birefringent crystals, and so forth, the Jones calculus provides a powerful method for the evaluation of such a resonator [15]. The Jones calculus is a 2×2 matrix method in which the state of polarization is represented by a two-component vector and each optical element is represented by a 2×2 matrix. The polarization state of the transmitted beam is computed by multiplying all the matrices. The overall matrix is then multiplied with the vector representing the input beam.

5.2 Longitudinal Modes

5.2.1 The Fabry–Perot Interferometer

The Fabry–Perot interferometer is based on multiple reflections between two plane and parallel surfaces. A basic understanding of its characteristics is important because the laser resonator is closely related to the Fabry–Perot interferome-

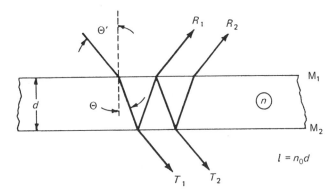

FIGURE 5.14. Interference of a plane wave in a plane-parallel plate.

ter, and the device is also used for longitudinal mode control, and as an instrument to measure line width. We will first consider a particularly simple version of the Fabry–Perot interferometer, called an etalon, which consists of a solid plate with plane and parallel surfaces (see Fig. 5.14).

Multiple reflections which occur between the surfaces cause individual components of the wave to interfere at M_1 and M_2. Constructive interference occurs at M_2 if all components leaving M_2 add in phase. For such preferred directions, the components reflected from M_1 destructively interfere and all the incident energy is transmitted by the etalon. For each member of either the reflected or the transmitted set of waves, the phase of the wave function differs from that of the preceding member by an amount which corresponds to a double traversal of the plate. This phase difference is

$$\varphi = \left(\frac{2\pi}{\lambda}\right) 2n_0 d \cos \Theta, \tag{5.34}$$

where $n_0 d$ is the optical thickness between the two reflecting surfaces, $\Theta = \Theta'/n_0$ is the beam angle in the material, and λ is the wavelength. The transmission of the Fabry–Perot resonator is

$$T = \left[1 + \frac{4R'}{(1 - R')^2} \sin^2\left(\frac{\varphi}{2}\right)\right]^{-1}, \tag{5.35}$$

where R' is the reflectivity of each of the two surfaces. The maximum value of the transmission, $T_{\max} = 1$, occurs when the path length differences between the transmitted beams are multiple numbers of the wavelength

$$2n_0 d \cos \Theta = m\lambda, \qquad m = 1, 2, 3, \ldots. \tag{5.36}$$

The reflectivity of the resonator can be expressed by

$$R = \left(1 + \frac{(1 - R')^2}{4R' \sin^2(\varphi/2)}\right)^{-1}. \tag{5.37}$$

The maximum value of the reflectivity

$$R_{max} = \frac{4R'}{(1 + R')^2} \tag{5.38}$$

is obtained when the pathlength difference of the light beam equals multiples of half-wavelength

$$2n_0 d \cos \Theta = \frac{m\lambda}{2}, \qquad m = 1, 3, 5, \dots . \tag{5.39}$$

Figure 5.15 illustrates the transmission and reflection properties of the etalon. In the absence of absorption losses, the transmission and reflectance of an etalon are complementary in the sense that $R + T = 1$. We define the ratio of the spacing between two adjacent passbands and the passband width as finesse $F = \Delta\lambda/\delta\lambda$:

$$F = \pi \left[2 \arcsin \left(\frac{2 + 4R'}{(1 - R')^2}\right)^{-1/2}\right]^{-1} \approx \frac{\pi (R')^{1/2}}{(1 - R')}. \tag{5.40}$$

The approximation can be used if $R' > 0.5$. For small values of R' the finesse approaches the value $F \approx 2$ and (5.37) is reduced to

$$R = R_{max} \sin^2 \frac{\varphi}{2}. \tag{5.41}$$

We see from (5.36) that the resonance condition of the etalon depends on the wavelength λ, the optical thickness nd, and the beam angle Θ. A variation of any of these quantities will shift the etalon passband. The separation of two passbands,

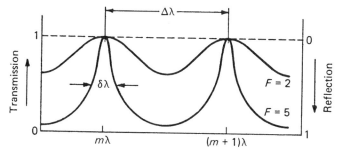

FIGURE 5.15. Intensity of the reflected and transmitted beams as a function of phase difference.

called the free spectral range, is given by

$$\Delta\lambda = \frac{\lambda_0^2}{2n_0 d \cos \Theta} \qquad \text{or} \qquad \Delta\nu = \frac{c}{2n_0 d \cos \Theta}. \qquad (5.42)$$

5.2.2   Laser Resonator

The spectral output from a laser is determined by the gain-bandwidth of the active material and the properties of the Fabry–Perot-type resonator. Typically, many reflectivity peaks of the resonator are within the spectral width of the laser material leading to many longitudinal modes. According to (5.42) the separation of the longitudinal modes in a laser cavity is given by

$$\Delta\lambda = \frac{\lambda_0^2}{2L} \qquad \text{or} \qquad \Delta\nu = \frac{c}{2L}, \qquad (5.43)$$

where L is the optical length of the resonator. For a resonator with $L = 50\,\text{cm}$, adjacent longitudinal modes are separated by $\Delta\nu = 300\,\text{MHz}$. For example, Nd : YAG lasers have a fluorescence linewidth of $120\,\text{GHz}$, and depending on the threshold and pump strength the gain–bandwidth is on the order of $12\,\text{GHz}$. Therefore, about 40 longitudinal modes can oscillate within the resonator described above.

If the reflectivities of the two surfaces of a resonator are different, as in a laser cavity, the reflectivity R' in (5.40) is the geometric mean reflectivity of the two mirrors: $R' = (R_1 R_2)^{1/2}$. Typical values for a laser cavity are $R_1 = 0.8$ and $R_2 = 1.0$ and a resonator length of 50 cm. Introducing these numbers into (5.40), (5.43), one obtains $F = 26.5$ and $\delta\nu = 11.3\,\text{MHz}$ for the finesse and the spectral width of the empty resonator (Fig. 5.16).

In the presence of several transverse modes, additional resonant frequencies occur in the laser cavity. The frequency separation of different TEM_{plq} modes in a laser resonator is given by [4]

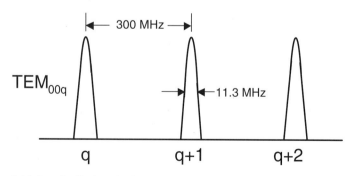

FIGURE 5.16. Longitudinal modes in an empty resonator composed of an 80% and a 100% reflective mirror separated by 50 cm.

$$\Delta v = \left(\frac{c}{2L}\right)\left[\Delta q + \left(\frac{1}{\pi}\right)\Delta(2p+l)\arccos\left(1-\frac{L}{R}\right)\right], \qquad (5.44)$$

where L is the length of the resonator and R is the radius of curvature for both mirrors. The term on the right containing Δq gives the frequency interval of the axial modes. The term with $\Delta(2p+l)$ describes the separation of the resonance frequencies of different transverse TEM_{pl} modes. Note that the resonant frequencies depend on $(2p+l)$, and not on p and l separately. Therefore, frequency degeneracies arise when $(2p+l)$ is equivalent for different modes. By replacing $(2p+l)$ with $(m+n)$, the cavity frequencies for transverse modes expressed in Cartesian coordinates are obtained. From (5.44) it follows that the frequency separation between transverse modes is not only a function of mirror separation, as is the case with the axial modes, but also depends on the curvature of the mirrors.

For a near plane-parallel or long-radius resonator $(L \ll R)$, the second term within the brackets of (5.44) becomes small compared to Δq. In this case the resonant-mode spectrum is composed of the relatively large axial mode spacing $c/2L$, with each axial mode surrounded by a set of transverse-mode resonances, as shown in Fig. 5.17. For example, the 50 cm long resonator mentioned before, if terminated by two mirrors with 5 m curvatures, will have a resonance spectrum of $\Delta v/\mathrm{MHZ} = 300\Delta q + 43(2p+l)$. Because of these additional resonance frequencies, which occur with higher order transverse modes, techniques for narrowing the spectral output of a laser (to be discussed in Section 5.2.3) require operation at the TEM_{00} mode.

If a laser is operated without any mode-selecting elements in the resonator, then the spectral output will consist of a large number of discrete frequencies determined by the transverse and axial modes. The linewidth of the laser transition limits the number of modes that have sufficient gain to oscillate. The situation is diagrammed schematically in Fig. 5.18, which shows the resonance frequencies of an optical resonator and the fluorescence line of the active material. Laser emission occurs at those wavelengths at which the product of the gain of the laser

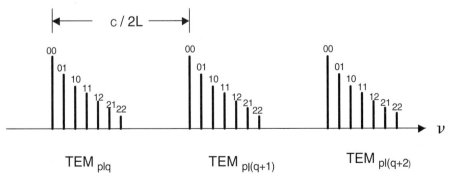

FIGURE 5.17. Resonance frequency of a resonator containing higher order transverse modes.

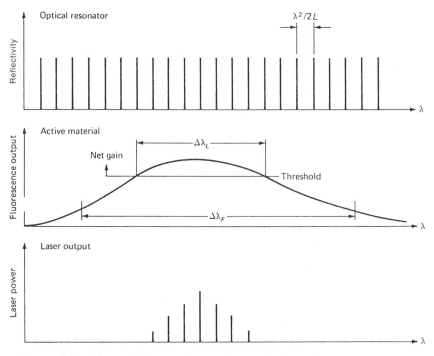

FIGURE 5.18. Schematic diagram of spectral output of a laser without mode selection.

transition and the reflectivity of the mirrors exceed unity. In the idealized example shown, the laser would oscillate at seven axial modes.

The beam emitted from a laser, which emits at a discrete number of integrally related wavelengths, is strongly modulated. The situation can be illustrated by considering the simplest case of two superimposed traveling waves whose wavelengths are specified by adjacent axial modes. This situation is shown schematically in Fig. 5.19. The two waves interfere with one another and produce traveling nodes that are found to be separated from one another in time by twice the cavity separation. For the resonator mentioned above the bandwidth of the laser is

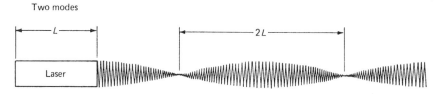

FIGURE 5.19. Amplitude modulation of a laser beam due to emission of two adjacent longitudinal modes.

$\Delta v = 300\,\text{MHz}$ and the coherence length, if defined by

$$l_c = c/2\pi\,\Delta v, \tag{5.45}$$

has a value of $l_c = 16\,\text{cm}$.

5.2.3 Longitudinal Mode Control

In contrast to transverse mode control, which is important for most laser applications, control of the longitudinal modes is generally only required for lasers employed in interferometric applications, coherent lidar, or holographic systems. A reduction of longitudinal modes can be achieved by insertion of a tilted etalon into the resonator. This will cause a strong amplitude modulation of the closely spaced reflectivity peaks of the basic laser resonator, and thereby prevent most modes from reaching threshold.

The etalon is inserted at a small angle in the laser resonator. The tilt effectively decouples the internal transmission etalon from the resonator; that is, no other resonances will be formed with other surfaces in the main resonator. If the etalon is sufficiently misaligned, it acts simply as a bandpass transmission filter. The tilted etalon has no reflection loss for frequencies corresponding to its Fabry–Perot transmission maxima. At other frequencies the reflections from this mode selector are lost from the cavity, and thus constitute a frequency-dependent loss mechanism. The transmission maxima can be tuned into the central region of the gain curve by changing the tilt angle.

The transmission peaks of an etalon with index of refraction $n = 1.5$ and thickness $d = 1$ cm are separated by 10 GHz according to (5.42). If, for example, the etalon has a finesse of $F = 10$, the spectral width is 1 GHz, which is considerably narrower than the 12 GHz gain-bandwidth of a typical Nd:YAG laser. The laser output will be composed of longitudinal modes, which fall inside the transmission band of the etalon, hence a narrowing of the laser linewidth is observed. In pulsed lasers longitudinal mode selection can be enhanced by a passive Q-switch. Longitudinal-mode selection in the laser takes place while the pulse is building up from noise. During this buildup time, modes that have a higher gain or a lower loss will increase in amplitude more rapidly than the other modes. Besides differences in gain or losses between the modes, there is one other important parameter which determines the spectral output of the laser. This parameter is the number of round trips it takes for the pulse to build up from noise. The difference in amplitude between two modes becomes larger if the number of round trips is increased. Therefore, for a given loss difference between the modes it is important for good mode selection to allow as many round trips as possible.

Since a passive Q-switch requires more round trips for the development of a pulse, as compared to a Pockels cell Q-switch, the former has better mode selection properties. While the above-mentioned techniques will narrow the line width of a laser, reliable single longitude mode output can only be achieved from a laser which meets the following design criteria: a very short resonator such that

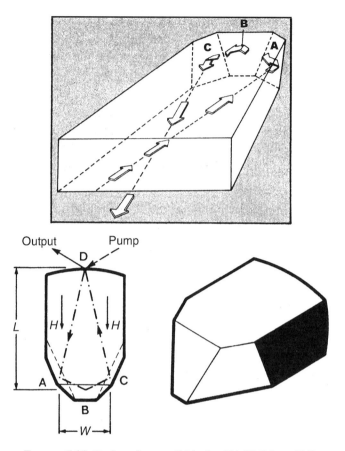

FIGURE 5.20. Design of a monolithic ring Nd:YAG laser [16].

the axial modes are spread so far apart that only one mode has sufficient gain; elimination of spatial hole burning caused by the standing wave pattern in the resonator; mechanical stability of the resonator elements; and a stable, low-noise pump source. These desirable features are combined, for example, in the diode-pumped, monolithic ring laser illustrated in Fig. 5.20.

The laser resonator is only a few millimeters long, therefore axial modes are spaced sufficiently far apart for axial mode control. The ring laser configuration avoids spatial hole burning, and end-pumping provides a small gain volume for efficient and stable TEM_{00} mode operation. Furthermore, the monolithic design assures good mechanical stability of the resonator. Spatial hole burning is a descriptive term for the fact that, along the axis of the resonator, "holes" in the inversion are burnt as a result of the standing wave nature of the mode. The fixed nodal planes on the other hand represent regions of untapped inversion which will preferentially contribute gain to other modes leading to mode-hopping. The tendency of a laser for mode-hopping can be eliminated by providing relative motion between the atoms in the active material and the electric field of the resonator. This task can be accomplished with a traveling wave ring structure.

The oscillators that we have discussed so far are characterized by standing waves in the resonator. In an oscillator consisting of a ring-like resonator utilizing three or four mirrors and a nonreciprocal optical gate, a traveling wave can be generated. The optical gate provides a high loss for one of the two countercirculating traveling waves. The wave with the high loss is suppressed, and unidirectional output from the laser is obtained.

In a discrete element ring laser the unidirectional gate is formed by a Faraday rotator, a half-wave plate, and a polarizer. The Farady device rotates the plane of polarization by a small angle $\pm\theta$, the sign being dependent on the propagation direction. The half-wave plate is oriented such that in one direction it compensates for the polarization rotation induced by the Faraday element. For this direction the beam passes the polarizer without loss. In the opposite direction the beam experiences a 2θ polarization rotation and suffers a loss at the polarizer.

The basic idea of a monolithic, diode-pumped unidirectional ring laser is to provide the equivalent of a discrete element design. The polarizer, half-wave plate equivalent, and the Faraday rotator are all embodied in the nonplanar ring Nd : YAG laser illustrated in Fig. 5.20. With a magnetic field H present in the direction shown, the YAG crystal itself acts as the Faraday rotator, the out-of-plane total internal reflection bounced (labeled A and C) acts as the half-wave plate, and the output coupler (mirror D) acts as a partial polarizer. Polarization selection results from nonnormal incidence at the dielectrically coated output mirror.

Such a device produces an extremely stable, narrow linewidth cw output of a few hundred milliwatts. For most applications the power must be increased by injecting the output from this laser source into a large slave oscillator. Both the injected single-mode radiation and spontaneous emission from the high-power Q-switch slave laser will be regeneratively amplified in the slave cavity during the pulse buildup period. If the injected signal has enough power on a slave cavity resonance, the corresponding single axial mode will eventually saturate the homogeneously broadened gain medium and prevent development of any other axial modes.

The linewidth of the output of a laser operated in a single axial mode is usually many orders of magnitude narrower than the linewidth of the empty or passive resonator. The theoretical limit of the laser linewidth is determined by spontaneously emitted radiation which mixes with the wave already present in the resonator and produces phase fluctuations. The photons in the resonator all correspond to in-phase waves except for the spontaneous emission, which occurs at the same frequency but with a random phase. It is this phase jitter that causes the finite laser linewidth $\Delta\nu_L$ which is given by the Schalow–Townes limit [17]

$$\Delta\nu_L = \frac{2\pi h\nu(\Delta\nu_c)^2}{P_{out}}, \qquad (5.46)$$

where $\Delta\nu_c$ is the linewidth of the passive resonator, $h\nu$ is the photon energy, and P_{out} is the laser output.

The limit can also be expressed as

$$\Delta \nu_L = \frac{\Delta \nu_c}{N_{\mathrm{ph}}} \tag{5.47}$$

which says that the theoretical minimum laser linewidth equals the linewidth of the empty resonator divided by the number of photons in the resonator. The above equation is derived if we note that the laser output power P_{out} equals the number of photons N_{ph} in the resonator times the energy per photon divided by the photon lifetime τ_c,

$$P_{\mathrm{out}} = \frac{N_{\mathrm{ph}} h \nu}{\tau_c}. \tag{5.48}$$

Furthermore, from Chapter 3, we recall that in a resonator the loss mechanism, besides limiting the lifetime of the oscillation, causes a broadening of the resonance frequency. The width $\Delta \nu$ of the resonance curve, at which the intensity has fallen off to half the maximum value, is

$$\Delta \nu = (2\pi \tau_c)^{-1}. \tag{5.49}$$

Introducing (5.48), (5.49) into (5.46) results in the expression (5.47).

We will now calculate the fundamental linewidth limits for such a laser and compare it with the values achieved in actual systems. The physical length of a minilaser is typically 5 mm and the refractive index is 1.82; therefore the optical length of the resonator is $2L = 1.82$ cm. The output coupling and internal losses for such a monolithic laser are $\delta + T = 0.01$. These values result in a lifetime of the photons in the resonator of $\tau_c = 6$ ns or a linewidth of the empty resonator of $\Delta \nu_c = 26$ MHz. The theoretical limit for the laser linewidth at a typical output power of 1 mW is, according to (5.46), $\Delta \nu_L = 1$ Hz. In practice, the laser linewidth is determined by temperature fluctuations and mechanical vibrations and instabilities which produce rapid changes in frequency. For example, the temperature-tuning coefficient for a monolithic Nd : YAG laser operating at 1064 nm is 3.1 GHz/$^\circ$ C. The major contribution is the temperature dependence of the refractive index. The design of narrow linewidth laser oscillators requires extremely good temperature control. A state-of-the-art diode-pumped monolithic Nd : YAG laser has achieved a linewidth of 300 Hz during a 5 ms observation time, 10 kHz in time intervals of 300 ms, and the laser exhibited a long-term drift of 300 kHz/min [18].

5.3 Unstable Resonators

As was discussed in Section 5.1, in a stable resonator the radiation is confined between the surfaces of the resonator mirrors and do not walk out past their edges. In order to produce a diffraction-limited output beam from a stable resonator, the Fresnel number must be on the order of unity or smaller (Section 5.1.5), otherwise sufficient discrimination against higher-order modes cannot be achieved. For practical resonator lengths, this usually limits the diameter of the TEM_{00} mode to a few millimeters or less.

If the resonator parameters g_1 and g_2 lie outside the shaded regions defined in Fig. 5.8, one obtains an unstable resonator. In these configurations, the beam is no longer confined between the mirrors. A beam in an "unconfining" or unstable resonator diverges away from the axis, as shown in Fig. 5.21, eventually radiation will spill around the edges of one or both mirrors. This fact can, however, be used to advantage if these walk-off losses are turned into useful output coupling.

The most useful property of an unstable resonator is the attainment of a large fundamental mode volume and good spatial mode selection at high Fresnel numbers. In other words, unstable resonators can produce output beams of low divergence in a short resonator structure that has a large cross section. Whether an unstable resonator has an advantage over a stable resonator for a particular system depends on the gain of the laser, since gain, mode volume, and sensitivity to misalignment are closely related in unstable resonators. Generally speaking, only in a high-gain laser can a large mode size be realized in an unstable resonator at a reasonable misalignment tolerance. Besides the issue of alignment stability, the cost and difficulty of fabrication of the output coupler is another consideration.

The unstable resonator first described by Siegman [19]–[21] has been studied extensively for applications in high-energy chemical and gas dynamic lasers. The most useful form of an unstable resonator is the confocal unstable resonator. A primary advantage of this configuration is that it automatically produces a collimated output beam.

Confocal configurations can be divided into positive-branch resonators that correspond to the case $g_1 g_2 > 1$, and negative-branch resonators for which $g_1 g_2 < 0$. Confocal unstable resonators of the positive or negative branches are shown in Fig. 5.22. These configurations are defined by the following relationships:

$$2L = R_1 + R_2 \tag{5.50}$$

or

$$L = f_1 + f_2,$$

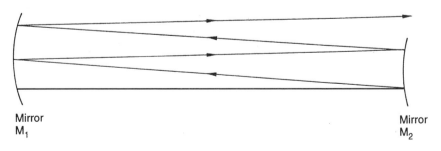

Mirror
M_1

Mirror
M_2

FIGURE 5.21. Light ray in an unstable resonator.

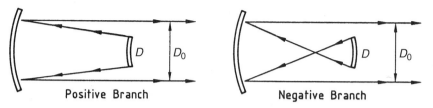

FIGURE 5.22. Positive- and negative-branch confocal unstable resonators.

where L is the optical length of the resonator and R_1, R_2, f_1, f_2 are the radii and focal lengths of the two mirrors, respectively. For concave mirrors, the sign for R and f is positive and for convex mirrors it is negative.

The confocal positive-branch unstable resonator is the most widely used form of the unstable resonator for solid-state lasers because it does not have an intra-cavity focal point that could lead to air breakdown or could cause damage to optical components.

Referring to Fig. 5.22 the annular output beam has an outer diameter of D_0 and inner diameter D, where D is also the diameter of the output coupler.

The resonator magnification

$$m = D_0/D \tag{5.51}$$

is the amount that the feedback beam is magnified when it travels one round trip in the resonator and becomes the output beam.

The magnification in the transverse beam dimensions results in a decrease of intensity by a factor $1/m^2$ in each round trip, and radiation spilling out around the edges of the output mirror. A beam that is just contained within the mirrors will, after one round trip, lose a fraction T of its energy.

The geometrical output coupling is related to the magnification by

$$T = 1 - \frac{1}{m^2}. \tag{5.52}$$

If we insert a gain medium between the mirrors, the loss in energy has to be made up by the gain of the laser. According to the laser threshold condition (3.7), we obtain

$$2gl = \delta + \ln m^2. \tag{5.53}$$

Ignoring internal resonator losses δ for a moment, the round-trip gain of the laser has to be

$$G = \exp(2gl) \geq m^2. \tag{5.54}$$

For a confocal resonator, the mirror radii are given by

$$R_1 = \frac{-2L}{m-1}, \quad \text{and} \quad R_2 = \frac{2mL}{m-1}, \tag{5.55}$$

where L is the length of the resonator and R_1 and R_2 are the output and back-cavity mirror curvatures. Note that the output mirror has a negative curvature and thus is convex, while the high-reflectivity mirror has positive curvature and is concave.

Actually, the relationship between m and the output coupling T for confocal unstable resonators is such that T is less than the geometrically predicted value $(1 - 1/m^2)$. This is because the intensity distribution, according to wave optics, tends to be more concentrated toward the beam axis than that predicted by geometrical optics.

The equivalent Fresnel number of a resonator characterizes the destructive or constructive interference of the mode at the center of the feedback mirror due to the outcoupling aperture. For a positive-branch, confocal resonator, the equivalent Fresnel number is

$$N_{eq} = \frac{(m-1)(D_0/2)^2}{2L\lambda}, \tag{5.56}$$

where D_0 is the diameter of the output mirror and L is the resonator length. Plots of the mode eigenvalue magnitudes versus N_{eq} show local maxima near half-integer N_{eq}'s.

Physically, the half-integer equivalent Fresnel numbers correspond to Fresnel diffraction peaks centered on the output coupler which leads to increased feedback into the resonator. Resonators should be designed to operate at half-integer equivalent Fresnel numbers (N_{eq}) to obtain best mode selectivity.

As a final step one has to take into account the effect of the focal length f of the active medium. One usually chooses an available mirror curvature R and calculates the rod focal length at the desired lamp input power required to achieve an effective mirror curvature R_{eff}. If the mirror to rod distance is less than the rod focal length, then

$$\frac{1}{f} + \frac{1}{R_{eff}} = \frac{1}{R}. \tag{5.57}$$

Essentially the focusing effect of the laser rod is compensated for by increasing the radius of curvature of the mirror.

The design of such a resonator is an iterative process. It starts with some knowledge of the optimum output coupling T or the saturated gain coefficient g obtained from operating the laser with a stable resonator. From either of these two parameters, we obtain from (5.52) or (5.53) the magnification m. Selecting a value of $0.5, 1.5, 2.5, \ldots$ for N_{eq} from (5.56) provides a relationship between D_0 and L. With (5.50), (5.51), (5.55) the pertinent resonator design parameters such as D, D_0, L, R_1, and R_2 can be calculated. The effect of thermal lensing is then accounted for by adjusting the radius of curvature according to (5.57). There are

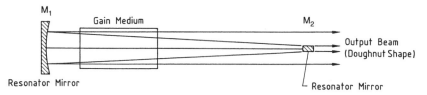

FIGURE 5.23. Small mirror out-coupling scheme in a confocal, positive branch unstable resonator.

basically two types of output couplers possible with solid-state lasers, the dot mirror and the radially variable reflectivity mirror.

The design depicted in Fig. 5.23 consists of a concave mirror M_1 and a convex output mirror M_2, both of which are totally reflecting. The dot mirror M_2 is a small circular dielectrically coated area of radius d centered on a glass substrate. The disadvantage of a dot mirror, namely the generation of an annular output beam containing diffraction rings and a hot spot in the center, can be eliminated by employing a partially transparent output coupler with a radially variable reflectivity profile. In such a mirror, reflectivity decreases radially from a peak in the center down to zero over a distance comparable to the diameter of the laser rod. The output is a smooth Gaussian-shaped beam. Design procedures for such mirrors are described in [22], [23].

Illustrated in Fig. 5.24 is a diode-pumped Nd : YAG slab laser with a positive-branch, confocal unstable resonator in a folded configuration. The output coupler is a dielectrically coated mirror with a variable reflectivity profile. The resonator in Fig. 5.24 has a magnification $M = 1.38$ and a cavity length of 58 cm. The laser generated 100 mJ of Q-switched output at a repetition rate of 100 Hz with a near diffraction limited beam.

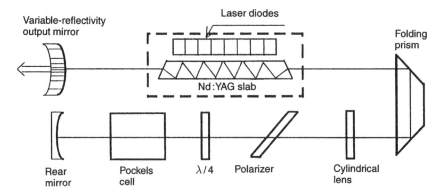

FIGURE 5.24. Diode-pumped Nd : YAG slab laser with positive branch unstable resonator and variable reflectivity output coupler [24].

Summary

In this chapter, the definition and characteristics of transverse and longitudinal modes of optical resonators are reviewed, and mode-selecting techniques as well as typical resonator configurations are discussed.

In resonators employed in microwave technology the cavity is totally enclosed and the dimensions are on the order of the wavelength. By virtue of the directional properties of light, a structure formed by two mirrors facing one another can sustain high Q-resonances and therefore constitutes a resonator in spite of the fact that it is an open structure. In an optical resonator the mirror size and the separation of the mirrors are very large compared to the wavelength. The modes of an open passive resonator are eigenfunctions of an integral equation that depend on the shape and spacing of the mirrors and on the beam diameter. The gain provided by the active material compensates for the resonator losses. The oscillator will lase at those modes that have the smallest losses.

Transverse modes determine beam divergence and power distribution within the cross section of the beam. The number of longitudinal modes that are oscillating in a resonator are responsible for the spectral line width of the output beam. The smallest beam divergence is obtained if the resonator is operated in the lowest transverse mode. Single transverse mode operation is obtained by restricting the lateral extent of the beam inside the resonator. This is usually associated with a utilization of a smaller active volume of the laser material and therefore leads to a reduction in output power. Since virtually all laser applications benefit from a low beam divergence, many ingenious concepts and designs have been developed to achieve maximum output at the TEM_{00} mode.

In applications requiring a narrow line width from the laser the number of longitudinal modes having sufficient gain to oscillate has to be restricted to just a single mode. This can be achieved by shortening the resonator (and thereby also the active material) to a point where the modes are spaced so far apart that only one mode has sufficient gain to oscillate. From these considerations it becomes clear that single transverse and longitudinal mode lasers have a very small volume.

Unstable resonators, also discussed in this chapter, are resonators where the $g_1 g_2$ product falls outside the range $0 < g_1 g_2 < 1$. The modes are not confined in the transverse direction, and the size is determined by some limiting aperture in the system.

References

[1] A.G. Fox and T. Li: *Bell Syst. Tech. J.* **40**, 453 (1961).

[2] G.D. Boyd and J.P. Gordon: *Bell Syst. Tech. J.* **40**, 489 (1961).

[3] G.D. Boyd and H. Kogelnik: *Bell Syst. Tech. J.* **41**, 1347 (1962).

[4] H. Kogelnik and T. Li: *Appl. Opt.* **5**, 1550 (1966); see also *Proc. IEEE* **54**, 1312 (1966).

[5] H. Kogelnik: in *Lasers* I (edited by A.K. Levine), (Dekker, New York), 1966, pp. 295–347.

[6] J.S. Kruger: *Electro-Opt. Syst. Designs* **12** (September 1972).

[7] R.J. Freiberg and A.S. Halsted: *Appl. Opt.* **8**, 335 (1969).

[8] T. Li: *Bell Syst. Tech. J.* **44**, 917 (1965).

[9] R. Phillips and L. Andrews: *Appl. Opt.* **22**, 643 (1983).

[10] W.H. Carter: *Appl. Opt.* **19**, 1027 (1980).

[11] D.E. McCumber: *Bell Syst. Tech. J.* **44**, 333 (1965).

[12] H. Kogelnik: *Bell Syst. Tech. J.* **44**, 455 (1965).

[13] J.E. Geusic, H.J. Levingstein, S. Singh, R.C. Smith, and L.G. Van Uitert: *Appl. Phys. Lett.* **12**, 306 (1968).

[14] H. Steffen, J.P. Lörtscher, and G. Herziger: *IEEE J. Quantum Electron.* **QE-8**, 239 (1972).

[15] A. Yariv: *Optical Electronics*, 4th ed. (Holt, Rinehart and Winston, Orlando, FL), 1991.

[16] T.J. Kane and R.L. Byer: *Opt. Lett.* **10**, 65 (1985).

[17] A. Yariv: *Quantum Electronics*, 3rd ed. (Wiley, New York), 1988, Sect. 21.1.

[18] B. Zhou, T.J. Kane, G.J. Dixon, and R.L. Byer: *Opt. Lett.* **10**, 62 (1985).

[19] A.E. Siegman: *Proc. IEEE* **53**, 227 (1965).

[20] A.E. Siegman and E. Arrathon: *IEEE J. Quantum Electron.* **QE-3**, 156 (1967).

[21] A.E. Siegman: *Appl. Opt.* **13**, 353 (1974); see also *Lasers* (University Science Books, Mill Valley, CA), 1986.

[22] N. McCarthy and P. Lavigne: *Appl. Opt.* **22**, 2704–2708 (1983).

[23] N. McCarthy and P. Lavigne: *Appl. Opt.* **23**, 3845–3850 (1984).

[24] E. Armandillo, C. Norrie, A. Cosentino, P. Laporta, P. Wazen, and P. Maine: *Opt. Lett.* **22**, 1168 (1997).

Exercises

1. When you work in modern optics you often have to find the values of the electric field amplitude in a beam of laser light. In the laboratory your instruments do not measure this directly. When working with a pulsed laser (e.g., a Q-switched solid-state laser) your instruments allow you to measure the total energy in the pulse, the temporal waveform of the pulse, and the spatial distribution of the beam. The first is done with what is essentially a calibrated absorber that determines the change in temperature of the absorber and computes the energy that was absorbed. With knowledge of the absorbance of the surface of the absorber the instrument can read out the number of joules of energy that struck it. Pulse waveform is measured with a fast photodetector and an oscilloscope where both are selected to have enough bandwidth to assure accurate waveform determination. The beam's spatial properties are measured with an imaging camera (e.g., a CCD camera) and electronics and software that display the intensity as a function of position normal to the beam axis on a computer screen.

Now suppose you are in the laboratory using these instruments and you measure an energy, U, a pulse duration (the duration as defined by the pulse full width at half-maximum, FWHM) of τ_p, and the beam spatial distribution. The latter is determined to be the lowest order Gaussian distribution with a full width at $1/e$ of the maximum intensity of w. Use the definitions of terms in Section 5.1.2 and your knowledge of the definition of intensity to calculate the value of the electric field on axis in terms of your measured quantities. Use proper units so that you give your answer in volts/m. (*Hint:* You might want to refer to an electricity and magnetism text to refresh yourself on the definition of the Poynting vector and to get the units straight.)

2. Is the resonator shown below stable or unstable for Gaussian mode propagation?

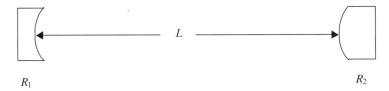

R_1 \hfill R_2

Here $L = 100\,\text{cm}$, $R_1 = 1000\,\text{cm}$, $R_2 = -1000\,\text{cm}$, and the laser wavelength $\lambda = 1\,\mu\text{m}$. If it is stable, locate the beam waist and find the beam radii at each mirror. Where would you put the laser rod in this resonator to maximally fill it, the gain medium, with the lowest order Gaussian mode?

3. Consider the laser shown below:

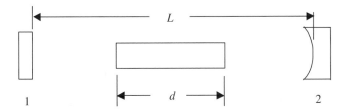

In this laser L is the length of the resonator; $d = 10\,\text{cm}$ is the length of the gain medium that has index of refraction of 1.8; the surfaces of the gain medium are antireflection coated so they do not contribute to cavity losses; mirror 1 has reflectivity of 98% and is flat, and mirror 2 has reflectivity of 80% and radius of curvature, $R_2 = 100\,\text{cm}$. The laser wavelength is $1.06\,\mu\text{m}$.

(a) Find L such that this resonator is hemispherical and give the distances between the ends of the centered rod and the mirrors.

(b) What are g_1 and g_2 for the hemispherical resonator?

(c) Find the mode size at each mirror and at each end of the laser rod in this hemispherical resonator.

(d) What is the frequency difference between two adjacent TEM$_{00}$ modes?

(e) Find the frequency difference in megahertz between the TEM_{00q} and TEM_{10q} modes.

(f) What is the photon lifetime in this cavity?

(g) What is the minimum gain coefficient for this laser to oscillate?

(h) What gain coefficient would be needed if there were no antireflection coatings on the laser rod?

6

Optical Pump Systems

6.1 Pump Sources
6.2 Pump Radiation Transfer Methods
    References
    Exercises

An efficient pump system utilizes a radiation source having a spectral output that closely matches the absorption bands of the gain medium, transfers the radiation with minimum losses from the source to the absorbing laser material, and creates an inversion which spatially matches the mode volume of the resonator mode. In this chapter we will review the most common pump sources and radiation transport systems employed in solid-state lasers.

6.1 Pump Sources

A number of different radiation sources have been employed over the years to pump solid-state lasers; today only flashlamps, cw arc lamps, and laser diodes depicted in Figs. 6.1 and 6.2 are of practical interest. Pump sources for solid-state lasers range in size and power from tiny laser diodes at the 1 W level, pumping a microchip laser, to flashlamp-pumped lasers employed in inertial confinement fusion research, which occupy large buildings and provide tens of megajoules of pump radiation.

6.1.1 Flashlamps

Lamp Design and Construction

Flashlamps used for laser pumping are essentially long arc devices designed so that the plasma completely fills the tube. A flashlamp consists of a linear quartz tube, two electrodes which are sealed into the envelope, and a gas fill. Standard

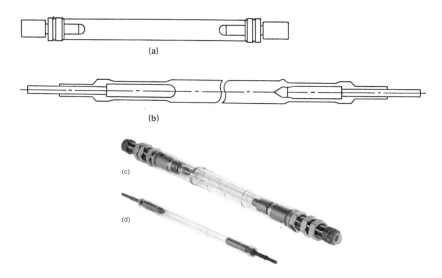

FIGURE 6.1. Outline drawing of (a) a flashlamp and (b) a cw krypton arc lamp and photograph of an arc lamp (c) with and (d) without cooling jacket.

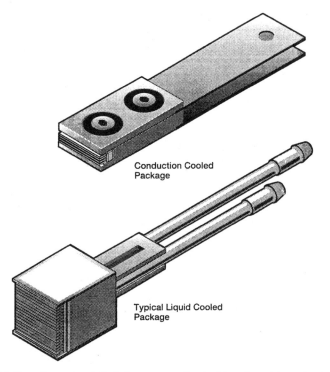

FIGURE 6.2. Two-dimensional diode laser arrays. Stacked 1 cm bars (a) conduction-cooled and (b) internally water-cooled.

linear lamps have wall thicknesses of 1 to 2 mm, bore diameters between 3 and 19 mm, and lengths from 5 cm to 1 m.

Flashlamps are typically filled with gas at a fill pressure of 300 to 700 torr at room temperature. Xenon is generally chosen as the gas fill for flashlamps because it yields a higher radiation output for a given electrical input energy than other gases. However, in a few special cases, such as small, low-energy lamps employed to pump Nd : YAG lasers, the low-atomic-weight gas krypton provides a better spectral match to the absorption bands of Nd : YAG.

The anode in flashlamps consists of pure tungsten or often thoriated tungsten because the alloy is easier to machine. The cathode is comprised of a compressed pellet of porous tungsten impregnated with barium calcium aluminate. This pellet is attached onto a tungsten heat sink. The surface area of the tip must be large enough to handle the peak current while the shape of the tip positions the arc during the trigger pulse. During the lamp operation, barium is transported to the cathode surface where it forms a monolayer with a work function of about 2 eV versus 4.5 eV for pure tungsten. A lower work function improves electron emission for a given temperature. For the same reason, a cathode material with a low work function makes it easier to trigger the lamp. Since in a standard lamp the cathode is more emissive than the anode, flashlamps are polarized and will pass current in only one direction without damage.

The envelope and the electrode-seal area of flashtubes are normally cooled by free or forced air, water, or water-alcohol mixtures. Liquid cooling permits operation of the lamps at a maximum inner-tube wall surface loading of 300 W/cm^2 of average power. Free-air convection cooling is limited to handling about 5 W/cm^2 of dissipation; forcing air across the flashtube envelope increases this value up to 40 W/cm^2. Liquid-cooled linear flashlamps are available with outer quartz jackets, which permit cooling of the lamps with a highly turbulent flow.

Optical Characteristics. The radiation output of a gas discharge lamp is composed of several different components, each corresponding to a different light-emission mechanism. The relative importance of each of these mechanisms depends strongly on the current density in the lamp, and so the low- and high-power optical output spectra are markedly different. The total radiation is made up of both line and continuum components. The line radiation corresponds to discrete transitions between the bound energy states of the gas atoms and ions (bound–bound transitions). The continuum is made up primarily of recombination radiation from gas ions capturing electrons into bound states (free–bound transitions) and of bremsstrahlung radiation from electrons accelerated during collisions with ions (free–free transitions). The spectral distribution of the emitted light depends in complex ways on electron and ion densities and temperatures.

In the continuous-power, wall-stabilized noble gas arc, current densities are such that there is a high number of bound–bound transitions and therefore the radiation spectrum is characteristic of the fill gas and is broadened by increases in pressure.

In high current density, pulsed laser applications, the spectral output of the lamp is dominated by continuum radiation, and the line structure is seen as a relatively

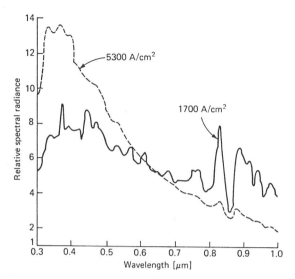

FIGURE 6.3. Spectral emission of a xenon flash tube (EG&G, model FX-47A) operated at high current densities. Lamp fill pressure is 0.4 atm. The spectrum at the two current densities can be approximated by blackbodies at 7000 and 9400 K, respectively [1].

minor element. Between these two cases, a pulsed-power region can exist where the pulsed-power level is such that discrete line radiation is still emitted and is superimposed on a strong background continuum.

In Fig. 6.3 the spectral emission of a xenon flashtube is plotted for two current densities. As a result of the high current densities, the line structure is in this case masked by a strong continuum. From Fig. 6.3 it also follows that a high current density shifts the spectral output toward the shorter wavelengths. One can relate the radiation characteristics of a flashlamp $R(\lambda, T)$ to the characteristics of a blackbody. The departure of any practical light source from the blackbody characteristic is accounted for in the emissivity, $\varepsilon(\lambda, T)$, which varies between zero and one, and is both wavelength and temperature dependent; thus $R(\lambda, T) = \varepsilon(\lambda, T)B(\lambda, T)$. In a flashlamp, $\varepsilon(\lambda, T)$ depends strongly on temperature and wavelength. If we assume local thermodynamic equilibrium in the plasma, the blackbody characteristic $B(\lambda, T)$ becomes the blackbody curve corresponding to the electron temperature in the plasma. In general, the emissivity of the flashlamp radiation at longer wavelengths is greater than the emissivity at shorter wavelengths. An increase in power density will result in a large increase in emissivity (and radiation) at short wavelengths, but only small changes at longer wavelengths where the emissivity is already close to unity.

The xenon lamp is a relatively efficient device as it converts 40 to 60% of the electrical input energy into radiation in the 0.2 to 1.0 μm region. As noted in Chapter 2, principal pump bands of Nd : YAG are located from 0.73 to 0.76 μm and 0.79 to 0.82 μm. The xenon spectrum has no major line radiation in these bands, so pumping is primarily due to continuum radiation. In general, the pumping efficiency of Nd : YAG increases as the pressure is increased. In the range 450 to 3000 torr no maxima are found. The limitation to further increases appears to be the fact that high-pressure lamps are difficult to trigger. The effect of an increase in gas fill pressure is to reduce the mean free path of the electrons and atoms in the discharge, and thus to increase their collision frequency. This leads to the production of more excited species in the discharge and the emission of more useful line radiation.

Electrical Characteristics. The impedance characteristics of a flashlamp determine the efficiency with which energy is transferred from the capacitor bank to the lamp. The impedance of the arc is a function of time and current density. Most triggering systems initiate the arc as a thin streamer which grows in diameter until it fills the whole tube. The expansion period is fast, of the order of 5 to 50 μs for tubes of bore diameter up to 1.3 cm. During the growth of the arc, lamp resistance is a decreasing function of time. The decreasing resistance arises in part from the increasing ionization of the gas and from the radial expansion of the plasma.

After the arc stabilizes in the high-current regime the voltage is proportional to the square root of the current [1, 2]

$$V = K_0 i^{1/2}, \tag{6.1}$$

where the flashlamp impedance K_0 depends on the arc length l and the bore diameter D of the flashlamp, and on the kind of gas and fill pressure p. For xenon the following relation holds

$$K_0 = 1.27 \left(\frac{p}{450}\right)^{0.2} \frac{l}{D}, \tag{6.2}$$

where p is the flashlamp pressure in torr.

Tube Failure

The flashtube end-of-life can occur in either of two modes, and we can distinguish between catastrophic and nondestructive failures. The factors that contribute to catastrophic failure are explosion of the walls due to the shock wave in the gas when the lamp is fired or overheating of the lamp envelope or seal and subsequent breakage because of excessive thermal loading of the lamp. The first failure mode is a function of pulse energy and pulse duration, whereas the second catastrophic-failure mode is determined by the average power dissipated in the lamp. When the flashtube is operated well below the rated maximum pulse energy and average power, the lamp usually does not fail abruptly. Rather, the tube will continue to flash with a gradual decrease in light output, which will eventually fall below a level necessary for the particular application. In the latter mode of failure, the reduced light output is caused by the erosion of the flashtube electrodes and of the quartz walls and by the gradual buildup of light-absorbing deposits within the flashtube envelope.

Flashtube Explosion

When a high-voltage trigger pulse is applied between the electrodes of a flash-tube, the gas breaks down generally near the axis of the tube and a conducting filament is established. As energy is released into the channel, heating of the ambient gas causes the filament to expand radially, forming cylindrical shock waves. The shock front and its associated plasma travel radially from the tube axis to its wall. The radial velocity of the plasma discharge and the shock amplitude are proportional to the input energy. The cylindrical shock wave and the associated plasma heat cause stress on the inside of the tube wall, which is axial in tension toward the electrodes. If the energy discharged exceeds the explosion limit for the lamp, the shock wave will be sufficiently intense to rupture the lamp walls and consequently destroy the lamp.

The explosion energy is directly related to the inside-wall surface area of the lamp and to the square root of the pulse duration. If l and D are measured in centimeters and t_p is in seconds, and if a critically damped single-mesh discharge circuit is assumed, then it has been found empirically that for xenon-filled quartz flashtubes [1, 3] the explosion energy in Joules is

$$E_{ex} = (1.2 \times 10^4) l D \, t_p^{1/2}. \tag{6.3}$$

Having determined the ultimate limit for the tube, the question of the life that the tube will have in a given application has to be addressed. It has been shown that lamplife can be related to the fraction of explosion energy at which the lamp is operated. The life in flashes for a xenon-filled quartz-envelope flashlamp as a function of the single-shot explosion energy is empirically given by [1, 4]

$$N \cong \left(\frac{E_{ex}}{E_{in}} \right)^{8.5}. \tag{6.4}$$

The extremely strong dependence of flashlamp lifetime on the input energy E_{in} is a great incentive to underrate the lamp.

In Fig. 6.4 life expectancy as a function of lamp input is shown, assuming that the interpulse interval is sufficiently long to permit the tube to cool between pulses. The lifetime expressed by (6.4) is defined as the number of shots at which the light output is reduced by 50% of its initial value. The life expectancy of a flashlamp begins to deteriorate rapidly when peak current exceeds 60% of explosion current. Above 60%, the lamp's life is usually limited to less than 100 discharges, even if all other parameters are optimized. At the 80% level, explosion can occur on any discharge. Most flashlamps, with the exception of lamps installed in single-shot lasers, such as, for example, employed in inertial confinement fusion, are operated at only a small fraction of the explosion energy. In this case, cathode degradation is the primary failure mode. The two dotted lines in Fig. 6.4 indicate where the cathode end of life begins to occur. As explained below, lamplife is considerably improved by operation in the simmer mode which

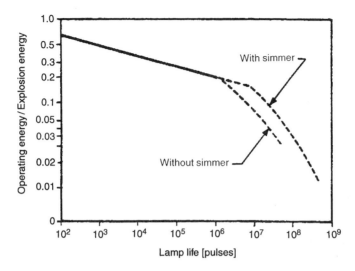

FIGURE 6.4. Lamp life as a function of operating energy [4].

keeps the lamp continuously ionized at a low-current level. Because of the increased lamplife most lamps are operated in the simmer mode today.

Examples of Typical Operating Conditions

In selecting a particular lamp the following steps are usually followed:

1. A flashlamp arc length and bore diameter is chosen that matches the rod size.
2. The explosion energy for the lamp chosen is calculated.
3. An input energy is selected that is consistent with the desired lamp life. If the lamp input required to achieve a certain laser output is not consistent with a desired minimum lamplife, the application of a multiple-lamp pump cavity has to be considered.
4. The average power loading of the lamp is calculated and compared with the limit specified by the lamp manufacturer.
5. In designing the pulse-forming network, care must be taken to avoid fast current rise times and/or reverse currents. To obtain the expected life, manufacturers recommend that the rise time of the flashlamp current be greater than $120\,\mu s$ for lamps discharging more than a few hundred joules. Both electrode erosion and wall vaporization are accelerated by fast-rising peak currents.

First we will consider a typical small linear flashlamp used in military equipment such as target designators or laser illuminators. These systems operate typically at a flashlamp input of 10 J with pulse repetition rates of 20 pps. The lasers typically use 63 mm long Nd : YAG laser rods of 6.3 mm diameter and have a Q-switched output of 100 to 200 mJ. A 4 mm diameter and 60 mm long lamp operated at 10 J input and a pulse width of $300\,\mu s$ is operated at only 2% of the explosion energy. The wall loading of the liquid-cooled lamp operated at 10 pps is $15\,W/cm^2$. From the extensive data that exist for this type of lamp operated in a closed coupled laser cavity, it follows that the useful lamplife (10% drop in laser performance) is in excess of 10^7 shots.

Figure 6.5 illustrates a circuit, which is widely used for flashlamp-pumped solid-state lasers. For clarity we will assume for the moment that the simmer power supply and the thyristor have been removed, and the flashlamp is directly connected to the inductor. The energy delivered to the lamp is stored in the capacitor C. The discharge of the stored energy into the flashlamp is initiated by a high-voltage trigger pulse generated by a pulse generator and a step-up transformer. The trigger pulse applied to a wire wrapped around the tube envelope creates an ionized spark streamer between the electrodes. The energy stored in the capacitor is discharged through the inductor L into the lamp. The pulse duration and pulse shapes are determined by the time constant of the LC circuit. The lamp extinguishes after the capacitor is discharged. Simplicity is the major advantage of this operation, which requires no high voltage and current switching device since the flashlamp serves as a switch.

It has been found that the lifetime of a flashlamp can be significantly increased, and pump efficiency can be improved if the lamp is preionized.

The improvement is mainly due to a decrease in cathode sputtering, a more uniform arc formation, and an associated smaller shock wave. In the so-called simmer mode a low-current discharge is maintained between the high current pulses. The simmer mode of operation requires a switching element between the flashlamp and the pulse-forming network (PFN). The flashlamp is initially ignited as described before by a trigger pulse on the order of 20 kV. A low-current discharge of a few hundred milliamps is maintained in the lamp by the simmer supply shown

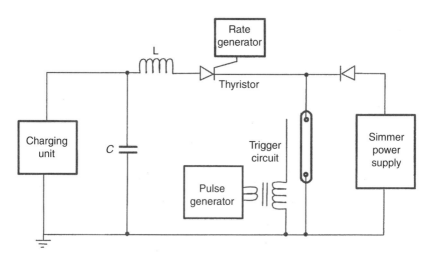

FIGURE 6.5. Typical power supply employed for the operation of flashlamps.

in Fig. 6.5. The energy storage capacitor is discharged via a switch such as an SCR, thyristor, a gas, or mercury-filled tube. The pulse shape is determined by the PFN, as before, and the switch, as well as the flashlamp, are turned off once the capacitor is discharged.

The dynamic behavior of the discharge circuit, which consists of the flashlamp impedance defined in (6.1), a capacitance C, and an inductance L, is described by a nonlinear differential equation. The impedance Z_0 and the time constant t_{LC} of the circuit is given by

$$Z_0 = (L/C)^{1/2} \quad \text{and} \quad t_{LC} = (LC)^{1/2}. \tag{6.5}$$

In addition, a damping factor α is defined, which determines the pulse shape of the current pulse. It is

$$\alpha = K_0 (V_0 Z_0)^{-1/2}, \tag{6.6}$$

where V_0 is the initial voltage of the capacitor. Solutions of the nonlinear differential equation reveal that the current waveform is critically damped for a value of $\alpha = 0.8$ [5, 6]. The current waveform and energy, as a function of time for a critically damped pulse-forming network, is shown in Fig. 6.6.

The current pulse duration t_p at the 10% points is approximately $3t_{LC}$; during this time about 97% of the energy has been delivered. If we substitute

$$\alpha = 0.8 \quad \text{and} \quad t_p = 3t_{LC} = 3(LC)^{1/2} \tag{6.7}$$

for a critically damped pulse into (6.5), (6.6), then we can determine for a given lamp type the relationship between energy input, pulse duration, pulse shape, inductance, capacitance, and operating voltage.

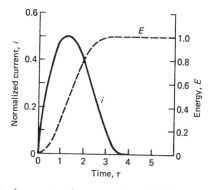

FIGURE 6.6. Normalized current and energy of a critically damped flashlamp discharge circuit. Pulse shape factor $\alpha = 0.8$.

The energy initially stored in the capacitor is

$$E_{in} = \tfrac{1}{2} C V_0^2.$$

(6.8)

Equation (6.8) is used to eliminate V_0 from (6.6), from which follows the value of the capacitor

$$C^3 = 0.09 \frac{E_{in} t_p^2}{K_0^4}.$$

(6.9)

The inductance can be calculated from (6.7)

$$L = \frac{t_p^2}{9C}.$$

(6.10)

We now have a set of three equations (6.8)–(6.10) that, given the specifications of the flashlamp parameter K_0, the desired input energy E_{in}, and pulse width t_p, provide explicit values of C, L, and V_0. If the lamp parameter K_0 is not specified by the lamp manufacturer, it can be calculated from (6.2). From Fig. 6.6 also follows the peak current for a critically damped current pulse

$$i_p = 0.5 \frac{V_0}{Z_0}.$$

(6.11)

The rise time to reach this peak value is

$$t_R = 1.25(LC)^{1/2} \cong 0.4 t_p.$$

(6.12)

6.1.2 Continuous Arc Lamps

Mechanical Design

Continuous arc lamps are similar in design to linear flashlamps, with the exception that the cathode has a pointed tip for arc stability (see Fig. 6.1).

Electrical Characteristics

Continuous arc lamps are ignited just like flashlamps. For example, a series injection trigger transformer supplying a 30 kV spike will lower the lamp impedance to a level that is within the range of a 300 V dc supply. The operating parameters and impedance depend on current density, arc length, bore size, and fill pressure. Typical krypton arc lamps are operated between 80 and 150 V and 20 to 50 A.

Optical Characteristics

Arc lamps filled with inert gas are rich in line structure in the near-infrared. Xenon has the highest overall conversion efficiency and is commonly used in arc lamps.

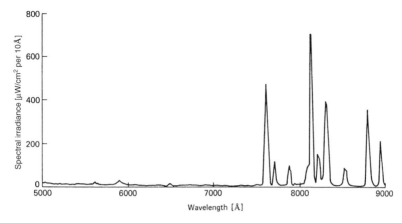

FIGURE 6.7. Emission spectrum of a typical cw-pumped krypton arc lamp (6 mm bore, 50 mm arc length, 4 atm fill pressure, 1.3 kW input power) (ILC Bulletin 3533).

However, the infrared line spectra of xenon misses all of the Nd : YAG pump bands, which are located at 0.73 to 0.76, 0.79 to 0.82, 0.86 to 0.89, and 0.57 to 0.60 μm. The line spectrum from krypton is a better match to Nd : YAG than the line spectrum of xenon since two of its strongest emission lines (7600 and 8110 Å) are strongly absorbed by the laser crystal (Fig. 6.7).

The laser output from a krypton-filled lamp is about twice that obtained from a xenon arc lamp operated at the same input power. Krypton arc-pumped Nd : YAG lasers are currently the highest continuous power solid-state lasers. A summary of the special data of a representative krypton arc lamp is given in Table 6.1. Note the large fraction of radiation between 0.7 and 0.9 μm.

TABLE 6.1. Spectral data for cw krypton arc lamps. Data are typical for lamps having a 6–13 mm bore, 7.5–25 cm arc length, 2–3 atm fill pressure, operated at 6–16 kW.

Quantity	Definition	Numerical data
Radiation efficiency	Radiation output/ electrical input	0.45
Spectral output	Fraction of radiation in spectral lines	0.40
	Fraction of radiation in continuum	0.60
Spectral power distribution	Fraction of total radiation below 0.7 μm	0.10
	Fraction of total radiation between 0.7 and 0.9 μm	0.60
	Fraction of total radiation between 0.9 and 1.4 μm	0.30

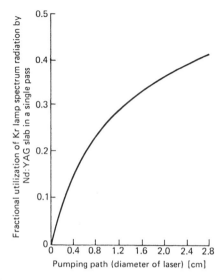

FIGURE 6.8. Fractional utilization of krypton lamp output by Nd : YAG.

Combining the absorption spectra of Nd : YAG with the emission spectra of the krypton lamp, the spectral utilization as a function of sample thickness has been calculated. The result is shown in Fig. 6.8. The curve illustrates rather dramatically the kind of improvement one can achieve by increasing the diameter of an Nd : YAG laser rod in a pumping cavity.

6.1.3   Laser Diodes

The high pump efficiency compared to flashlamps stems from the excellent spectral match between the laser diode emission and the Nd absorption bands at 808 nm. Actually, flashlamps have a radiation efficiency (40–60%) comparable to laser diodes (35–50%); however, only a small fraction of the radiation match by the laser material absorption bands. In contrast, the output wavelength of laser diodes can be chosen to fall completely within an absorption band of a particular solid-state laser. This has also the added benefit of reduced thermal loading of the laser crystal, which leads to better beam quality. The directionality of the diode output and the small emitting area provides great flexibility of shaping and transferring the pump radiation from the source to the laser medium that created whole new design architectures, such as end-pumped systems, microchip lasers, and fiber lasers. Diode laser pumps are also an enabling technology for new laser materials. The most prominent laser materials that are pumped with diode pump sources can also be pumped with flashlamps. However, a number of very useful materials such as Nd : YVO$_4$, Yb : YAG, and Tm : YAG have only reached prominence as a result of diode laser pumps.

Internal Structure of Laser Diodes

The most basic form of a diode laser consists of a semiconductor in which one part is doped with electron donors to form n-type material and the other part is doped with electron acceptors (holes) to produce p-type material. Application of a negative voltage to the n-material and a positive voltage to the p-material drives electrons and holes into the junction region between the n- and p-doped material. This is called forward biasing the pn-junction. In the junction electron–hole recombination takes place. In this process, electrons from a higher-energy state (conduction band) transfer to a lower-energy state (valence band) and this releases energy in the form of photons. In order to produce stimulated emission in the junction area, population inversion and optical feedback are required. Passing a high current through the junction area provides a population inversion. A Fabry–Perot resonator, formed by cleaving the chip along parallel crystal planes, provides optical feedback. Gain in laser diodes is very high; therefore only a small length of active material is needed. Typical edge emitting laser chips have active layers about 500 μm long. In order to increase efficiency and output power, actual laser diodes contain a number of different layers with the active region sandwiched in between. Semiconductor-laser technology has produced an amazing variety of new device structures made possible by sophisticated growth techniques such as MetallOrganic Chemical–Vapor Deposition (MOCVD) and molecular beam epitaxy (MBE) [7]. Most techniques differ in the way confinement is achieved regarding the width and depth of the gain region and mode volume of the optical radiation.

Figure 6.9 shows the dimensions of a typical 1 cm monolithic linear array (top) and a cross section of the device (bottom) that is made up of a stack of several semiconductor layers with varying thickness and doping levels. The thickest layer on the bottom is the n-GaAs substrate onto which the other layers are deposited. The top layer is p-GaAs. Sandwiched between several other layers is the very thin active region of the pn-junction. The top and bottom surfaces of the whole stack are coated with metallic contacts for electrical connection to the drive electronics. The active region is divided into many parallel and narrow stripes by restricting the current in the lateral direction. The width of the active region has to be restricted to avoid the onset of amplified spontaneous emission which would propagate in off-axis directions and compete with the laser beam for the available population inversion. In the gain guided design illustrated in Fig. 6.9(a) particular emitter width is achieved by deposition of regions of high resistivity material beneath the contact area. Therefore the current is channeled through narrow low-resistance areas of the p-GaAs surface layer (white rectangular area in Fig. 6.9(b)). Electrical current injected through the forward biased pn-junction creates the population inversion and gain in the active region. The distribution of the charge-carriers injected in the active region, and therefore the gain distribution, is determined by the width of the current channel. The voltage across the junction is on the order of 2 V for GaAs/AlGaAs materials, and the slope efficiency is about 1 W/A. Laser emission, which is between 790 and 860 nm for

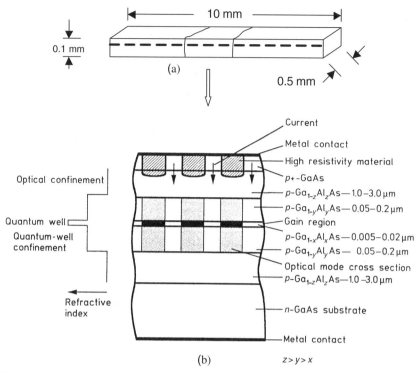

FIGURE 6.9. Gain-guided, single quantum-well separate confinement heterostructure stripe laser [8]. (a) Typical dimensions of a 1 cm bar and (b) detail of internal structure. (Thickness of layers is greatly exaggerated).

GaA/AlGaAs laser diodes, is perpendicular to the planar front facet of the laser diode bar shown in Fig. 6.9.

A 1 cm bar contains 10 to 50 independent laser emitters, all radiating in the same direction perpendicular to the planar front of the structure. The width of each emitter (white rectangle in Fig. 6.9(b)) is between 50 to 200 μm. The spacing between adjacent emitters ranges from 200 μm to 1 mm. In so-called broad area emitters each emitter consists of a single stripe. In other designs the total width is subdivided in subarrays of 10 to 20 closely spaced narrow-stripe emitters for better lateral mode control. The width of, and spacing between, individual emitters is dictated by the average power, peak power, or duty cycle requirements and the type of cooling applied to the bar.

Radiation in the vertical direction of the *pn*-junction is confined by a wave guide structure comprised of several layers of differently doped materials. The active layer shown in Fig. 6.9 is sandwiched between two pairs of layers having a different concentration of aluminum. The main purpose of the innermost pair of layers is to confine the carriers to the active region, whereas the purpose of the outer layers is to confine the optical beam. In a conventional double-heterostructure pn-junction the active region has a thickness around 0.1–0.3 μm. The quantum-well active region illustrated in Fig. 6.9(b) is an order of magnitude thinner. Extremely high gains on the order of 3 to 5×10^4 cm^{-1} and low thresholds can be achieved in such a thin layer.

A quantum well is a thin layer of semiconductor located between two layers with a larger bandgap. Electrons in the quantum-well layer lack the energy to escape and cannot tunnel through the thicker surrounding layers. The quantum-well layer in Fig. 6.9(b) has a composition of gallium and aluminum indicated by x, which defines the emission wavelength. A higher aluminum concentration increases the bandgap and shifts the output toward shorter wavelength. The quantum-well structure is sandwiched between two thick layers of a composition which contains a higher concentration of aluminum ($y > x$). The higher aluminum concentration increases the bandgap, thereby defining the quantum well, and the large thickness of the layer prevents tunneling of the carriers out of the quantum well. Instead of a single quantum-well active layer, several quantum wells can be sandwiched one above the other to generate multiple quantum-well stacks.

The thin active region incorporating a quantum-well active layer structure provides low threshold and high electrical-to-optical efficiency. However, such a very small emitting surface poses one problem since the power from a diode laser is limited by the peak flux at the output facet. One can increase the output from these devices by spreading the beam over an area that is larger than the active layer or gain region. The standard approach is to deposit layers next to the active layer, each of which has a slightly lower refractive index than the active layer, thus making a wave guide of the active layer. In the separate carrier and optical confinement heterostructure (SCH) design shown in Fig. 6.9(b), the refractive index boundary is abrupt, at the layer boundary. Alternatively, the refractive indices of the surrounding layers may be graded, forming a Graded-Index and Separate carrier and optical Confinement Heterstructure (GRINSCH).

In the single quantum-well structure depicted in Fig. 6.9(b), the effective aperture of the laser output is 0.1 to 0.4 μm vertical to the pn-junction, which is about a factor 20 larger than the gain region. This substantially reduces the energy density at the output facet and enhances reliability by minimizing catastrophic facet damage.

The small white rectangle in Fig. 6.9(b) indicates the gain region, whereas the large rectangle surrounding it represents the optical aperture of one emitter. The drawing in Fig. 6.9 exaggerates the vertical dimension by several orders of magnitude. The emitting aperture is, in reality, a wide and very narrow horizontal line with a width up to 200 μm and a thickness of no more than 0.4 μm.

As described above, the beam cross section of a single emitter is given by the width of the current channel in the plane of the pn-junction and by the waveguide structure in the vertical direction. The front and rear facets, separated by about 500 μm, define the length of the Fabry–Perot resonator. The facets, which provide optical feedback and output coupling, are created by cleaving the bars from the processed semiconductor wafer. Cleaving along directions dictated by the crystal structure generates extremely planar and parallel surfaces. The facets are dielectrically coated to provide a high reflectivity at the rear facet and an appropriate output transmission at the front facet.

Diode Pump Source Configuration

Depending on the output-power requirement of the laser and the particular pump configuration chosen, one can select a diode pump source from a number of standard commercial designs. In accordance with the average power capability we can classify diode pumps into four groups:

(a) Laser diodes with emitting apertures between 100 and 500 μm and output powers up to 5 W.
(b) Single 1 cm long bars yielding a cw output up to 40 W.
(c) Stacked 1 cm bars, forming two-dimensional arrays that are conductively cooled and mounted to a common heat sink.
(d) Stacked bar two-dimensional arrays with individual, internal liquid cooling for each bar.

Depending on the number of bars and the type of cooling employed, several kilowatts of cw output and tens of kilowatts of peak power in quasi-cw mode can be obtained from two-dimensional arrays.

For small solid-state lasers a typical pump source is a laser diode as illustrated in Fig. 6.10. The diode has a total output of 2 W in the cw mode. The output is emitted from an area of $200 \times 1.0 \mu$m with a beam divergence of 40° by 10°. The device can be coupled to an optical fiber or mounted on a thermoelectric cooler. In the latter configuration shown in Fig. 6.10 the array is mounted on a pyrami-

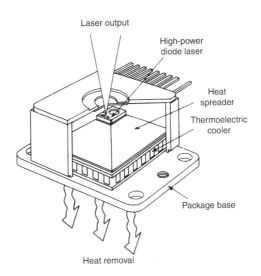

FIGURE 6.10. Diode laser mounted on a thermoelectric cooler (SDL, Inc.).

dal heat sink that connects to a thermoelectric cooler. Heat dissipates through the base, while the radiation emerges from the top. The thermoelectric cooler permits temperature tuning of the output to match the absorption lines of the solid-state laser. In a GaAlAs diode laser, the wavelength changes with temperature according to 0.3 nm/° C. Combining several of the 200 μm wide emitters up to 5 W of output can be achieved in similar packages.

Linear Arrays

The most common pump source is a 1 cm diode bar, as illustrated in Figs. 6.11 and 6.12. It consists of a monolithic chip of linear diode arrays that is bonded to an alloy submount. An insulated wire bond plate is soldered to the submount behind the laser diode array. Electrically the individual diodes in the array are

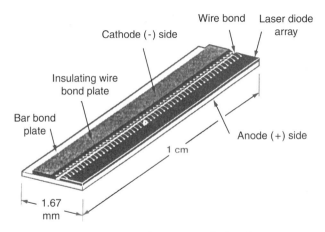

FIGURE 6.11. Basic structure of a 1 cm bar.

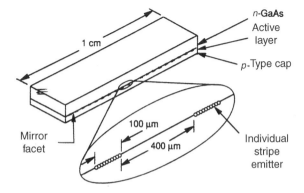

FIGURE 6.12. Monolithic 1 cm bar for cw operation [9].

all connected in parallel to a wire bond plate. The bonding of the bar to the heat sink is made through the thinner p-doped side to minimize the thermal resistance between the gain medium and the heat sink.

In the fabrication of laser bars 10 to 50 independent laser emitters are serially repeated on a single substrate to form a monolithic bar structure. Practical processing considerations limit the length of such bars to 1 cm. In the example shown in Fig. 6.12, the device is composed of 20 emitters spaced on 500 μm centers and produces an output of 20 W in the cw mode. Recent devices are capable of up to 40 W output power in a 1 cm bar. In the particular design shown in Fig. 6.12 each emitter is further subdivided into a number of narrow stripes. The spacing of the emitters, which amounts to a fill factor of 20% for the device, is determined by thermal limitations.

The use of diode lasers for pumping pulsed neodymium lasers requires a pulse length on the order of several hundred microseconds. One limit of the output of diode lasers is the temperature of the junction and associated thermal runaway condition that raises the local temperature of the facet to the melting temperature. Because of the small volume of material involved, this temperature rise occurs in less than 1 μs. Therefore, the power limit of diodes in the long-pulse mode is the same as for cw operation. Thus, long-pulse operation ($\gg$ 1 μs) is called quasi-cw operation. The peak power obtained from a 1 cm bar can be increased over the cw value by increasing the number of stripe lasers; this is achieved by increasing the fill factor to close to 100%. Of course, the duty cycle has to be limited such that the average power from the bar is about the same as for a cw bar. Operated in the quasi-cw mode, a single bar typically generates 60 W peak power, or 12 mJ in a 200 μs long pulse, although devices up to 100 W are available at reduced lifetime. Depending on cooling conditions, duty cycles on the order of 30% are possible. Therefore, the maximum average output power for cw and quasi-cw devices is 20 to 40 W for a 1 cm bar. Microlenses can be mounted in front of a bar to concentrate pump radiation into a fiber. Typically a 20 W cw bar can be coupled to a 400 μm fiber with 16 W available at the output end of the fiber. The ability to mount the pump source away from the laser onto a heat dissipating structure is very useful in some applications.

Two-Dimensional Arrays

A number of 1 cm bars can be combined to form a two-dimensional array. The key issues are the mounting technique and the efficient removal of waste heat from the laser emitter. In the so-called rack-and-stack assembly technique, laser diode bars are first mounted on individual submounts which are then stacked to produce two-dimensional arrays. The spacing (pitch) between adjacent bars depends on the average output power and on the cooling method.

For quasi-cw arrays one common design approach uses thin copper heat sink spacers between the bars. The assembly of sandwiched bars and spacers is mounted onto a common heat sink. Such a device is shown in Fig. 6.2(a). This internal passively cooled diode stack is attached via the two mounting holes to a larger cold plate or water-cooled heat sink. Electrically the bars are connected in series, the two large copper leads extending from the back of the device in Fig. 6.2(a) are for connection to a power supply. The thickness of the copper heat spreader, and therefore the stacking density, depends on the duty cycle of operation. Higher average power operation requires thicker heat sinks. For low-duty-cycle operation around 3%, the bars can be densely packaged with only 0.4 mm space in between; this allows a power density of 1.5 kW/cm². At a 6% duty cycle, the spacing increases to 0.8 mm and 800 W/cm² can be achieved. Operation of a 10 stack array, with 200 μs pulses at 100 Hz, produces typically a peak power of 600 W at 40% efficiency.

Multibar arrays for cw or high-duty cycle-pulsed operation require active cooling close to the junction of the individual emitters. This can be achieved by using microchannel cooling technology whereby each diode bar has its own liquid heat exchange submount. In one approach individual bars are separated by cooling plates apprxoimately 1 mm thick, through which cooling water is channeled. Using this approach, bars can be stacked with a 2 mm bar-to-bar pitch. Figures 6.2(b) and 6.13 depict internally liquid-cooled two-dimensional arrays. Arrays comprised of 20 cw bars have achieved 1 kW output in a 4 cm² emitting area. Pulsed arrays can be operated up to 30% duty cycle with internal active cooling. The most powerful multibar pump modules generate peak powers in excess of 30 kW and cw powers of several kilowatts.

Spatial Profile

The output beam emerging from the facet of a diode laser is in the form of a light cone that is oval in cross section, as illustrated in Fig. 6.14. In the vertical direction with respect to the *pn*-junction the beam is diffraction limited and has a divergence of about 40° FWHM. The Gaussian beam appears to emerge from a 1 μm aperture. If one introduces in (5.6) a wavelength of $\lambda = 0.8\,\mu$m and a spot size for a Gaussian beam of 1 μm the beam divergence at the $1/e^2$ points is $2\theta = 57°$.

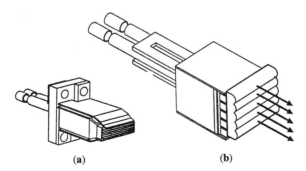

(a)   (b)

FIGURE 6.13. High-duty cycle internally liquid-cooled stacked bars (SDL, Inc.); (a) without and (b) with microlens array.

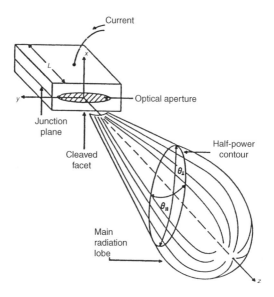

FIGURE 6.14. Radiation pattern from a laser diode [10].

Collimation of such a highly divergent beam requires optics with a numerical aperture of NA $= \sin(2\theta/2) = 0.5$ or higher. Optical systems with such a high numerical aperture are said to be fast. From this it follows that the direction of the output beam from a laser diode in the vertical direction of the emitting stripe is termed the fast axis.

In the plane parallel to the pn-junction, the beam divergence is much smaller and is about 10° FWHM. However considering the width of the emitter which is between 50 and 200 μm, this is many times diffraction limited. For example, a Gaussian beam emerging from an aperture of 150 μm would have a divergence of 0.4° at the $1/e^2$ points. The output consists of many high-order spatial modes, termed lateral modes in the literature. The number of lateral modes depends on the design of the emitter (i.e., broad area or subdivided into narrow stripes), cavity length, and various material parameters. Compared to the vertical direction, the smaller divergence of the beam in the plane of the pn-junction allows the use of optics with a smaller numerical aperture. Therefore the beam in this direction is referred to as the slow axis.

Reformatting the highly astigmatic pump beam emitted from a line source into a circular beam, which is required for end-pumped lasers, is a challenging task. End-pumped lasers require focusing of the pump radiation to a small spot size for mode matching or for coupling into an optical fiber. In Section 6.2.3 a number of different optical schemes that accomplish this task will be discussed.

Diode Laser Lifetime

The long lifetime of laser diodes as compared to arc lamps is an important consideration for pumping solid-state lasers with these devices. We will distinguish between catastrophic failures and normal degradation during operation. Laser diodes, like all semiconductors, are very susceptible to damage by electrostatic discharges or high-voltage transients. Therefore handling of these components requires electric discharge control such as wearing wrist straps and grounded work surfaces. Also great care has to go into power supply and driver design to protect this very expensive pump source from transients, current surges, and high voltage. Operation at excessive currents will lead to catastrophic mirror facet damage. Also, insufficient cooling combined with high currents can melt the solder joints or vaporize the wire bonds within the diode structure.

Under proper operating conditions laser diodes degrade in a fairly predictable manner which results in a decrease in output over time. The higher the operating current and/or the operating temperature, the faster the degradation. The decrease in output over time can be caused by structural defects, which spread through the laser diode forming light absorbing clusters. Dark lines or spots in the output beam are a manifestation of this damage. The degradation of AlGaAs laser diodes, the most common pump source, is usually attributed to oxidation and migration of aluminum under high-power operation. Structural defects can also be caused by thermal stress due to different thermal expansion coefficients of semiconductor and heat sink material.

Facet damage due to the environment or high-radiation flux is another source of degradation, so is diffusion of solder into the semiconductor lattice over time caused by the high-current density. After-screening of the laser diodes during a burn-in period performed by the manufacturer has eliminated early failures due to production problems, and very long operational lifetimes can be achieved. Lifetimes of cw devices are typically on the order of 10^4 hr. In most commercial systems requiring long life, diode arrays are typically operated at 75% of their rated output. Quasi-cw bars and stacked arrays achieve lifetimes of several billion shots.

Spectral Properties

The spectral properties of laser-diode arrays that are most critical for the pumping of solid-state lasers are the center wavelength and the spectral width of the emission and the wavelength shift with temperature.

The wavelength of a laser diode is inversely proportional to the energy difference ΔE between the conduction and valence band

$$\lambda = hc/\Delta E, \tag{6.13}$$

where h is the Planck constant and c is the speed of light. The bandgap depends on the crystalline structure and chemical composition of the semiconductor. A binary compound such as GaAs has a fixed wavelength. Adding a third element, by substituting a fraction of gallium with aluminum in GaAs, changes the bandgap.

In a compound semiconductor, such as GaAlAs, the wavelength of the emitted light can be tailored over a wide range by changing the ratio of aluminum and gallium. Adding aluminum increases the bandgap, producing lasers with shorter wavelengths. Most commercial GaAlAs lasers have wavelengths of 750 to 850 nm. Neodymium ions have substantial absorption in the vicinity of 808 nm, which is the emission wavelength of diode lasers with $Ga_{0.91}Al_{0.09}As$ active regions. As far as the pumping of neodymium lasers is concerned, the output wavelength can be tailored to the peak absorption by adjustments of the aluminum concentrations according to [11]

$$\Delta E = 1.424 + 1.247 \cdot x \quad (eV), \tag{6.14}$$

where x is the aluminum concentration.

Compositional changes and temperature gradients within an array lead to a broader spectral output for the whole array as compared to a single device. A typical spectral width of a single bar or small stacked array is on the order of 4 to 5 nm. State-of-the-art performance is about 2.2 nm for a 10 bar array. The bandwidth of the Nd : YAG absorption line at 808 nm is 2 nm for an absorption coefficient larger than $3.8 \, cm^{-1}$. Similarly, in Nd : YLF, the absorption line is about 2 nm wide for absorption coefficients larger than $2 \, cm^{-1}$. The excellent spectral overlap between the absorption line of Nd : YAG at 808 nm and the emission of a two-dimensional array of GaAlAs laser diodes is illustrated in Fig. 6.15.

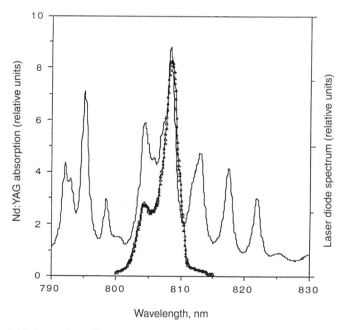

FIGURE 6.15. Spectral overlap between Nd : YAG absorption and emission spectrum of a 10 bar array [12].

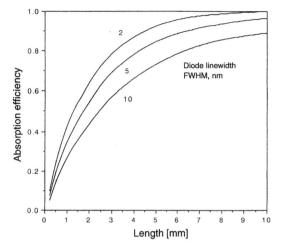

FIGURE 6.16. Absorption efficiency of Nd : YAG as a function of absorption length for a pump wavelength of 808 nm [13].

In diode-pumped lasers that have only a short absorption path, it is important to have a narrow spectral emission from the diode array in order to absorb most of the pump radiation. In optically thick materials, such as large diode-pumped lasers with rod or slab dimensions on the order of 10–15 mm, spectral width becomes less important because eventually all the pump radiation gets absorbed. In Fig. 6.16 the absorption of diode-pumped radiation for Nd : YAG is plotted versus optical thickness with spectral width as a parameter.

For laser materials with an absorption band between 750 and 850 nm, pumping with GaAlAs diodes is the obvious choice because of the availability and high state of development of these devices. Lasers that can be pumped with GaAlAs diodes include neodymium doped materials and Tm : YAG, as listed in Table 6.2. For pumping Yb : YAG or yttrbium-sensitized lasers such as the Er : Yb : glass, a

TABLE 6.2. Important lasers and diode pump sources.

Laser material	Pump wavelength [nm]	Semiconductor laser material
Nd : YAG	808	GaAlAs
Nd : YLF	798,792	GaAlAs
Nd : YVO$_4$	809	GaAlAs
Tm : YAG	805 (wing)	GaAlAs
	785 (peak)	GaAlAs
Yb : YAG	941	InGaAs
Er : glass	980	InGaAs
Cr : LiSAF	670 (peak)	AlGaInP
	760 (wing)	GaAlAs

pump source around 0.98 μm is needed. For this wavelength, so-called strained-layer superlattice lasers based upon InGaAs on GaAs substrates have been developed with emission wavelengths between 0.95 and 0.98 μm [14]. Traditional diode lasers are restricted to materials that can be lattice-matched to GaAs or InP substrates because unmatched materials are prone to defects and quickly degrade. However, it was discovered that very thin layers—below a critical thickness of a few tens of nanometers—can accommodate the strain of a small lattice mismatch. This discovery led to the development of a single, thin strained layer used as a quantum well which has an emission wavelength not achievable with GaAlAs or InGaAsP laser diodes.

On the short-wavelength end of available diode pump sources are the visible AlGaInP diode lasers operating in the 640–680 nm region. These devices have been used to pump chromium-doped lasers, such as $Cr:LiCaAlF_6$ (LiCAF) and $Cr:LiSrAlF_6$ (LiSAF).

Input Versus Output Characteristics

The optical losses in the diode structure determines the threshold current i_s that has to be exceeded before amplification can take place. A diode laser that is operating above threshold will exhibit a linear relationship between output power and electrical current as shown in Fig. 6.17. The optical output power as a function of current input can be expressed by

$$P_{out} = \eta_d (\Delta E / e)(i - i_s), \qquad (6.15a)$$

where η_d is the differential quantum efficiency characterized by the number of photons emitted per injected electrons, ΔE is the bandgap of the recombination region, and e is the electron charge.

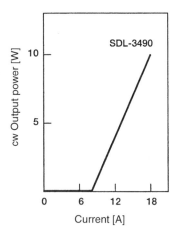

FIGURE 6.17. The cw output power versus input current for a 1 cm bar.

The slope efficiency σ_s, determined from the slope of the output power versus input current, follows from (6.15a)

$$\sigma_s = \eta_d (\Delta E / e). \tag{6.15b}$$

Figure 6.17 depicts the output power versus current for a 10 W cw bar. The device has a slope efficiency of $\sigma_s = 1$ W/A and with $\Delta E = 1.5$ eV follows from (6.15b) a differential quantum efficiency of $\eta_d = 0.66$. The applied forward voltage for diode lasers is

$$V_0 = (\Delta E / e) + i R_d, \tag{6.15c}$$

where R_d is the series resistance of the diode due to the resistivity of the epilayers and the resistivity of the metal contacts. For typical devices the voltage drop across the junction is 1.5 eV and one obtains $V_0 = 1.5V + i R_d$.

The energy conversion efficiency of the diode laser, defined as the ratio of optical output power to electrical input power, which we call the pump efficiency η_p, follows from the above equations

$$\eta_p = \frac{\sigma_s (i - i_s)}{i[(\Delta E / e) + i R_d]}. \tag{6.15d}$$

The device characterized in Fig. 6.17 has an overall efficiency of $\eta_p = 0.32$ at the maximum output based on a series resistance of $R_d = 0.015\,\Omega$ and $\Delta E / e = 1.5$ eV.

If quasi-cw bars are stacked into two-dimensional arrays, the bars are connected electrically in series to avoid high currents and the associated resistive losses. This is illustrated in Fig. 6.18. The drive current for this 10 bar stack is identical to the current of a single bar. The nonlinear behavior of the overall efficiency

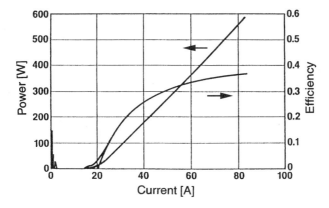

FIGURE 6.18. Performance of a quasi-cw 10 bar array (Spectra Diode Laboratories).

with increasing current occurs because the ohmic losses in the material increase as the square of the current (see (6.15d)), whereas the output power is linear in current.

Thermal Management

Diode-array pumping offers dramatic improvements in efficiency of solid-state laser systems. However, the need to maintain the operating temperature within a relatively narrow range requires a more elaborate thermal management system as compared to flashlamp-pumped lasers.

In end-pumped systems, the diode wavelength is usually temperature tuned to the peak absorption line of the laser and maintained at that wavelength by controlling the array temperature with a thermoelectric cooler. This approach works well for small lasers, and systems including thermoelectric coolers can work over a large range of ambient temperatures.

In large systems, the power consumption of thermoelectric coolers is usually prohibitive. For these systems, a liquid cooling loop with a refrigeration stage can be employed that maintains the temperature of the coolant independent of the environment. However, the coefficient of performance for refrigerators is typically no greater than one. Therefore, the electrical input power requirement is equal to the heat load to be controlled.

Recently, the development of diode arrays which can operate at high junction temperatures has eliminated the need for refrigerated cooling loops for special cases such as military systems. Diode arrays have been operated at junction temperatures as high as $75°$ C, which is higher than ambient in most situations. A simple liquid-to-air cooling system provides the most efficient thermal control system for the laser because only the power consumption of a pump and possibly a fan is added to the total electrical requirements.

The power dissipation of the various subsystems of a pulsed diode-pumped laser is listed in Table 6.3. The system has a repetition rate of 200 Hz and an output pulse energy of 0.8 J/pulse. By far the greatest heat dissipation occurs in the diode pump arrays, and accounts for approximately 65% of the system input power. The solid-state laser medium itself dissipates heat at a rate approximately equal to the output power from the laser system. The electronics dissipate on the order of 19% of the input power.

TABLE 6.3. Power dissipation from a diode-array-pumped solid-state laser.

System output	160 W
Diode array	1300 W
Laser crystal	160 W
Electronics/other components	380 W
Total electrical input	2000 W

Operation of Laser–Diode Arrays

Compared to arc lamps, the electrical operating conditions of laser diodes are much more benign. The operating voltages are low and a trigger circuit is not required. Essentially, only a low-voltage dc power supply of the appropriate voltage and current range is required to operate the cw devices. Despite the very straightforward electronics, there are several challenges for operating laser diodes efficiently and reliably. The series resistance of laser diodes is very low; therefore, the internal resistance of the power supply must be minimized for efficient energy transfer.

Second, compared to flashlamps, laser diodes are electrically much more vulnerable. Therefore, overvoltage, reverse bias, and current protection are essential features in the design of power sources. The power supply of a diode-pumped solid-state laser is the subsystem where most of the weight and volume savings occur compared to a flashlamp-based laser. High efficiency of a diode-pumped laser, low-voltage operation of the diodes, and the use of electrolytic capacitors for energy storage, allow the design of very small and compact power supplies. In a pulsed diode-pumped solid-state laser system, the diode-array drivers replace the pulse-forming network in a conventional flashlamp-pumped laser system. The diode-array driver is a low-voltage switching network which supplies a constant-current pulse to the diode arrays. The low-voltage requirements for the diode arrays allow the use of power MOSFETs as switching elements. Electrical energy stored in electrolytic capacitors is transferred to the diode arrays on each laser pulse.

6.2   Pump Radiation Transfer Methods

The efficiency in the transfer of radiation from the source to the laser element determines, to a large extent, the overall efficiency of the laser system. The optical design, besides providing good coupling between the source and the absorbing active material, is also responsible for the pump power distribution in the laser element which influences the uniformity, divergence, and optical distortions of the output beam.

In the development of solid-state lasers, many different optical designs have been employed to transfer the radiation from a light source to the active material. Depending on the shape of the active material and the type of pump source used, one can broadly divide pumping geometries into systems in which the active material is side-pumped, end-pumped, face-pumped, or edge-pumped. Side- or transverse-, and end- or longitudinal-pumping refers to the orientation of the pump beam relative to the direction of the laser beam in the gain medium. Face- and edge-pumped slabs and disks differ in the directions in which waste heat is removed, and in the resulting directions and magnitudes of the thermal gradients relative to pump and output laser beam directions.

6.2.1 Side-Pumping with Lamps

The most widely used arrangement to collect and concentrate radiation emitted by
an arc lamp into the laser medium is an enclosure comprising a highly reflective
elliptical cylinder with the laser rod and pump lamp at each focus. The ellip-
tic configuration is based on the geometrical theorem that rays originating from
one focus of an ellipse are reflected into the other focus. Therefore an elliptical
cylinder transfers energy from a linear source placed at one focal line to a linear absorber placed along the second focal line. The elliptical cylinder is closed
by two plane-parallel and highly reflecting end plates. This makes the cylinder
optically infinitely long.

An elliptical cylinder can have a cross section with a large or small eccentricity.
In the former case, the laser rod and lamp are separated by a fairly large distance,
in the second case they are close together. If the elliptical cylinder closely surrounds the lamp and rod, then one speaks of a close-coupled elliptical geometry
(see Fig. 6.19(a)). This geometry usually results in the most efficient cavity and
it has the further advantage that it minimizes the weight and size of the laser
heads. Since the energy delivered to a discharge lamp is limited, schemes to focus
the energy from several lamps onto a single crystal are attractive. Figure 6.19(b)
shows two partial elliptical cylinders having one common axis at which the crystal
is placed. Multilamp close-coupled cavities provide a higher degree of pumping
uniformity in the laser rod than single-lamp designs.

Figure 6.20 exhibits examples of close-coupled nonfocusing pump cavities.
The lamp and rod are placed as close together as possible, and a reflector
closely surrounds them. The reflector can be circular (Fig. 6.20(a)) or oval

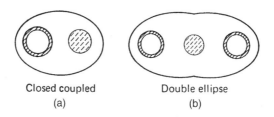

Closed coupled   Double ellipse
(a)   (b)

FIGURE 6.19. (a) Single and (b) double elliptical pump cavity.

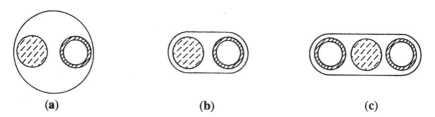

(a)   (b)   (c)

FIGURE 6.20. Nonfocusing pump cavities; (a) circular cylinder, (b) single, and (c) double-
lamp closed-coupled configuration.

(Fig. 6.20(b),(c)). The latter type is often used in laboratory setups of low-repetition-rate pulsed lasers. The reflector consists simply of silver or aluminum foil wrapped around the flash lamp and laser rod.

The fraction of pump energy that is transferred from the source to the laser rod can be expressed by

$$\eta_t = \eta_{tg} \times \eta_{to}, \tag{6.16}$$

where η_{tg} is the geometrical cavity transfer coefficient. It is the calculated fraction of rays leaving the source which reach the laser material either directly or after reflection from the walls. The parameter η_{to} expresses the optical efficiency of the cavity and essentially includes all the losses in the system. This parameter can be expressed as

$$\eta_{to} = R_w(1 - R_r)(1 - \delta)[1 - (A_n/A_{tot})], \tag{6.17}$$

where R_w is the reflectivity of the cavity walls at the pump bands, R_r is the reflection at the laser-rod surface or at the glass envelopes of the cooling jackets and Fresnel losses of any filters inserted in the cavity, δ is the absorption loss in the optical medium between the lamp and the laser rod such as coolant liquid, filters, etcetera, and (A_n/A_{tot}) is the ratio of the nonreflecting area of the cavity A_n to the total inside area A_{tot}. This factor accounts for losses due to openings in the reflector which are required, for example, to insert the laser rod and pump lamp. Equations (6.16), (6.17) are rough approximations based on the assumption that all the lamp radiation undergoes just one reflection. If one were to include direct radiation and multiple reflections from the cavity walls, these equations would have to be developed in series expressions.

The pump source is usually assumed to be a cylindrical radiator having a Lambertian radiation pattern. This implies that it appears as a source having constant brightness across its diameter when viewed from any point. In Fig. 6.21 trajecto-

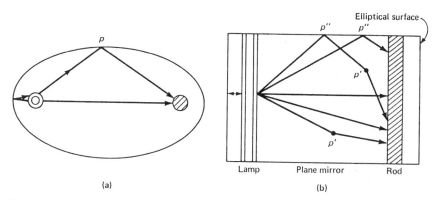

FIGURE 6.21. Trajectories of photons emitted from the pump source; (a) cross section and (b) top view of elliptical pump cavity.

ries of photons that originate from a volume element dv of the source are shown. In Fig. 6.21(a) the rays leave normal to the surface of the source and, therefore, remain in a cross-sectional plane. Figure 6.21(b) depicts trajectories of photons that leave the source at an angle with respect to a cross-sectional plane of the ellipse. In this case the photons can be reflected off the end-plate reflectors as well as undergo a reflection at the elliptical cylinder. In an elliptical pump cavity, all rays emanating from one point of the source are transformed into parallel lines at the laser rod. Each line corresponds to rays leaving the source inclined at the same angle with respect to the major axis of the ellipse, but at different angles with respect to the cross-sectional plane.

Image formation, in its usual sense, is meaningless for the elliptical cylinder since rays emanating in different directions from a point at the source converge after reflection from the cavity at altogether different points. In a pump cavity, all that is actually desired is the transfer of radiant energy from the source to the laser rod. From the foregoing considerations, it is obvious that the condition of an infinitely long elliptical pump cavity can be satisfied by enclosing the ends of the cylinder with highly reflective plane mirrors. Therefore, in the analysis of this arrangement the two-dimensional case can be treated by considering the light distribution in a plane perpendicular to the longitudinal axes of the cylinder.

In the theoretical expressions obtained for the efficiency of elliptical configurations, the pump system is usually characterized by the ratio of the rod and lamp radii r_R/r_L, the ratio of the lamp radius to the semimajor axis of the ellipse r_L/a, and the eccentricity $e = c/a$, where $2c$ is the focal separation. Consider any point P on the surface of the cavity with a distance l_R from the crystal and distance l_L from the lamp as shown in Fig. 6.22. Suppose that the lamp has a radius r_L, then, as a consequence of the preservation of angles, upon reflection the image will have the radius $r'_L = r_L l_R/l_L$. This means that the portion of the elliptical reflector nearer the lamp forms a magnified image at the laser rod, while that

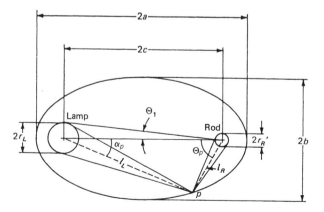

FIGURE 6.22. Cross section of elliptical pump cavity. Eccentricity $e = c/a$; focal point separation $c = (a^2 - b^2)^{1/2}$.

portion nearer the crystal forms a reduced image of the lamp. A point P, with corresponding angles α_p and Θ_p measured from the lamp and rod axis, respectively, may be defined dividing these two regions. At this point, the ellipse generates an image of the lamp which exactly fills the crystal diameter. We must allow for this effect of magnification and demagnification when determining how much energy is captured by the crystal. From the properties of an ellipse, and noting that at P, $l_R/l_L = r_R/r_L$, we obtain

$$\cos \alpha_p = \frac{1}{e}\left[1 - \frac{1-e^2}{2}\left(1 + \frac{r_R}{r_L} \right) \right], \tag{6.18}$$

and

$$\sin \Theta_p = \left(\frac{r_L}{r_R} \right) \sin \alpha_p. \tag{6.19}$$

The geometrical cavity transfer coefficient can be calculated by considering what fraction of the energy radiated by the lamp into an angle $\Delta\alpha$ is trapped by the crystal. Integration over all angles leads to [15, 16]

$$\eta_{tg} = \frac{1}{\pi}\left[\alpha_p + \left(\frac{r_R}{r_L} \right)\Theta_p \right]. \tag{6.20}$$

This expression is plotted in Fig. 6.23. A certain portion of the reflecting surface behind the lamp is screened from the crystal by the lamp itself. In flashlamps, which resemble a blackbody source, the radiation reflected back into the lamp will be absorbed. On the other hand, the plasma in cw arc lamps is optically thin, and the reflected pump radiation is transmitted through the plasma. If we assume the radiation reflected back to the pump source to be lost, we must reduce the angle Θ_p in the above formula by Θ_1, where

$$\sin \Theta_1 = \frac{r_L}{4ae}, \tag{6.21}$$

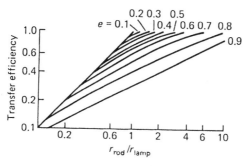

FIGURE 6.23. Dependence of transfer efficiency of an elliptical reflector on the quantity r_R/r_L and eccentricity e [17].

and

$$\eta'_{tg} = \frac{1}{\pi} \left[\alpha_p + \left(\frac{r_R}{r_L} \right) (\Theta_p - \Theta_1) \right]. \tag{6.22}$$

It is seen from Fig. 6.23 that the efficiency increases with an increase in the ratio r_R/r_L and with a decrease in the eccentricity e. This result stems from the fact that the magnification of the pump source increases with the eccentricity of the ellipse. A superior efficiency is therefore obtained for an almost circular cavity and as small a pump source as possible.

It should be noted that in these calculations the crystal is assumed to absorb the radiation which falls upon it; no second traverses of light through the crystal are considered. Furthermore, no allowance was made for the reflection loss at the cavity walls or losses due to Fresnel reflection at the surface of the laser rod. Multiple reflections can become significant if the absorption coefficient of the crystal is small compared to the inverse of the crystal diameter.

Table 6.4 summarizes the energy balance in a gold-plated double-elliptical pump cavity which contained an Nd:YAG rod pumped by two cw krypton arc lamps. In a pump cavity the electrical input power supplied to the lamp is dissipated as heat by the lamp envelope and electrodes or emitted as radiation, and a portion of the radiation will be absorbed by the metal surfaces of the pump cavity. The radiation reflected from the walls will either be absorbed by the lasing medium or will return to the lamp. The light that is absorbed by the lamp will add energy to the radiation process in the same way as the electrical power does, and the returned light will be radiated with the same efficiency as the power supplied electrically. One consequence of the reabsorption is that a lamp, when enclosed in the pumping cavity, is operated under a higher thermal loading, resulting in shorter lamplife than when operated in the open for the same electrical input power. The heat dissipated by the laser rod assembly listed in Table 6.4 contains not only heat rejected by the laser crystal, but also energy absorbed by the flow tube and cooling water. The pump power absorbed by the laser rod causes stimu-

TABLE 6.4. Energy transfer in a cw krypton arc lamp, pumped Nd:YAG laser.

Heat dissipation of lamps		55%
Heat dissipation of pump reflectors		30%
Power absorbed by coolant and flow tubes		7%
Heat dissipation by rod	5%	
Laser output	2%	
Fluorescence output	0.4%	
Optical losses	0.6%	
Power absorbed by laser rod		8%
Electrical input to lamps		100%

lated emission and fluorescence at the laser wavelength and other main emission bands. The remainder is dissipated as heat by the laser material.

In elliptical-pump cavities, the laser rod and lamp are often liquid cooled by circulating the coolant in flow tubes that surround these elements. The inside of the pumping chamber itself is dry. However, in most cases the body of the reflector contains cooling chambers through which the coolant fluid passes.

In the so-called flooded cavity approach, the whole inside of the pump cavity is immersed in cooling fluid. The absence of flow tubes and separate cooling chambers for the reflector makes this type of cavity very compact and simple in design. For example, only one inlet and outlet are required for the cooling loop, whereas in an elliptical cavity with flow tubes, one pair of coolant ports with the associated fittings, tubing, and so on, is required for each reflector half, lamp, and laser rod. Also, in the flooded cavity design, lamp and laser rod can be brought very close together, and no reflection losses from additional glass surfaces are encountered. Figure 3.18 presents a photograph of a commercial cw-pumped Nd : YAG laser featuring a single elliptical reflector in a flooded cavity.

Most liquid-cooled military-type Nd : YAG lasers and most commercial cw-pumped Nd : YAG lasers feature this design because of its compactness and simplicity. Figure 6.24 shows a photograph of a liquid-immersed pump cavity used for a high-repetition-rate military Nd : YAG laser. The cavity is sealed by one large O-ring in the top cover. The reflector inserts are machined from aluminum which is nickel-plated, polished, and silver-plated. The laser head is machined from aluminum that is hard-anodized to prevent corrosion.

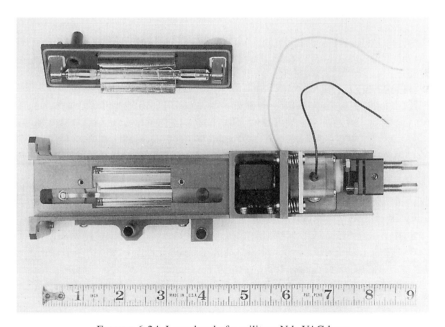

FIGURE 6.24. Laser head of a military Nd : YAG laser.

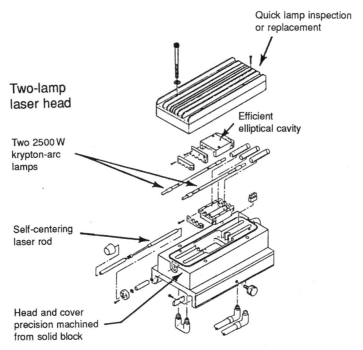

FIGURE 6.25. Exploded view of a single- and double-elliptical pump cavity of a cw-pumped Nd : YAG laser.

Figure 6.25 illustrates an exploded view of a double-elliptical pump cylinder using immersion cooling. The laser head is machined from a solid block of acrylic. This material alleviates the problem of arcing and eliminates the need to electrically insulate the electrodes from the laser head. The reflector inserts are machined from copper and are gold-plated.

The planar pump and cooling geometry achieved with a slab provides the possibility for one-dimensional heat flow. The design of flashlamp-pumped slab lasers has been extensively studied, and the potential advantages and deficiencies of such systems will be discussed in Chapter 7.

In Fig. 6.26 a single lamp is placed along the axis of the slab on each side. Appropriate reflectors are employed to distribute the pump light evenly over the slab surface area. Uniform pumping and cooling of the optical surfaces is absolutely crucial for good performance of this type of laser as will be discussed in Chapter 7.

6.2.2   Side-Pumping with Diodes

In contrast to flashlamps, the emission from laser diodes is highly directional; therefore many arrangements are possible for transferring the pump radiation to the solid-state laser material. Since the output beams of laser diodes can be shaped

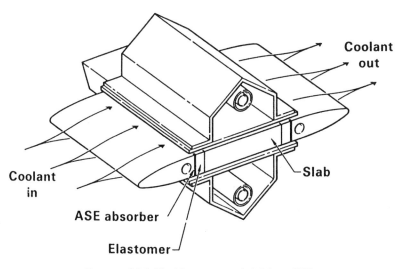

FIGURE 6.26. Flashlamp-pumped slab laser [18].

and focused, a major consideration is the design of a pump geometry which maximizes the overlap of the pumped volume with the volume occupied by a low-order resonator mode. The optimization of the overlap is referred to as mode-matching.

In side-pumping the diode arrays are placed along the length of the gain material and pumping is transverse to the direction of the laser beam. Scalability to higher power is the advantage of this design approach. As more power is required, more diode arrays can be added along and around the laser rod. There are three approaches to couple the radiation emitted by the diode lasers to the rod:

(a) *Direct coupling.* From a design standpoint this is by far the most desirable approach. However, this arrangement does not allow for much flexibility in shaping the pump radiation inside the laser rod.

(b) *Intervening optics between source and absorber*: In this case, the pump distribution can be peaked at the center of the rod allowing for a better match with the resonator modes. Optical coupling can be achieved by using imaging optics such as lenses or elliptical and parabolic mirrors or by nonimaging optics such as reflective or refractive flux concentrators.

(c) *Fiber optic coupling.* Due to the coupling losses combined with the increased manufacturing cost of fiber-coupled diode arrays, this technique is not very attractive for side-pumping.

Whether or not to use intervening optics between the arrays and the laser rod depends mainly on the desired beam diameter and optical density of the active medium. Generally speaking, oscillator pump heads usually employ optical elements in order to concentrate the pump radiation in the center for efficient extractions at the TEM_{00} mode, whereas large amplifier rods are direct-pumped.

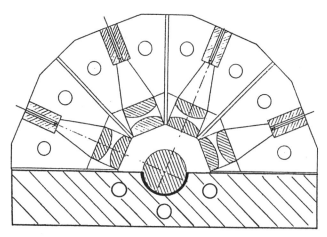

FIGURE 6.27. Side-pumping of a cylindrical laser rod employing a hemispherical pump geometry.

An example of a side-pumped Nd : YAG rode is shown in Fig. 6.27, the laser rod is conductively cooled by a copper heat sink. The pumping cavity consists of a number of linear arrays symmetrically located around the rod. The arrays are mounted on long heat-sink structures and contain cylindrical optics. The optics allow the arrays to stand-off from the laser crystal and the arrays. The output apertures of the diode bars are imaged near the center of the rod by using a symmetrical doublet of plano-convex cylindrical lenses. The rod surface in contact with the heat sink has a reflective coating that returns the pump light for a second pass through the rod.

For optimization of the pump distribution within the gain medium, a ray trace program is very useful. Figure 6.28 presents the result of such a calculation for the

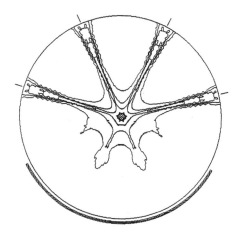

FIGURE 6.28. Contour plot of pump intensity distribution.

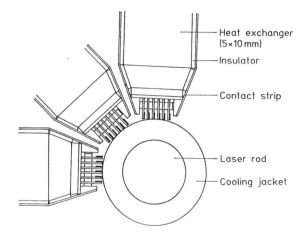

Heat exchanger
(5×10 mm)

Insulator

Contact strip

Laser rod

Cooling jacket

FIGURE 6.29. Arrangement of stacked diode array bars around a laser rod.

pump arrangement of Fig. 6.27. The code calculates the spatial pump distribution in the laser material as a function of spectral absorption, and the spectral and spatial properties of the radiation source as modified by the optical surfaces between the source and absorber. The strong concentration of pump radiation at the center of the rod was responsible for the good TEM$_{00}$ performance of the laser.

A pump configuration of a high-power amplifier head which does not use coupling optics is illustrated in Fig. 6.29. The laser rod is surrounded by eight stacked arrays, each containing six bars. Since the bars are 1 cm long, and each bar produces 12 mJ in a 200 μs pump pulse, the rod can be pumped at 576 mJ per centimeter rod length.

Figure 6.30 displays a photograph of a pump head containing four rings of stacked diode arrays. In this design, 4 cm of the 2 in long rod are pumped by the diode arrays. Large diode-pumped lasers employ liquid-cooled laser rods or slabs similar to flashlamp-pumped systems. The laser-diode arrays are directly mounted on liquid-cooled heat sinks. Extremely compact and lightweight structures have emerged which take full advantage of the high packaging density which can be achieved in the diode arrays.

Figure 6.31 presents a photograph of an assembled pump head for a side-pumped Nd: YAG rod pumped by 16 diode arrays. Each array is 1 cm long, and the devices are arranged in four pairs symmetrically around the rod. The adjacent four pairs are off-set 45° in order to achieve an eight-fold symmetry around the laser crystal. The active pump area is 4 cm long and the complete assembly, including the end plates, has a length of 7 cm.

Each diode array contains five bars, that is, the laser crystal is pumped by 80 bars at a pulse energy of 800 mJ in a 200 μs-long pulse. Electrical input is 2 J per pulse that at a repetition rate of 100 Hz amounts to about 180 W of heat, which has to be carried away by the laser head. To carry off heat from the diodes and laser rod, liquid flows into the pump head through a fitting that screws into the

FIGURE 6.30. Photograph of diode-array-pumped laser head (background) and stacked diode array bars mounted on copper heat sink (foreground).

end plate. Directed by the manifolds, the coolant flows back and forth through passages in the triangular structures, and also through the annulus between the rod surface and sapphire sleeve. The flat areas of the triangular bars provide the mounting surface for the diodes and the thermal interface. Eventually, the coolant exits the pump head through a second fitting at the end plate. With the diode arrays mounted on the internally water-cooled bars, the coolant, as it moves through the structure, is close to the heat source, thus facilitating heat transfer. For maximum

FIGURE 6.31. Liquid-cooled laser heads for diode-pumped Nd : YAG lasers containing 16 diode arrays.

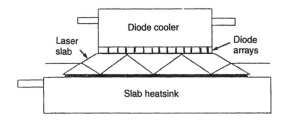

FIGURE 6.32. Side-pumped conductively cooled slab laser.

heat transfer, a thin layer of indium is applied between the mounting surface and the diode arrays.

Side-Pumped Slab Laser

Slab lasers are usually pumped by large densely packed two-dimensional arrays. In a slab laser, the face of the crystal and the emitting surface of the laser diodes are in close proximity and no optic is employed (Fig. 6.32).

The slab is usually pumped from one face only. The opposite face is bonded to a copper heat sink containing a reflective coating to return unused pump radiation back into the slab for a second pass. An antireflection coating on the pump face is used to reduce coupling losses (the diode array is not in contact with the active material). Liquid cooling is employed to remove heat from the crystal and diode heat sink. The purpose of the zigzag optical path shown in Fig. 6.32 is to mitigate the effects of thermal lensing caused by thermal gradients in the slab, as will be explained in Chapter 7.

A photograph and details of a side-pumped and conductively cooled slab laser are shown in Fig. 6.33. The zigzag slab Nd : YAG laser is air-cooled and battery-powered. At a repetition rate of 10 pps, the system produces 270 mJ per pulse at the frequency doubled wavelength of 532 nm. The slotted compartment shown in the photograph houses the pump module consisting of the Nd : YAG slab, diode arrays, heat sink, and cooling fans. The zigzag slab is pumped by twelve 18 bar arrays with a total optical output of 2.6 J at a pulsewidth of 200 μs.

A cross section of the pump module is shown in Fig. 6.33(b). The slab was bonded to a thermal expansion-matched heat sink which is clamped to the bridge assembly. This clamping method was employed to reduce mounting and thermal cycling stresses in the laser crystal. Zigzag slab lasers can be scaled to the kilowatt level employing diode arrays. The large face area available in a slab geometry allows stacking of a sufficient number of diode arrays to achieve a high-output power. The large flat surfaces eliminate the need for intervening optics between the pump source and the laser crystal. This facilitates scaling to large systems.

The pump module of a high-power diode laser is shown in Fig. 6.34. A 20 $\times$ 2 $\times$ 0.7 cm Nd : YAG slab is pumped by 14 diode arrays on each side. Each array is comprised of 16 bars with a peak output of 50 W at a 30% duty cycle. The total optical pump power to the slab is 6.72 kW at an electrical input of 15 kW. The

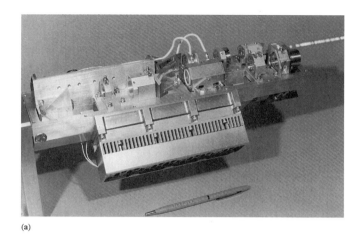

(a)

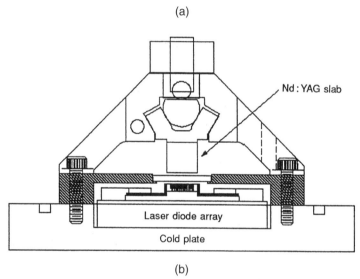

(b)

FIGURE 6.33. (a) Photograph of an air-cooled and battery-operated zigzag slab Nd : YAG laser and (b) cross section of pump module [19].

system was designed for a pulse repetition rate of 2 kHz and a pump pulse length of 150 μs. With a 2 m long resonator, the pump head generated 740 W of average Q-switched power in a 7 × diffraction limited beam. Peak power was 1 MW.

Depending on the way heat is extracted from the slab we can distinguish the three design configurations illustrated in Fig. 6.35. If the slab is water-cooled, pumping and heat extraction can occur through the broad faces of the rectangular slab as shown in Fig. 6.35(a). Space and military applications require conductively cooled slabs. One approach, shown in Fig. 6.35(b), is to conductively cool one face of the slab and pump through the opposite side. More powerful lasers can be

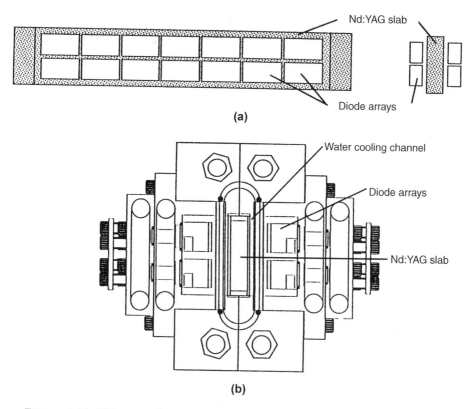

FIGURE 6.34. High-power diode-pumped Nd : YAG slab laser; shown is (a) the arrangement of the diode arrays and (b) a cross section of the pump head [20].

built by conduction cooling of both large surface areas of the slab. Pump radiation is introduced through the two narrow sides of the slab as shown in Fig. 6.35(c). Therefore this pumping geometry is termed edge-pumping [21]. The concentration of pump radiation, which is possible with diode sources, makes this approach feasible. In all three cases depicted in Fig. 6.35 pump radiation enters into the slab transverse to the resonator axis and the laser output is extracted from the ends of the slab. Also the plane of the zigzag beam is collinear with the direction of the thermal gradients in the slab.

Pump Configurations Based on Internal Reflections of the Laser Beam

Besides the classical side-pumped configurations discussed in this section, there is an almost unlimited number of pump-configuration variations possible. Most of the more unusual design configurations utilize internal reflections of the laser beam. In Fig. 6.36 designs are illustrated which contain one, or many internal reflections of the laser beam. An extension of this concept is the monolithic ring laser which was discussed in Section 5.2.3.

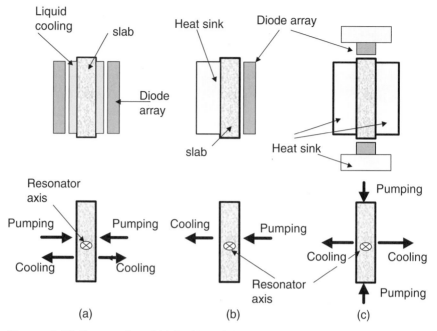

FIGURE 6.35. Cross section of (a) liquid-cooled slab, (b) one-sided conductively cooled, and (c) two-sided conductively cooled slab.

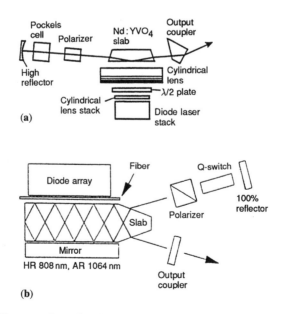

FIGURE 6.36. Pump configurations based on internal reflections of the laser beam. (a) Slab with grazing angle at the pump face [22], and (b) slab with folded zigzag path [23].

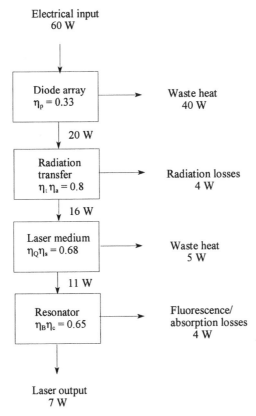

Electrical input
60 W

Diode array
$\eta_p = 0.33$ → Waste heat
40 W

20 W

Radiation
transfer
$\eta_t \eta_a = 0.8$ → Radiation losses
4 W

16 W

Laser medium
$\eta_Q \eta_s = 0.68$ → Waste heat
5 W

11 W

Resonator
$\eta_B \eta_c = 0.65$ → Fluorescence/
absorption losses
4 W

Laser output
7 W

FIGURE 6.37. Energy flow in a typical diode-pumped Nd : YAG laser. (Efficiencies are defined in Section 3.4.1.)

In the design shown in Fig. 6.36(a) side-pumping through a polished face of the gain medium is used to obtain high gain. The resonator mode makes a single grazing-incidence total internal reflection at the pump face, thus remaining in the region of the highest gain throughout its passage through the slab. This design has been used for small Nd : YVO$_4$ and Er : YAG lasers. Figure 6.36(b) exhibits a miniature slab laser with a folded zigzag path of the laser beam. The entrance and exit faces are cut at Brewster's angle. The beam path in this particular Nd : YAG slab consisted of five bounces. The pump radiation is collimated in the plane of the zigzag path by a fiber lens to maximize the overlap between pump beam and resonator mode.

The excellent spectral match of the diode output and the absorption of the solid-state material, in conjunction with a high spatial overlap between pump radiation and resonator mode volume, are responsible for the high overall efficiencies that can be achieved in diode-pumped lasers. Figure 6.37 illustrates the energy flow in a typical Nd : YAG laser, pumped with a 20 W diode bar at 808 nm. Laser diodes have efficiencies of $\eta_p = 0.3$–0.45, with quasi-cw diodes at the upper range since they are more efficient pump sources as compared to cw diodes. Transfer of diode radiation and absorption by the lasing medium is expressed by the transfer and absorption efficiencies that in most systems are around $\eta_t\eta_a = 0.7$–0.8. In end-pumped configurations, diode radiation is, in most cases, not all collected by the optical system and reflection losses further diminish pump radiation. On the other hand, the diode radiation is almost completely absorbed in the longitudinally pumped active medium. In side-pumped configurations, transfer efficiency is close to one, but absorption of the pump radiation is often incomplete because a portion of the diode output is transmitted through a small rod or thin slab.

The absorbed pump radiation produces heat in the crystal. The fractional thermal loading, that is, the ratio of heat generated to absorbed power, is 32% in Nd : YAG pumped at 808 nm. These losses are accounted for by the Stokes shift and quantum efficiency $\eta_s\eta_Q = 0.68$. Mode-matching and optical losses in the resonator are expressed by the beam overlap and coupling efficiency which typically are $\eta_B\eta_c = 0.6$–0.8. The generic laser postulated in Fig. 6.37 has an overall efficiency of 11.6% and an optical-to-optical efficiency of 35%. In reviewing the energy-flow diagram, it is clear that the laser designer has only control over Steps 2 and 4, namely the design of the pump radiation transfer, pump geometry, and mode volume.

6.2.3   End-Pumped Lasers

In the end-pumped technique, which is unique to laser diode pump sources, the pump beam is circularized, and focused into the gain medium coaxial with the resonator beam. There are two requirements that must be met for an efficient TEM$_{00}$ end-pumped laser. First, the gain medium must be long enough to absorb all, or at least a large fraction, of the pump radiation. Second, the pump beam spot size must be less than or equal to the spot size of the fundamental mode. If these requirements are met and if we further assume that the pump beam and laser resonator mode are TEM$_{00}$ Gaussian beams, and the resonator length is equal to the gain medium, a closed-form expression for laser threshold and slope efficiency can be derived for a four-level system [24]. From the rate equations we obtain the upper laser level population n at threshold

$$\frac{dn}{dt} = W_p r_p(r, z) - \frac{n(r, z)}{\tau_f} = 0, \tag{6.23}$$

where W_p is the pump rate and $r_p(r, z)$ is the normalized pump intensity distribution, τ_f is the upper state lifetime, and $n(r, z)/\tau_f$ describes the losses due to spontaneous emission. The pump rate is defined by

$$W_p = \eta_Q P_a / h\nu_p, \tag{6.24}$$

where P_a is the absorbed pump power, $h\nu_p$ is the pump photon energy, and η_Q is the quantum efficiency.

From the threshold condition (3.7) and with (3.12) we obtain

$$2n\sigma l s_0(r, z) = T + \delta, \tag{6.25}$$

the term $s_0(r, z)$ has been added to describe the normalized resonator mode intensity distribution. From (6.23)–(6.25) follows

$$P_a = \frac{h\nu_p(T + \delta)}{\tau_f \eta_Q 2\sigma l r_p(r, z) s_0(r, z)}. \tag{6.26}$$

The spatial distribution of the pump volume $r_p(r, z)$ and resonator mode $s_0(r, z)$ is taken into account by calculating an overlap integral that averages the upper laser level population over the laser resonator mode. With the assumption of TEM$_{00}$ Gaussian beams, the normalized expressions for the pump and resonator modes are given by

$$r_p(r, z) = \frac{2\alpha_0 \exp(-\alpha_0 z) \exp(-2r^2/w_p^2)}{\pi w_p^2[1 - \exp(-\alpha_0 l)]} \tag{6.27a}$$

and

$$s_0(r, z) = \frac{2 \exp(-2r^2/w_0^2)}{\pi w_0^2 l}, \tag{6.27b}$$

where

$$\iiint r_p(r, z)\, dV = \iiint s_0(r, z)\, dV = 1, \tag{6.27c}$$

α_0 is the absorption coefficient at the pump wavelength, w_p, w_0 is the Gaussian beam radius of the pump beam and resonator modes, respectively, and L is the length of the gain medium and resonator. With (6.27) the overlap integral can be calculated

$$\iiint s_0(r, z) r_p(r, z)\, dV = 2/l(w_0^2 + w_p^2)\pi. \tag{6.28}$$

With (6.28) following from (6.26) for the absorbed pump power at threshold

$$P_a = \frac{\pi h\nu_L}{4\sigma \eta_Q \eta_s \tau_f}(w_0^2 + w_p^2)(T + \delta), \tag{6.29}$$

where $h\nu_p$ has been replaced by the laser photon energy $h\nu_L$, and η_s is the quantum defect efficiency.

From (6.29) it follows that a low-laser threshold requires small spot sizes and low losses. The smallest mode dimensions are determined by diffraction and by the length l of the resonator. As explained in Chapter 5, the cross-sectional area of a mode is proportional to resonator length. Efficient absorption of pump radiation requires that l should be on the order of or greater than $1/\alpha_0$, where α_0 is the absorption coefficient for the pump radiation. With l determined by the pump absorption length, the smallest values of w_m and w_p that can be achieved are based on the propagation of Gaussian beams.

From (6.29) it follows that for a given laser mode volume the lowest threshold is achieved for the smallest pump volume. The spot size $w_p(z)$ of the pump beam inside the gain medium follows from (5.5):

$$
w_p(z) = w_{0p} \left[1 + \left(\frac{\lambda_p(z - l)}{\pi w_{0p}^2 n_0} \right)^2 \right]^{1/2} , \tag{6.30}
$$

where n_0 is the refractive index of the gain medium and the beam waist w_{0p} is assumed to be at the exit face $z = l$ of the laser crystal. The smallest pump volume is achieved if the mean value $\overline{w}_p^2$ is minimal over the crystal length. Integration of (6.30) yields the smallest pump radius, which is theoretically possible for the crystal of length

$$
w_{p,min} = \left(\frac{l\lambda_p}{\pi n_0} \right)^{1/2} . \tag{6.31}
$$

The power density in the resonator above threshold is obtained from the rate equations. From this analysis the slope efficiency, defined as the ratio of laser output to optical pump power for the case of $w_p < w_0$, is given by [24]

$$
\sigma = \eta_Q \eta_s \frac{T}{(T + \delta)}. \tag{6.32}
$$

As in all lasers, low losses lead to a high slope efficiency. The above expression is the same as (3.64) for the case where the pump radiation is completely absorbed by the laser medium ($\eta_a = 1$), and perfect overlap between the inversion and the resonator mode volume is assumed ($\eta_B = 1$). Since (6.32) gives the optical to optical efficiency, the pump efficiency η_P and the throughput of the optical transport system η_T, which occur in (3.64), are not included in this expression.

The practical realization of a complete spatial overlap between the longitudinal pump beam and the laser mode depends very much on the degree with which the strongly astigmatic and highly divergent radiation emanating from the pump source can be reshaped into a beam with circular symmetry and a small beam waist. Focusing the output from a 1 cm bar or a stack of bars into a laser crystal requires collimation of the emitted radiation, compensation of asigmatism and, finally, focusing the beam to a circular spot.

Common techniques of coupling radiation from laser diode bars into the gain medium of an end-pumped laser typically involve the use of:

- Two orthogonally mounted micro lens arrays for collimation of the output from each emitter followed by a focusing lens (Fig. 6.38(a)).

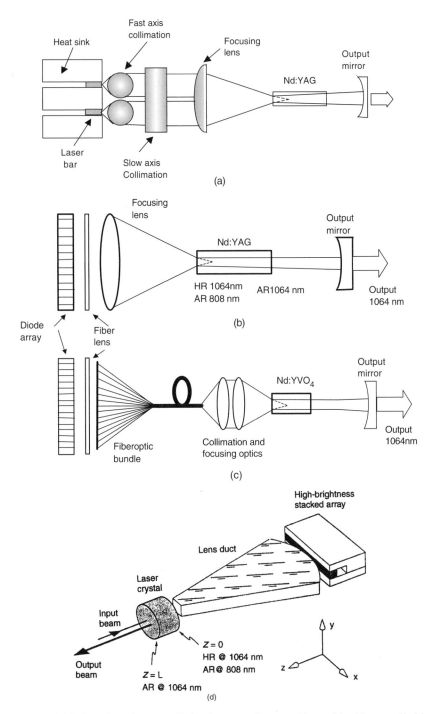

FIGURE 6.38. Focusing of pump radiation into an end-pumped laser: (a) with two cylindrical micro lens arrays and a focusing lens, (b) a single micro lens array and a focusing lens, (c) a micro lens array followed by a fiber bundle and focusing optics, and (d) a nonimaging light concentrator [25].

- A single cylindrical micro lens array for collimation of the fast axes of the bars and a macro lens for focusing (Fig. 6.38(b)).
- A single cylindrical micro lens array for collimation of the fast axes of the bars and a fiber bundle to circularize the beam followed by focusing optics (Fig. 6.38(c)).
- A nonimaging light concentrator (Fig. 6.38(d).
- A macro lens system (Fig. 6.39).

A single laser bar or a stack of bars represent a pump source with an overall cross section of up to a few centimeters squared, which consists of a multitude of independent emitters. The smallest focused spot size from a single bar or two-dimensional bar can be achieved if the radiation from each emitter is individually collimated and corrected for astigmatism. This is usually done with micro lenses or fiber bundles. After collimation, the multitude of beamlets are focused with a bulk lens or lens assembly into the gain medium. This approach requires extreme precision in positioning and aligning the optical components with the emitters.

Designs based on ordinary macro lens assemblies or nonimaging light concentrators treat the radiation emerging from a stack of bars as coming from a large extended source. The designs are relatively simple but the minimum spot size which can be achieved is at least an order of magnitude larger than can be obtained by collimating the output from each emitter.

A compromise between the above mentioned approaches are designs whereby the fast axis of each bar is collimated with a cylindrical micro lens and the emerging beams are focused with a conventional lens or nonimaging light concentrator.

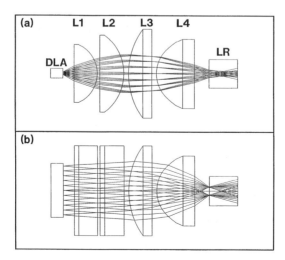

FIGURE 6.39. Optical system for end-pumping an Nd : YAG crystal with a stack of three 1 cm laser diode bars. Ray tracing plot (a) parallel and (b) perpendicular to the bars [27].

Lens Transfer System

A very effective method of collimating the output from two-dimensional bars is by means of two orthogonally oriented arrays of micro lenses. The first set of lenses are typically cylindrical micro lenses similar to short pieces of optical fibers. Each lens is oriented along the length of a bar close to the facets of the emitters. These fiber lenses, mounted in a rigid holder assembly, collimate the fast axis of each bar within the stacked array. The number of micro lenses required is thus given by the number of bars in the array. These fiber lenses are 1 cm long and have a maximum diameter determined by the pitch of the array (typically 0.4 to 1 mm). The distance from the facets is determined by the back focal length which is, for a micro lens with a circular cross section and uniform refractive index, $D(2 - n_0)/4(n_0 - 1)$. For a fiber lens with a diameter $D = 800\,\mu$m, made from fused silica with $n = 1.51$, the distance between the micro lens and the facet is therefore $200\,\mu$m.

Instead of simple step-index fibers with a circular cross section, lenses with noncircular cross section (aspherics) or graded index (GRIN) lenses can be used to reduce spherical aberration. Aspheric lenses can be fabricated in large volume from drawing of a shaped-glass preform. It is also common practice to make micro lens arrays monolithic rather than from discrete segments of optical fibers. Monolithic arrays consist of a glass plate in which parallel curved surfaces that optically act the same way as individual fiber lenses are shaped (see Fig. 6.13(b)).

Mounted orthogonally in front of the first array of micro lenses is another set of micro lenses for collimation of the slow axis or each emitter. The number of micro lenses is given in this case by the number of emitters in each bar. The diameter of the micro lenses cannot exceed the spacing between adjacent emitters (0.2 to 1 mm). Because of the stand-off distance from the facets, which is largely determined by the intervening horizontal lens array, the back focal length of the second lens array must be significantly larger than the lens diameter. This rules out micro lenses with a circular cross section since these optical elements have a very short back focal length. Either aspheric or plano-convex cylindrical micro lenses are normally used for the collimation of the slow axis of the emitters.

As an example, we consider a typical 1 cm cw bar which contains 20 emitters each $100\,\mu$m wide and spaced $400\,\mu$m apart as shown in Fig. 6.12. If we assume a 5-bar internally cooled stack with a pitch of 1 mm, a collimator could be designed that consists of a matrix of five horizontal cylindrical micro lenses with a circular diameter of $800\,\mu$m and a back focal length of $200\,\mu$m and 20 vertical plano-convex lenses with a diameter of $500\,\mu$m and a back focal length of 2.5 mm. With the approach discussed so far, the output from a stacked bar can be focused to a spot with a diameter of a few hundred micrometers.

If larger spot sizes can be tolerated, a compromise between complexity and performance can be reached if only one array of micro lenses is employed followed by an ordinary lens system. In a typical design, a set of fiber lenses reduces the beam divergence of the fast axis of each bar to match the divergence of the slow axis and a bulk lens focuses the beam into the active medium. Figure 6.38(b) illustrates such an approach.

In the absence of micro lenses or fiber-optic bundles one can attempt to focus the radiation from a single bar or stacked bar array with conventional optics. As already mentioned, the power density at the focal point will be far below that which can be achieved with micro-optics. To illustrate this point and to provide a comparison the following example is given.

Figure 6.39 depicts an imaging optical system for focusing pump radiation from a stack of three 1 cm bars into an Nd : YAG crystal. The optic consists of two cylindrical lenses, L1 and L2, with 12.7 and 19 mm focal length, respectively, followed by a 30 mm focal length plano-convex lens L3 and a 12 mm focal length aspheric L4. The rays traced in the plane perpendicular to the bars span a total angular field of 60° at the source and those traced in the plane parallel to the bars cover a total field of 10°. The corresponding width at the pump end of the laser rod is ≈ 2 mm.

In order to at least approximately match such a large pump profile with the resonator mode, a near hemispherical resonator was chosen. The resonator length was adjusted until the TEM_{00} mode at the curved mirror and the front surface of the Nd : YAG was expanded to 1.8 mm. At the flat output coupler the $1/e^2$ beam diameter was 120 μm. The expansion of the mode at the pump end of the resonator is achieved at the expense of alignment stability. A photograph of the system is shown in Fig. 3.27.

Fiber Coupling

Using fiber optics to deliver the diode output to the laser crystal has a number of advantages: the beam at the fiber output has a circular distribution and the ability to remove heat from the diode remote from the optical components of the laser is an attractive feature. Also the pump source can be replaced relatively easily without disturbing the alignment of the laser. Typically one of the following three techniques is implemented in coupling the output from a diode bar or stacked bar into an optical fiber.

The first approach is identical to the technique already described for end-pumping a laser. Two sets of micro lenses and a condensor lens are used for focusing the radiation into a multimode fiber with a diameter of 200 to 400 μm. In another design, the output from a diode bar is first collimated using a fiber lens and subsequently imaged with a combination of lenses into a two-mirror beam-shaping device. With this focusing scheme the output from a 1 cm bar has been focused into a circular spot of less than 200 μm [26].

In the third approach a fiber bundle is employed to transform a line source, which a single 1 cm bar represents, into a circular beam. A micro lens reduces the divergence of the fast axis of the emitters before the radiation is collected by a flat bundle of fibers (see Fig. 6.38(c)). The line-to-bundle converting fiber-optic coupler consists of a large number of small fibers spread out in a linear array which faces the emitting diodes of the laser bar. At the other end, the fibers are arranged into a round fiber bundle.

The technique of using a fiber-optic bundle to circularize the output from a 1 cm diode bar is implemented in the laser illustrated in Fig. 6.40. Shown is an Nd : YVO$_4$ laser crystal pumped from both ends by two fiber-coupled 20 W diode bars. For separation of the pump and resonator beam, two dichroic beam splitters are inserted into the optical train, which results in a z-configuration of the optical resonator. The radiation from each bar is coupled into a flat fiber bundle by a micro lens, which reduces the divergence in the fast axis of the diode output. At

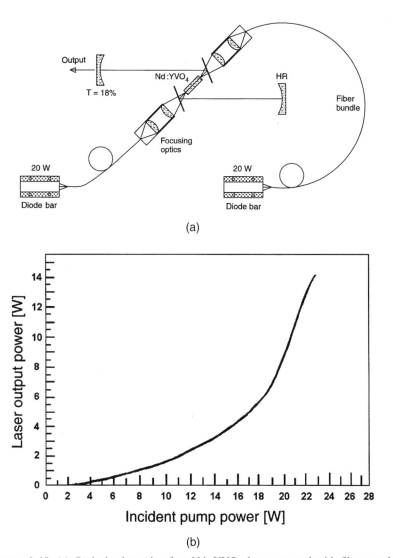

(a)

(b)

FIGURE 6.40. (a) Optical schematic of an Nd : YVO$_4$ laser pumped with fiber-coupled diode bars (Spectra-Physics Lasers, Inc.) and (b) laser output versus optical pump power incident on the crystal.

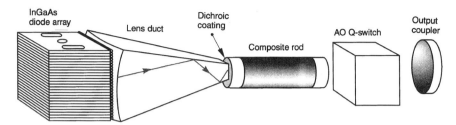

FIGURE 6.41. End-pumping with a nonimaging light concentrator [28].

the pump end, the fibers form a round bundle, and a collimating and focusing lens transfer the pump radiation into the crystal through a dichroic fold mirror. The high efficiency that can be achieved with diode end-pumped solid-state lasers is illustrated in Fig. 6.40(b). With a fiber coupling efficiency of 85% and some reduction in the rated power level of the diode to preserve lifetime, each side of the Nd : YVO$_4$ crystal is pumped with 13 W. Pump radiation at the crystal is converted with over 50% conversion efficiency into a polarized TEM$_{00}$ mode.

Maximum output of the system is 13 W. An Nd : YAG laser used in the same configuration produced 11 W of polarized cw output.

Nonimaging Light Concentrators

Figure 6.41 illustrates an end-pumping technique based on a nonimaging devices. In this approach the output from a stack of bars is first collimated in the fast axis by cylindrical micro lenses mounted in front of each bar. The collimated radiation is directed into a lens duct that combines focusing at its curved input face and total internal reflection at the planar faces to channel the diode pump radiation to the laser rod. Powerful Nd : YAG, Tm : YAG, and Yb : YAG lasers have been pumped with this technique and outputs in excess of 100 W have been reached in these lasers. A slightly different version of a light concentrator that has parallel walls in the vertical direction is shown in Fig. 6.38(c). Concentrators are an attractive solution for large stacks because they are relatively easy to fabricate. The degree of concentration is governed by the radiance theorem stating that the radiance of the light distribution produced by an imaging system cannot be greater than the original source radiance.

6.2.4   Face-Pumped Disks

In a disk laser amplifier the surfaces are set at Brewster's angle, as shown in Fig. 6.42. The slab faces have a high-quality optical finish to minimize scattering loss. The beam is linearly polarized in the p-plane to avoid reflection loss. Flashlamps are used for pumping, with lamp radiation incident on the disk faces and transverse to the beam direction. Thus, the disks are said to be face-pumped. Nd : glass amplifiers containing glass disks have mainly been developed at labo-

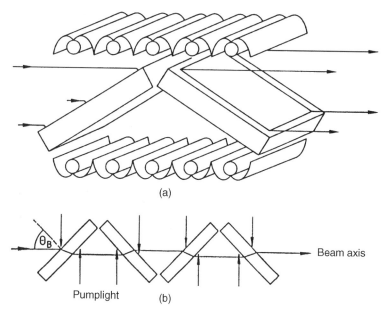

(a)

(b)

Beam axis

Pumplight

FIGURE 6.42. Schematic diagram of (a) a disk amplifier and (b) alternate arrangement of disks.

ratories engaged in laser fusion studies. Disk amplifiers with apertures as large as 74 cm have been built and operated.

Generally, an even number of disks is used in an amplifier since beam translation occurs due to refraction in a single disk. Beam steering can be minimized if the disks are placed in an alternate rather than parallel arrangement, as shown in Fig. 6.42(b).

The concept of face-pumping, as opposed to pumping through the edges of a disk or slab, is important in that the pump radiation can be distributed, in an easily controlled manner, uniformly over the faces of the slab. Pumping in this way yields uniform gain over the aperture for a laser beam, but more important, the slab heating is uniform laterally, thus allowing transverse temperature gradients to be minimized. Such temperature gradients give rise to distortion in the slab just as in the rod geometry.

The design of the flashlamp reflector housing requires a careful design to provide uniform illumination of the slabs. The cusp-shaped reflectors direct radiation to the slabs and shield each flashlamp from the radiation from adjacent lamps. Table 6.5 lists the results of an energy-transfer analysis performed on a large Nd : glass disk amplifier. About 1% of the total electrical input is available as stored energy at the laser level. Of course, only a fraction of this energy can be extracted during the amplification of a laser pulse.

The other extreme of a face-pumped laser, in terms of size, is the thin-disk approach employed for pumping Yb : YAG. As illustrated in Fig. 6.43 a disk about

TABLE 6.5. Energy transfer in an
Nd : glass disk amplifier [29].

Circuit losses	8%
Lamp heat	50%
Heating of pump cavity walls	30%
Ultraviolet absorption	7%
Heating of glass disks	2%
Fluorescence decay	2%
Useful laser energy	1%
Electrical input to lamps	100%

300 μm thick is pumped in a small spot of a few millimeters in diameter at a small angle about the normal of the disk face. The 941 nm pumps radiation reflected back from the disk is refocused by a mirror for a second path through the crystal. The backside of the disk, which is in contact with a heat sink, is coated to form one of the resonator mirrors. The other resonator mirror is the curved output coupler shown in Fig. 6.43.

In the thin-disk laser, waste heat generated in the disk is extracted through the back high-reflectivity mirror and to an actively cooled submount that holds the laser disk. The key feature of the thin-disk design approach is that this cooling geometry constrains resulting thermal gradients to lie essentially parallel to the laser resonator axis. Consequently, thermal focusing by the gain disk is greatly reduced, facilitating the achievement of high beam quality.

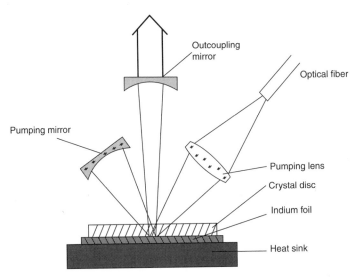

FIGURE 6.43. Thin-disk-pumping geometry of a Yb : YAG laser [30].

Summary

This chapter deals with the design and characteristics of the pump sources employed in solid state lasers and with the techniques and methods that have been developed for transferring the radiation from the pump source to the laser medium. The designs of an optical pump system for a solid state laser is governed by the following considerations: the spectral and spatial emission characteristics of the pump source, transfer of pump radiation to the gain medium, absorption of the pump radiation, and spatial overlap with the mode volume in the resonator.

The pump sources for solid-state lasers are flash lamps, continuous arc lamps, and laser diode arrays. In flash lamps that are typically operated at high current densities, the spectral line structure of the particular fill gas is masked by a strong continuum. The output spectrum resembles a blackbody radiator. In continuous arc lamps, the continuum emission spectrum contains sharp emission lines that are characteristic of the particular fill gas. The output from flash lamps and arc lamps is broadband, and only a small fraction of the total output spectrum falls within the absorption bands of the active medium. Diode lasers, on the other hand, are narrowband emitters and by adjusting the composition of the semiconductor material, the output wavelength can be shifted to match an absorption band of a particular solid-state laser material.

The transfer of radiation from the pump source to the laser medium should be highly efficient, and ideally the pump radiation should create an inversion that spatially overlaps the volume of the resonator mode. The total absorption of pump radiation depends on the absorption coefficient and thickness of the laser medium. However, mode structure and thereby beam uniformity and divergence are strongly affected by the pump power distribution inside the active material. In flash lamps and arc-lamp pumped lasers, the active medium and the radiation source are contained in highly reflective enclosures. Typical pump cavities consist of reflective elliptical cylinders with the lamp and active medium at each focal axis, or so-called closed coupled cavities where the lamp and laser rod are in close proximity surrounded by reflective metal or white ceramic walls.

Since radiation from diode lasers is directional, reflective enclosures are not needed. In the so-called side-pumped configuration, the laser diode arrays are placed along the length of cylindrical laser rods or rectangular slabs and the pump radiation propagates perpendicular to the axis of the resonator. In this configuration laser diode arrays are essentially used as replacements for flashlamps. Other designs exploit more fully the spatial properties of laser diodes by employing intervening optics to shape the pump beam to achieve good overlap with the resonator fundamental mode. The most efficient pump method is end-pumping, whereby pump and laser radiation propagate collinear inside the active material. End pumping has the advantage that total absorption of the pump radiation is obtained, and an excellent overlap between the pump volume and the laser mode can be achieved.

References

[1] J.H. Goncz: *Instr. Soc. Am. Trans.* **5**, 1 (1966).

[2] J.H. Goncz: *J. Appl. Phys.* **36**, 742 (1965).

[3] H.E. Edgerton, J.H. Goncz, and J. Jameson: Xenon flashlamps, limits of operation. *Proc. 6th Int'l Congr. on High Speed Photography*, Haarlem, Netherlands, 1963, p. 143.

[4] F. Schuda: in *Flashlamp-Pumped Laser Technology* (edited by F. Schuda). *SPIE Proc.* **609**, 177 (1986).

[5] J.P. Markiewicz and J.L. Emmett: *IEEE J. Quantum Electron.* **QE-2**, 707 (1966).

[6] J.F. Holzrichter and J.L. Emmett: *Appl. Opt.* **8**, 1459 (1969).

[7] J. Hecht: *The Laser Guidebook*, 2nd ed. (McGraw-Hill, New York), 1991.

[8] W. Streifer, D.R. Scifres, G.L. Harnagel, D.F. Welch, J. Berger, and M. Sakamoto: *IEEE J. Quantum Electron.* **QE-24**, 883 (1988).

[9] R.S. Geels, D.F. Welch, D.R. Scifres, D.P. Bour, D.W. Treat, and R.D. Bringans: CLEO 93, *Tech. Digest*, Paper CThQ3 (1993), 478.

[10] M. Ettenberg: *Laser Focus* **22**, 86 (May 1985).

[11] E. Kapon: Semiconductor Diode Lasers: Theory and Techniques, in *Handbook of Solid-State Lasers*, (edited by P.K. Cheo), (Marcel Dekker, New York), 1989.

[12] R. Burnham, Fibertek, Inc. Private communication.

[13] N.P. Barnes, M.E. Storm, P.L. Cross, and M.W. Skolaut: *IEEE J. Quantum Electron.* **QE-26**, 558 (1990).

[14] G.C. Osbourn, P.L. Gourley, I.J. Fritz, R.M. Biefield, L.R. Dawson, and T.E. Zipperian: Principles and applications of strained-layer superlattices, in *Semiconductors and Semimetals*, Vol. 24, *Applications of Multiple Quantum Wells, Doping, and Superlattices* (edited by R. Dingle) (Academic Press, San Diego, CA), 1987, p. 459.

[15] S.B. Schuldt and R.L. Aagard: *Appl. Opt.* **2**, 509 (1963).

[16] C. Bowness: *Appl. Opt.* **4**, 103 (1965).

[17] Yu.A. Kalinin and A.A. Mak: *Opt. Tech.* **37**, 129 (1970).

[18] Lawrence Livermore National Lab., *Medium Average Power Solid State Laser Technical Information Seminar*, CA (October 1985).

[19] A.D. Hays, G. Witt, N. Martin, D. DiBiase, and R. Burnham: *UV and Visible Lasers and Laser Crystal Growth* (edited by R. Scheps, M.R. Kokta). *SPIE Proc.* **2380**, 88 (1995).

[20] R. Burnham, G. Moule, J. Unternahrer, M. McLaughlin, M. Kukla, M. Rhoades, D. DiBiase, and W. Koechner: *Laser '95* (Munich), Paper K9.

[21] T.S. Rutherford, W.M. Tulloch, E.K. Gustafson, and R.L. Byer: *IEEE J. Quantum Electron.* **QE-36**, 205 (2000).

[22] J.E. Bernard and A.J. Alcock: *Opt. Lett.* **19**, 1861 (1994); also **18**, 968 (1993).

[23] J. Richards and A. McInnes: *Opt. Lett.* **20**, 371 (1995).

[24] T.Y. Fan and R.L. Byer: *IEEE J. Quantum Electron.* **QE-24**, 895 (1988).

[25] A. Brignon, G. Feugnet, J.P. Huignard, and J.P. Pocholle: *IEEE J. Quantum Electron.* QE-**34**, 577 (1998)

[26] W.A. Clarkson and D.C. Hanna: *Opt. Lett* **21**, 869 (1996).

[27] H.R. Verdun and T. Chuang: *Opt. Lett.* **17**, 1000 (1992).

[28] E.C. Holnea, C.A. Ebbers, R.J. Beach, J.A. Speth, J.A. Skidmore, M.A. Emanuel, and S.A. Payne: *Opt. Lett.* **23**, 1203 (1998).

[29] J. Trenholme: Optimizing the design of a kilojoule laser amplifier chain. *Laser Fusion Program*, Semiannual Report, Lawrence Livermore Laboratory, 1973, p. 60.

[30] Datasheet, *Jenoptik Laser*, Jena, Germany.

Exercises

1. In the text an example of a small, xenon-filled, linear flashlamp used in military equipment is given. It indicates that such a lamp is operated at 10 J per pulse and at 20 pulses per second. The lamp's inner diameter is 4 mm and its discharge length is 60 mm long. If the pulse width is 300 μs long calculate the explosion limit and the expected useful lifetime (number of pulses) for 10 J pulses using (6.3) and (6.4). Does your result agree with that in the example and why? (There may be more than the obvious reason.)

2. It seems clear that you would prefer to deliver all the pump energy you could to the gain medium before fluorescence losses start to significantly deplete the inversion that you are trying to build. This comment is particularly important when you are building Q-switched lasers since you are trying to hold off lasing until the inversion gets large and then to let it lase. You would prefer not to waste any pump energy by pumping too slowly. Therefore, for a 6.0 mm diameter and 100.0 mm long Nd : YAG laser rod, use the equations derived in this chapter to select a reasonable xenon-filled flashlamp capable of delivering up to 20 J per pulse at a repetition rate of 10 Hz. "Reasonable" means one that delivers a pulse shorter than the fluorescent lifetime (you can look the lifetime up in the text) and which demands reasonable size capacitors and inductors, reasonable voltage, and operates with a critically damped pulse. The choice of pressure and lamp dimensions will tell you something of the lamp's electrical properties. From this you can find the circuit parameters needed for its operation. Be careful to make the pulse duration short, but not so short that the lamp's useful lifetime is less than 10^7 shots.

3. Calculate the heat flux in watts per square centimeter of a diode laser at 808 nm that emits 2 W of laser power and is 0.1 mm wide by 1.0 mm long. Assume it is 40% efficient and that what does not appear as light appears as heat in the diode. Compare this to the heat flux into the heat removal plates that this pump light would produce if it were completely absorbed in a solid-state laser medium in a cylinder of diameter 200 μm and length 1 mm. This microlaser gain medium is a small chip 1 mm thick by 1 mm in diameter held between two copper heat removal plates with 500 μm diameter holes in them as sketched below.

4. Derive equation (6.28).

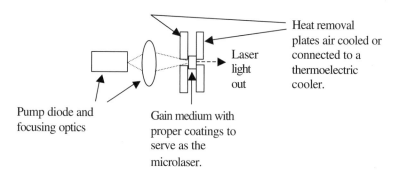

Heat removal
plates air cooled or
connected to a
thermoelectric
cooler.

Laser
light
out

Pump diode and
focusing optics

Gain medium with
proper coatings to
serve as the
microlaser.

7

Thermo-Optic Effects

7.1 Cylindrical Geometry
7.2 Slab and Disk Geometries
7.3 End-Pumped Configurations
     References
     Exercises

The optical pumping process in a solid-state laser material is associated with the generation of heat for a number of reasons:

(a) the energy difference of the photons between the pump band and the upper laser level is lost as heat to the host lattice; similarly, the energy difference between the lower laser level and the ground state is thermalized. The difference between the pump and laser photon energies, termed quantum defect heating, is the major source of heating in solid-state lasers.

(b) In addition, nonradiative relaxation from the upper laser level to the ground state, due to concentration quenching, and nonradiative relaxation from the pump band to the ground state will generate heat in the active medium.

(c) In flashlamp-pumped systems, the broad spectral distribution of the pump source causes a certain amount of background absorption by the laser host material, particularly in the ultraviolet and infrared regions of the lamp spectrum. Absorption of lamp radiation by impurity atoms and color centers can further increase heating.

The temperature gradients set-up in the gain material as a result of heating can lead to stress fracture, which represents the ultimate limit in average power obtainable from a laser material. Below the stress fracture limit thermal lensing and birefringence adversely affect the output beam quality. Also, due to thermal lensing, the operating point of the resonator within the stability diagram becomes a function of input power. Therefore the output beam quality and mode structure are power dependent because the thermal lens can only be compensated for one input power level.

Efficient heat removal, and the reduction of the thermal effects that are caused by the temperature gradients across the active area of the laser medium usually dominate design considerations for high-average-power systems.

The ability to predict the amount of heat generated in a laser material is important to obtain an indication of the thermo-optic effects that can be expected in a particular rod or slab laser. In the laser literature several parameters have been introduced to quantify the amount of heat dissipated by the active material. The fractional thermal loading η_h is defined as the ratio of heat produced to absorbed pump power or energy. A different way to quantify heat generation is expressed by the parameter χ, defined as the ratio of heat deposited to energy stored in the upper laser level.

The theoretical lower limit of η_h for ND: YAG lasing at $1.06\,\mu$m and pumped with a laser diode pump source at 808 nm is $\eta_h = 1 - \lambda_p/\lambda_L = 0.24$, where $\eta_S = \lambda_p/\lambda_L$ is the Stokes factor discussed in Section 3.4.1 and λ_p and λ_L are the wavelengths of the pump and laser beam, respectively.

A Yb: YAG crystal lases at $1.03\,\mu$m and can be diode-pumped at 943 nm. Owing to the smaller difference between pump and laser photon energy the theoretical thermal loading is only $\eta_h = 0.09$.

In flashlamps and cw krypton arc lamps optical excitation is less efficient compared to laser diode pumping because of the comparatively high content of short wavelength radiation in the spectrum. Therefore the difference between pump and laser photon energies is larger and results in sizeable quantum defect heating.

In actual lasers there are also a number of nonradiative decay processes that generate heat but do not contribute to laser output. Concentration quenching, so named because it increases strongly with neodymium-doping levels, causes a nonradiative decay of excited ions from the upper laser level to the ground state. Concentration quenching reduces the radiative quantum efficiency to less than unity. So-called "dark neodymium ions" or "dead sites" are ions that absorb pump photons but do not contribute to inversion. This leads to a pump quantum efficiency of less than unity.

The fraction of absorbed pump radiation converted into heat was measured to be around 30 to 32% for different Nd: YAG lasers pumped at 808 nm [1]. Estimated values of η_h derived from thermal measurements without lasing also suggest a value of $\eta_h = 0.32$ for Nd: YAG, and $\eta_h = 0.11$ for Yb: YAG pumped at 943 nm [2].

In Section 3.4.2 we have combined the losses due to nonradiative processes in the quantum efficiency η_Q. For Nd: YAG we use $\eta_Q = 0.9$ (see Fig. 6.37), which gives a heat load of $\eta_h = 1 - \eta_Q \eta_s = 0.32$ for diode pumping, consistent with the measured value.

The parameter χ is usually determined by measuring the small signal gain with a probe beam at the end of the pump pulse. The stored energy is then calculated from the gain, and the heat deposited is deduced from the observed temperature rise of the material. In the energy storage mode the value of χ depends on the pump pulse length because of the fluorescence losses (see (3.53). Operated in the energy storage mode, a value of $\chi = 1.1$ was measured for a $200\,\mu s$ long pump pulse [3].

In flashlamp-pumped lasers, the quantum defect heating is larger as already mentioned, and in addition to the nonradiative processes described above there is additional heating because of the background absorption of the pump radiation by the host, and additional absorption because of impurity atoms and color centers. Measured values for χ range from 2.9 to 3.3 for Nd : YAG and from 1.6 to 3.0 for Nd : phosphate glass doped at 3 to 6% neodymium [4, 5].

Comparing the value of χ measured for flashlamp and diode pumping of an Nd : YAG laser suggests that diode pumping produces only one-third of the heat in the crystal compared to flashlamp-pumped systems.

From the magnitude of η_h and χ one can obtain an estimate of the amount of heat generated in the laser material relative to the output of the laser system. Absorbed power in the laser element is channeled into heat, stimulated emission, and fluorescence. In the presence of a strongly saturating stimulated emission the fluorescence becomes insignificant. Only the excited ions that are outside the volume occupied by the resonator modes will relax as fluorescence to the ground state.

For example, from the definition of η_h it follows that the ratio of heat to inversion power density available for extraction is $\eta_h/(1 - \eta_h)$. If we account for the resonator losses expressed by the coupling efficiency η_c and for the less than unity overlap of the gain region with the resonator modes expressed by η_B (see Section 3.4.1) we obtain, for the ratio of heat load to output power,

$$\frac{P_h}{P_{out}} = \frac{\eta_h}{\eta_c \eta_B (1 - \eta_h)}. \tag{7.1}$$

This expression is a lower limit because it does not include the portion of the resonator losses which appear as heat in the active medium due to absorption of intracavity laser radiation by the laser material.

Typical examples of heating generated in the laser process are provided in Section 6.2. In the laser diode-pumped systems analyzed in Fig. 6.37 an output of 7 W is associated with 5 W of heating due to the quantum defect and nonradiative processes. This is expressed by $\eta_h = 0.32$. With the coupling efficiency and beam overlap efficiency of $\eta_B \eta_c = 0.65$ given in Fig. 6.37, a ratio of $P_h/P_{out} = 0.72$ is obtained. In addition, a fraction of the absorption losses also contributed to the heating of the laser crystal. Let us assume that the 7 W output is produced by a 5 cm long Nd : YAG rod with an absorption coefficient of $10^{-3}\,cm^{-1}$. If the resonator has an output coupler with a reflectivity of 90%, then the intracavity power is 133 W which produces about 0.7 W of absorption losses in the Nd : YAG rod. Therefore, in the example given in Fig. 6.37, the total heat load for the crystal is around 5.7 W.

In general one can assume a ratio of

$$P_h/P_{out} = 0.8 - 1.1 \tag{7.2}$$

for a diode-pumped Nd : YAG laser, depending on the crystal quality and the overlap of the resonator and pump region. In flashlamp- or krypton arc lamp-pumped Nd : YAG lasers the thermal load of the crystal is about three times higher

$$P_h / P_{out} = 2.5 - 3.3. \qquad (7.3)$$

In the example of the energy flow in an arc lamp system summarized in Table 6.4, the measured ratio of heat load to laser output was about 2.5 for a highly multimode beam. Any measures to improve the beam quality will invariably result in a reduced overlap between resonator mode and pump region and, therefore, the ratio of heat load to laser output will increase.

7.1 Cylindrical Geometry

The combination of volumetric heating of the laser material by the absorbed pump radiation and surface cooling required for heat extraction leads to a nonuniform temperature distribution in the rod. This results in a distortion of the laser beam due to a temperature- and stress-dependent variation of the index of refraction. The thermal effects which occur in the laser material are thermal lensing and thermal stress-induced birefringence.

An additional issue associated with thermal loading is stress fracture of the laser material. Stress fracture occurs when the stress induced by temperature gradients in the laser material exceeds the tensile strength of the material. The stress-fracture limit is given in terms of the maximum power per unit length dissipated as heat in the laser medium.

The particular temperature profile which exists in the laser material depends, to a large degree, on the mode of operation, that is, cw- pumped, single-shot, or repetitively pulse-pumped. In the case of cw operation, a long cylindrical laser rod with uniform internal heat generation and constant surface temperature assumes a quadratic radial temperature dependence. This leads to a similar dependence in both the index of refraction and the thermal-strain distributions.

In a pulse-pumped system, laser action occurs only during the pump pulse or shortly thereafter in the case of Q-switching. Theoretical and experimental investigations have shown that heat transport during the pump pulse, which usually has a duration between 0.2 and 5 ms, can be neglected. Therefore, in single-shot operation, optical distortions are the result of thermal gradients generated by nonuniform pump-light absorption.

In repetitively pulse-pumped systems, distortions will occur from the cumulative effects of nonuniform pump processes and thermal gradients due to cooling. Which effects dominate depend on the ratio of the pulse interval time to the thermal relaxation time constant of the rod. At repetition intervals, which are short compared to the thermal relaxation time of the laser rod, a quasi-thermal steady state will be reached where the distortions from pumping become secondary to the distortions produced by the removal of heat from the laser material.

7.1.1   Temperature Distribution

We consider the case where the heat generated within the laser rod by pump-light absorption is removed by a coolant flowing along the cylindrical-rod surface. With the assumption of uniform internal heat generation and cooling along the cylindrical surface of an infinitely long rod, the heat flow is strictly radial, and end effects and the small variation of coolant temperature in the axial direction can be neglected. The radial temperature distribution in a cylindrical rod with the thermal conductivity K, in which heat is uniformly generated at a rate Q per unit volume, is obtained from the one-dimensional heat conduction equation [6]

$$\frac{d^2T}{dr^2} + \left(\frac{1}{r}\right)\left(\frac{dT}{dr}\right) + \frac{Q}{K} = 0. \tag{7.4}$$

The solution of this differential equation gives the steady-state temperature at any point along a radius of length r. With the boundary condition $T(r_0)$ for $r = r_0$, where $T(r_0)$ is the temperature at the rod surface and r_0 is the radius of the rod, it follows that

$$T(r) = T(r_0) + \left(\frac{Q}{4K}\right)(r_0^2 - r^2). \tag{7.5}$$

The temperature profile is parabolic, with the highest temperature at the center of the rod. The temperature gradients inside the rod are not a function of the surface temperature $T(r_0)$ of the rod. The heat generated per unit volume can be expressed as

$$Q = \frac{P_h}{\pi r_0^2 l}, \tag{7.6}$$

where P_h is the total heat dissipated by the rod and l is the length of the rod. The temperature difference between the rod surface and the center is

$$T(0) - T(r_0) = \frac{P_h}{4\pi K l}. \tag{7.7}$$

The transfer of heat between the rod and the flowing liquid creates a temperature difference between the rod surface and the coolant. A steady state will be reached when the internal dissipation P_h is equal to the heat removed from the surface by the coolant

$$P_h = 2\pi r_0 l h [T(r_0) - T_F], \tag{7.8}$$

where h is the surface heat transfer coefficient and T_F is the coolant temperature. With $A = 2\pi r_0 l$ being the surface area of the rod, it follows that

$$T(r_0) - T_F = \frac{P_h}{Ah}. \tag{7.9}$$

Combining (7.7) and (7.9) one obtains, for the temperature at the center of the rod,

$$T(0) = T_F + P_h \left(\frac{1}{4\pi Kl} + \frac{1}{Ah} \right). \tag{7.10}$$

Thus, from the geometry, and the appropriate system and material parameters, the thermal profile of the crystal can be determined, except that h must be evaluated. This coefficient is obtained from a rather complex expression involving the thermal properties of the coolant, the mass flow rate of the coolant, the Reynolds, Prandtl, and Grashof numbers, and the geometry of the cooling channel [7].

The boundary conditions for the heat transfer coefficient are a thermally insulated laser rod ($h = 0$), or unrestricted heat flow from the rod surface to a heat sink ($h = \infty$). For cases of practical interest the heat transfer coefficient is typically around $h = 0.5$–$2\,\mathrm{W\,cm^{-2}\,C^{-1}}$.

Figure 7.1 shows as an example the radial temperature profile in an Nd:YAG rod calculated from (7.10). The laser, which delivered between 200 to 250 W of output, was pumped at 12 kW of input power. The following parameters have been used in the numerical calculations of the temperature profile in the crystal: rod length $l = 7.5$ cm; rod radius $r_0 = 0.32$ cm; flow-tube inside radius $r_F = 0.7$ cm; power dissipated by the Nd:YAG rod $P_h = 600$ W; mass flow rate of the coolant $m^* = 142$ g/s; and fluid temperature entering the cavity $T_F = 20°$ C. As can be seen from this figure, the maximum temperature of the crystal occurring at the center is 114° C. The large temperature gradient of 57° C between the center of the crystal and the surface is responsible for the high stresses present in the material.

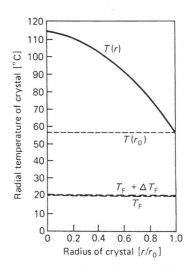

FIGURE 7.1. Radial temperature distribution within an Nd:YAG crystal as a function of radius. T_F is the temperature of coolant entering the flow-tube assembly, ΔT_F is the axial temperature gradient, and $T(r_0)$ is the rod surface temperature [7].

7.1.2   Thermal Stresses

The temperature gradients generate mechanical stresses in the laser rod since the hotter inside area is constrained from expansion by the cooler outer zone. The stresses in a cylindrical rod, caused by a temperature distribution $T(r)$, can be calculated from the equations given in [8]. Equations (7.11)–(7.13) describe the radial σ_r, tangential σ_ϕ, and axial σ_z stress in an isotropic rod with free ends and a temperature distribution according to (7.5),

$$\sigma_r(r) = QS(r^2 - r_0^2), \tag{7.11}$$

$$\sigma_\phi(r) = QS(3r^2 - r_0^2), \tag{7.12}$$

$$\sigma_z(z) = 2QS(2r^2 - r_0^2), \tag{7.13}$$

where the factor $S = \alpha E[16K(1 - \nu)]^{-1}$ contains the material parameters, E is Young's modulus, ν is Poisson's ratio, and α is the thermal coefficient of expansion. The stress components, σ_r, σ_ϕ, σ_z represent compression of the material when they are negative and tension when they are positive. We notice that the stress distributions also have a parabolic dependence on r. The center of the rod is under compression. The radial component of the stress goes to zero at the rod surface, but the tangential and axial components are in tension on the rod surface by virtue of the larger bulk expansion in the rod's center compared to the circumference.

Figure 7.2 gives the stresses as a function of radius inside the Nd:YAG rod whose temperature profile was shown in Fig. 7.1. From these curves it follows that the highest stresses occur at the center and at the surface. Since the tensile

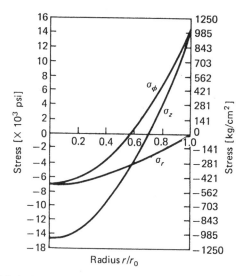

FIGURE 7.2. Radial (σ_r), tangential (σ_ϕ), and axial (σ_z) stress components within an Nd:YAG crystal as a function of radius [7].

strength of Nd : YAG is 1800 to 2100 kg/cm^2, the rod is stressed about 70% of its ultimate strength. As the power dissipation is increased, the tension on the rod surface increases and may exceed the tensile strength of the rod, thereby causing fracture. It is of interest to determine at what power level this will occur. Using (7.12) we find, for the hoop stress,

$$\sigma_\phi = \frac{\alpha E}{8\pi K (1 - \nu)} \frac{P_h}{l}. \tag{7.14}$$

The total surface stress $\sigma_{\max}$ is the vector sum of σ_ϕ and σ_z, i.e., $\sigma_{\max} = 2^{1/2}\sigma_\phi$.

We note from (7.14) that the tension on the surface of a laser rod depends on the physical constants of the laser material and on the power dissipated per unit length of the material, but does not depend on the cross section of the rod. Upon substitution of $\sigma_{\max}$ with the value of the tensile strength of Nd : YAG, we find that with 150 W dissipated as heat per centimeter length of Nd : YAG rod, the tension on the surface of the rod equals the tensile strength of the material. The actual rupture stress of a laser rod is very much a function of the surface finish of the rod, and can vary by almost a factor of 3 in Nd : YAG.

Stress Fracture Limit

The mechanical properties of the laser host material determine the maximum surface stress that can be tolerated prior to fracture. If there were no other constraints, such as stress-induced focusing and birefringence, the thermal loading and thus average output power of a rod laser could be increased until stress fracture occurred. If $\sigma_{\max}$ is the maximum surface stress at which fracture occurs, then we can rewrite (7.14) as follows:

$$\frac{P_h}{l} = 8\pi R_s, \tag{7.15}$$

where

$$R_s = \frac{K(1 - \nu)}{\alpha E}\sigma_{\max} \tag{7.16}$$

is a "thermal shock parameter." A larger R_s indicates a higher permissible thermal loading before fracture occurs. Table 7.1 lists typical values for a number of laser materials if we assume a standard surface treatment.

TABLE 7.1. Thermal shock parameter for different materials.

Material	Glass	GSGG	YAG	Al$_2$O$_3$
Thermal shock parameter R_s [W/cm]	1	6.5	7.9	100

7.1.3 Photoelastic Effects

The stresses calculated in the previous subsection generate thermal strains in the rod that, in turn, produce refractive index variations via the photoelastic effect. The refractive index of a medium is specified by the indicatrix that, in its most general case, is an ellipsoid. A change of refractive index due to strain is given by a small change in shape, size, and orientation of the indicatrix [9]. The change is specified by small changes in the coefficients B_{ij},

$$B_{ij} = P_{ijkl}\varepsilon_{kl} \qquad (i, j, k, l = 1, 2, 3), \tag{7.17}$$

where P_{ijkl} is a fourth-rank tensor giving the photoelastic effect. The elements of this tensor are the elasto-optical coefficients. ε_{kl} is a second-rank strain tensor.

We will confine our calculation to the case of an Nd:YAG laser rod. The method, however, is applicable to any material if the proper photoelastic matrix is used. Since Nd:YAG is a cubic crystal, the indicatrix is a sphere. Under stress the indicatrix becomes an ellipsoid. Nd:YAG rods are grown with the cylindrical axes along the [111]-direction. The light propagates in this direction, and thus the change of the refractive index along the [111] is of interest.

Since the transverse stresses are in the radial and tangential directions—relative to the coordinate system shown in Fig. 7.3—the local indicatrix also orients its axis in these directions. In a cylindrical coordinate system the photoelastic changes in the refractive index for the r and ϕ polarizations are given by [10]

$$\Delta n_r = -\tfrac{1}{2} n_0^3 \, \Delta B_r \tag{7.18}$$

and

$$\Delta n_\phi = -\tfrac{1}{2} n_0^3 \, \Delta B_\phi. \tag{7.19}$$

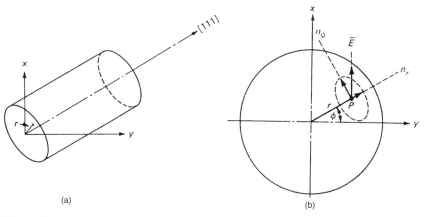

FIGURE 7.3. (a) Crystal orientation for an Nd:YAG rod and (b) orientation of the indicatrix of the thermally stressed Nd:YAG rod in a plane perpendicular to the rod axis.

A considerable amount of tensor calculation is required to determine the co-efficients ΔB_r and ΔB_ϕ in a plane perpendicular to the [111]-direction of the Nd : YAG crystal. The technique used in these calculations consists of introduc-ing the elasto-optic coefficients for Nd : YAG into the matrix form P_{mn} of the photoelastic tensor. The published values of these coefficients are given in a [100]-oriented Cartesian coordinate frame: $P_{11} = -0.029$; $P_{12} = +0.0091$; and $P_{44} = -0.0615$. The strain tensor is obtained from the stresses calculated in (7.11)–(7.13). After a coordinate transformation to bring P_{mn} and ε_{kl} into the same coor-dinate system, the tensor operation according to (7.17) can be performed. Intro-ducing the expression for ΔB_r and ΔB_ϕ into (7.18), (7.19), the refractive-index changes are given by [11], [12]

$$\Delta n_r = -\tfrac{1}{2}n_0^3 \frac{\alpha Q}{K} C_r r^2, \tag{7.20}$$

$$\Delta n_\phi = -\tfrac{1}{2}n_0^3 \frac{\alpha Q}{K} C_\phi r^2, \tag{7.21}$$

where C_r and C_ϕ are functions of the elasto-optical coefficients of Nd : YAG,

$$C_r = \frac{(17\nu - 7)P_{11} + (31\nu - 17)P_{12} + 8(\nu + 1)P_{44}}{48(\nu - 1)},$$

$$C_\phi = \frac{(10\nu - 6)P_{11} + 2(11\nu - 5)P_{12}}{32(\nu - 1)}.$$

The induced birefringence is determined from (7.20), (7.21)

$$\Delta n_r - \Delta n_\phi = n_0^3 \frac{\alpha Q}{K} C_B r^2, \tag{7.22}$$

where

$$C_B = \frac{1 + \nu}{48(1 - \nu)}(P_{11} - P_{12} + 4P_{44}).$$

Inserting the values of the photoelastic coefficients and the material parameters of Nd : YAG, $\alpha = 7.5 \times 10^{-6}/^\circ\text{C}$, $K = 0.14\,\text{W/cm}\,^\circ\text{C}$, $\nu = 0.25$, $n_0 = 1.82$ into (7.20)–(7.22), one obtains $C_r = 0.017$, $C_\phi = -0.0025$, $C_B = -0.0099$, and

$$\Delta n_r = (-2.8 \times 10^{-6})Q r^2,$$

$$\Delta n_\phi = (+0.4 \times 10^{-6})Q r^2,$$

$$\Delta n_r - \Delta n_\phi = (-3.2 \times 10^{-6})Q r^2,$$

where Q has the dimension of watts per cubic centimeter and r is measured in centimeters.

7.1.4   Thermal Lensing

Having explored the stresses in the laser rod, we now turn to the optical distortions which are a result of both temperature gradients and stresses. The change of the refractive index can be separated into a temperature- and a stress-dependent variation. Hence

$$n(r) = n_0 + \Delta n(r)_T + \Delta n(r)_\varepsilon, \tag{7.23}$$

where $n(r)$ is the radial variation of the refractive index, n_0 is the refractive index at the center of the rod, and $\Delta n(r)_T$, $\Delta n(r)_\varepsilon$ are the temperature- and stress-dependent changes of the refractive index, respectively.

The temperature-dependent change of the refractive index can be expressed as

$$\Delta n(r)_T = [T(r) - T(0)] \left(\frac{dn}{dT} \right). \tag{7.24}$$

With the aid of (7.5), (7.7) we obtain

$$\Delta n(r)_T = -\frac{Q}{4K} \frac{dn}{dT} r^2. \tag{7.25}$$

As can be seen from (7.20), (7.21), and (7.25) the refractive index in a laser rod shows a quadratic variation with radius r. An optical beam propagating along the rod axis suffers a quadratic spatial phase variation. This perturbation is equivalent to the effect of a spherical lens. The focal length of a lens-like medium, where the refractive index is assumed to vary according to

$$n(r) = n_0 \left(1 - \frac{2r^2}{q^2} \right) \tag{7.26}$$

is given by

$$f \cong \frac{q^2}{4 n_0 l}. \tag{7.27}$$

This expression is an approximation where it was assumed that the focal length is very long in comparison to the rod length. The distance f is measured from the end of the rod to the focal point.

The total variation of the refractive index is obtained by introducing (7.20), (7.21), and (7.25) into (7.23):

$$n(r) = n_0 \left[1 - \frac{Q}{2K} \left(\frac{1}{2 n_0} \frac{dn}{dT} + n_0^2 \alpha C_{r,\phi} \right) r^2 \right]. \tag{7.28}$$

As was discussed in the previous subsection, the change of refractive index due to thermal strain is dependent on the polarization of light; therefore the photo-elastic coefficient $C_{r,\phi}$ has two values, one for the radial and one for the tangential component of the polarized light. Comparing (7.28) with (7.26) yields

$$f' = \frac{K}{Ql} \left(\frac{1}{2} \frac{dn}{dT} + \alpha C_{r,\phi} n_0^3 \right)^{-1}. \tag{7.29}$$

In our final expression for the focal length of an Nd:YAG rod, we will include the contributions caused by end effects. Perturbations of the principal thermal distortion pattern occur in laser rods near the ends, where the free surface alters the stress character. The so-called end effects account for the physical distortion of the flatness of the rod ends. Self-equilibrating stresses causing a distortion of flatness were found to occur within a region of approximately one diameter from the ends of Nd:glass and one radius from the end for Nd:YAG. The deviation from flatness of the rod ends is obtained from

$$l(r) = \alpha l_0[T(r) - T(0)], \tag{7.30}$$

where l_0 is the length of the end section of the rod over which expansion occurs. With $l_0 = r_0$ and (7.5) we obtain

$$l(r) = -\alpha r_0 \frac{Q r^2}{4K}. \tag{7.31}$$

The focal length of the rod caused by an end-face curvature is obtained from the thick-lens formula of geometric optics

$$f'' = \frac{R}{2(n_0 - 1)}, \tag{7.32}$$

where the radius of the end-face curvature is $R = -(d^2l/dr^2)^{-1}$. From these expressions follows the focal length of the rod caused by a physical distortion of the flat ends

$$f'' = K[\alpha Q r_0(n_0 - 1)]^{-1}. \tag{7.33}$$

The combined effects of the temperature- and stress-dependent variation of the refractive index and the distortion of the end-face curvature of the rod lead to the following expression for the focal length

$$f = \frac{KA}{P_h} \left(\frac{1}{2} \frac{dn}{dT} + \alpha C_{r,\phi} n_0^3 + \frac{\alpha r_0(n_0 - 1)}{l} \right)^{-1}, \tag{7.34}$$

where A is the rod cross-sectional area and P_h is the total heat dissipated in the rod. If one introduces the appropriate materials parameters for Nd:YAG into (7.34) then one finds that the temperature-dependent variation of the refractive index constitutes the major contribution of the thermal lensing. The stress-dependent variation of the refractive index modifies the focal length about

20%. The effect of end-face curvature caused by an elongation of the rod is less than 6%.

Ignoring the end effects, we notice that the focal length is proportional to a material constant and the cross section A of the rod and is inversely proportional to the power P_h dissipated as heat in the rod. At first, it may be surprising that the length of the rod does not enter the equations. However, in a longer rod, for example, the reduction in power dissipation per unit length is offset by a longer path length.

We see from (7.34) that we have little flexibility in influencing the focal length. The material constants are determined when we choose the laser material: the dissipated power P_h is determined by the application (even though we may be able to reduce the heat load by avoiding unusable pump radiation); thus the only remaining design parameter is the rod cross section. The focal length can be increased by increasing A, but this is usually not a practical way of solving the problem, since a larger crystal reduces pump power density and therefore leads to lower gain.

According to (7.34), the focal length of a cylindrical laser rod, where heat is generated uniformly within the bulk material, can be written as

$$f = M_{r,\phi} P_{in}^{-1}, \tag{7.35}$$

where M contains all the material parameters of the laser rod and an efficiency factor η which relates the electrical input power to the power dissipated as heat in the rod ($P_h = \eta P_{in}$).

We can introduce a laser-rod sensitivity defined as

$$\frac{d(1/f)}{d P_{in}} = M_{r,\phi}^{-1}. \tag{7.36}$$

The sensitivity factor describes how much the optical power $1/f$ of a laser rod changes with input power. For an Nd : YAG rod 0.63 cm in diameter, and assuming that 5% of the electrical input power to the lamp is dissipated as heat, we obtain a change of focusing power of 0.5×10^{-3} diopters per watt of lamp input variation.

Returning now to (7.34), we can see that the rod acts as a bifocal lens with different focal lengths for radiation with radial and tangential polarization. Since a linear polarized wave or a nonpolarized wave incident on the crystal will always have components in radial and tangential directions, two focal points are obtained. For Nd : YAG one finds a theoretical value of $f_\phi/f_r = 1.2$. A difference in focal length for different polarizations means that a resonator designed to compensate for the lensing of the radial polarization cannot also compensate for the lensing of tangentially polarized light.

In Fig. 7.4 theoretical and measured thermally induced back focal lengths of various laser rods are plotted as a function of lamp input. Experimentally, the focal length was determined by projecting an HeNe laser beam through the rod and measuring the position of the beam diameter minimum. The measured curves have been observed for the same rod size in different pump cavities. The focal length changes faster than predicted, probably as a result of nonuniform pumping. The

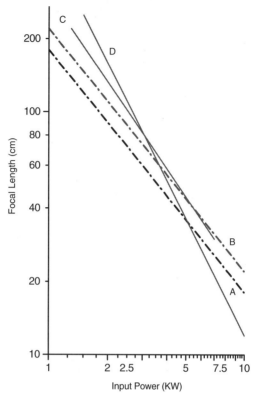

FIGURE 7.4. Thermally induced back focal length as a function of lamp input power for Nd : YAG rods. Shown are calculated curves for (A) the radially and (B) tangentially polarized beam components, and (C, D) measurements of average focal length for different rods and pump cavities [14].

calculations for the theoretical back focal length assumed that 5% of the electrical input power will be dissipated as heat in a 7.5×0.63 cm diameter crystal. The value of M for the radially polarized beam (A) is $M_r = 18 \times 10^4$ Wcm, and $M_\phi = 22 \times 10^4$ Wcm for the tangentially polarized beam (B).

7.1.5 Stress Birefringence

We will now investigate the influence of thermally induced birefringence on the performance of a solid-state laser. Taking Nd : YAG as an example, it was shown in (7.20), (7.21) that the principal axes of the induced birefringence are radially and tangentially directed at each point in the rod cross section and that the magnitude of the birefringence increases quadratically with radius r. As a consequence, a linearly polarized beam passing through the laser rod will experience a substantial depolarization. We refer to Fig. 7.3, where we have chosen a point $P(r, \phi)$

in a plane perpendicular to the rod axes. At this point we have a radial refractive index component n_r, which is inclined at an angle ϕ with respect to the y-axis and a tangential component n_ϕ perpendicular to n_r. Assume that $\overline{E}$ is the polarization vector for incident radiation. Radiation incident at point P must be resolved into two components, one parallel to n_r and the other parallel to n_ϕ. Since $\Delta n_r \neq \Delta n_\phi$, there will be a phase difference between the two components and the light will emerge elliptically polarized. This will occur for all points of the rod cross section with the exception for points located along the x- and y-axes in Fig. 7.3. Radiation incident along the y-axis will see only one refractive index, n_ϕ, while along the x-axis, n_r will be the only refractive index.

Birefringence effects in pumped laser rods can be studied in a polariscopic arrangement in which the expanded and collimated light beam from an HeNe laser serves as an illuminator for the observation of the rod between crossed polarizers. Because of thermally induced birefringence, the probe light suffers depolarization and is partially transmitted by the analyzer. The transmitted light forms the so-called isogyres, which display the geometrical loci of constant phase difference. Photographs of conoscopic patterns for various pump powers of an Nd : LaSOAP rod between crossed polarizers are shown in Fig. 7.5.

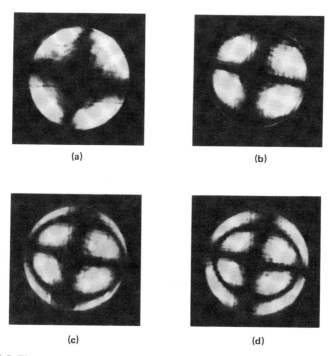

(a)   (b)

(c)   (d)

FIGURE 7.5. Thermal stresses in a 7.5-cm long and 0.63-cm-diameter Nd : LaSOAP crystal. The rod was pumped at a repetition rate of 40 pps by a single xenon flashlamp in an elliptical pump cavity. Input power (a) 115 W, (b) 450 W, (c) 590 W, and (d) 880 W [14].

The isogyre pattern exhibits a cross and ring structure, where the arms of the cross are parallel and orthogonal to the incident polarization. As mentioned before, the crosses correspond to those regions of the crystal where an induced (radial or tangential) axis is along a polarizer axis, so that the induced birefringence results only in a phase delay and not in a polarization rotation. The dark rings correspond to an integral number of full waves of retardation.

A number of operations of solid-state lasers, such as, for example, electro-optical Q-switching, frequency doubling, and external modulation of the beam, require a linearly polarized beam. An optically isotropic material, such as Nd : YAG, must be forced to emit a linearly polarized beam by the introduction of a polarizer in the resonant cavity. In the absence of birefringence, no loss in output power would be expected. However, the thermally induced birefringence causes a significant decrease in output power and a marked change in beam shape.

For a system with an intracavity polarizer, the effect of depolarization involves two phenomena: coupling of power into the orthogonal state of polarization followed by subsequent removal of that component by the polarizer and modification of the main beam by the depolarization process leading to a distortion of beam shape. When a birefringent crystal is placed between a polarizer and analyzer that are parallel, the transmitted intensity is given by [9]

$$\frac{I_{out}}{I_{in}} = 1 - \sin^2(2\phi) \sin^2\left(\frac{\varphi}{2}\right), \tag{7.37}$$

where ϕ is the angle between the polarizer and one of the principal birefringence axes and φ is the polarization phase shift of the light emerging from the crystal. The index difference, $\Delta n_\phi - \Delta n_r$, leads to a phase difference

$$\varphi = \frac{2\pi}{\lambda} l(\Delta n_\phi - \Delta n_r). \tag{7.38}$$

To illustrate the magnitude of stress birefringence in an Nd : YAG rod, the difference in optical path length normalized to the wavelength

$$\frac{\Delta l}{\lambda} = \frac{l(\Delta n_\phi - \Delta n_r)}{\lambda} \tag{7.39}$$

is plotted in Fig. 7.6 as a function of pump power. The following constants were used: $P_h = 0.05 P_{in}$, $l = 7.5\,\text{cm}$, $r_0 = 0.31\,\text{cm}$, and $\lambda = 1.06 \times 10^{-4}\,\text{cm}$. As can be seen from this figure, at maximum lamp input power of 12 kW, the path-length difference is approximately six wavelengths.

Assuming a plane wave, we can calculate the total transmitted intensity from (7.37), (7.38) by integrating over the cross-sectional area of the rod. The following integral must be evaluated

$$\left(\frac{I_{out}}{I_{in}}\right)_T = \left(\frac{1}{\pi r_0^2}\right) \int_{\phi=0}^{2\pi} \int_{r=0}^{r_0} \left[1 - \sin^2(2\phi) \sin^2 \frac{\varphi}{2}\right] r\, dr\, d\phi. \tag{7.40}$$

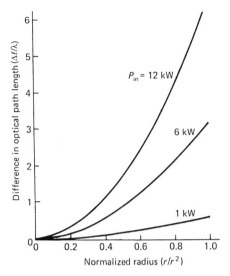

FIGURE 7.6. Differences in optical path length as a function of normalized rod radius in an Nd : YAG rod. Parameter is the lamp input power.

with $\varphi = C_T P_a (r/r_0)^2$ where $C_T = 2n_0^3 \alpha C_B / \lambda K$. Integration of (7.40) yields

$$\frac{I_{\text{out}}}{I_{\text{in}}} = \frac{3}{4} + \frac{\sin (C_T P_a)}{4 C_T P_a}. \tag{7.41}$$

A laser resonator containing a polarizing element is optically equivalent to a laser rod of twice the actual length located between a polarizer and analyzer. Subtracting $I_{\text{out}}/I_{\text{in}}$ from unity and multiplying the phase difference φ by a factor of 2 yields the fraction of the intracavity power which is polarized orthogonal to the polarizer. This beam, caused entirely by birefringence, will actually be ejected from the cavity and represents the depolarization loss of the resonator

$$\delta_{\text{depol}} = 0.25[1 - \text{sinc}(2C_T P_a)]. \tag{7.42}$$

This loss is plotted in Fig. 7.7, where the same constants as for Fig. 7.6 have been used. Note that this curve represents the round-trip loss in the cavity for a plane-parallel beam which was used as an approximation for a highly multimode laser beam.

Also shown is the depolarization loss for a TEM_{00} mode for which it was assumed that the beam radius w is $w = r_0/2$. A similar calculation as the one carried out for a plane wave yields the loss factor

$$\delta_{\text{depol}} = 0.25 \left(1 + \frac{16}{C_T^2 P_h^2} \right)^{-1}. \tag{7.43}$$

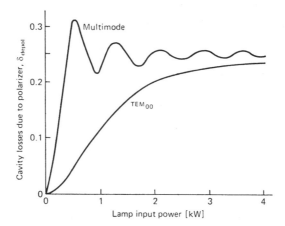

FIGURE 7.7. Calculated resonator loss caused by the combination of thermally induced birefringence in an Nd:YAG rod and an intracavity polarizer.

For the same lamp input power the losses for the TEM_{00} mode are less than for a highly multimode beam. This is expected since the energy in the TEM_{00} mode is concentrated nearer the center of the rod, where the induced birefringence is smaller.

The interaction of a linearly polarized beam with a birefringent laser rod and a polarizer not only leads to a substantial loss in power, but also a severe distortion of the beam shape. Figure 7.8 shows the output beam shape obtained from a cw-pumped Nd:YAG laser with and without an intracavity polarizer. The output is obtained in the form of a cross with a bright center. As discussed before, the

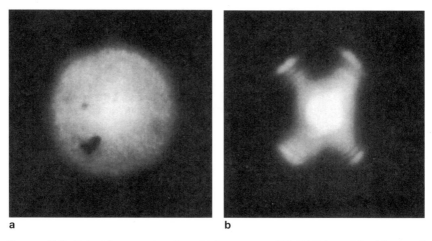

a   b

FIGURE 7.8. Output beam pattern for a high-power cw Nd:YAG laser (a) without and (b) with a Brewster plate in the cavity.

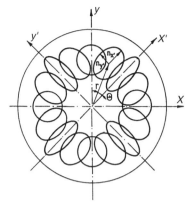

FIGURE 7.9. Orientation of the indicatrix of a thermally stressed rod [15].

depolarization losses are smallest in the rod center and in directions parallel and orthogonal to the preferred direction of the polarizer. Areas of high depolarization losses are removed from the cavity by the polarizer and are, therefore, missing in the output beam.

The $\sin(2\phi)$ dependence of birefringence on the azimuth angle ϕ is very nicely illustrated in Fig. 7.9. The plane of polarization is assumed to be in the y-direction and, consistent with observation, the largest depolarization occurs at 45° with respect to this direction.

7.1.6 Compensation of Thermally Induced Optical Distortions

Complete compensation of the thermal aberrations produced by a laser rod is difficult because:

(a) The focal length depends on the operating conditions of the laser and changes with pump power and repetition rate.

(b) The thermal lens is bifocal due to the stress-dependent variation of the refractive index.

(c) Nonuniform pumping leads to nonspherical aberrations.

In many pump configurations, with flashlamps as well as diode pump sources, pump radiation is more intense at the center than at the periphery of the rod. The focal length of a given area in the rod is inversely proportional to the intensity of the absorbed pumped radiation. Therefore, in the case of on-axis focusing of pump light, the focal length at the center of the rod is shorter than at the edges. Or expressed differently, the thermally induced refractive index profile contains terms that are higher than quadratic. A negative lens will remove the quadratic term; however higher-order effects cannot be compensated. The objective of a first-order compensation is to negate the positive lensing of the rod with a diverging optical component. The most common approaches are the insertion of a negative lens in the resonator, or a resonator featuring an internal telescope, or a design where one of the mirrors has a convex surface. The latter approach was discussed in Section 5.1.8.

A greater reduction of thermally induced optical distortions can be achieved if thermal lensing compensation is combined with birefringence compensation. In this case bifocusing is eliminated and depolarization losses are minimized in resonators containing polarized beams. The objective of birefringence compensation is to achieve equal phase retardation at each point of the rod's cross section for radially and tangentially polarized radiation. This can be accomplished by rotating the polarizations, either between two identical laser rods or in the same rod on successive passes, such that the radial and tangential components of the polarizations are exchanged.

On passing once through a rod with a radial temperature distribution, a phase difference $[\Delta P_r(r) - \Delta P_\theta(r)]$ is introduced between the radially polarized ray and the tangentially polarized ray at the same radius r. Now, if the two rays pass through a 90° polarization rotator, the radially polarized ray will be converted to tangential polarization, and vice versa. If the two rays are passed again through the same rod, or through an identical rod, the phase difference between the two rays will be removed.

For example, birefringence compensation in an oscillator containing two identical laser heads can be achieved by inserting a 90° quartz rotator between the laser rods. This rotator can be of crystalline quartz, for example, cut perpendicular to the optic axis. The rotator produces a 90° rotation of every component of the electric field of the laser beam. The part of a mode that is radially polarized in the first rod, is tangentially polarized in the second rod. Since each part of the beam passes through nearly identical regions of the two rods, the retardation induced by one rod is reversed by the other.

Figure 7.10 shows an example of a thermally compensated resonator designed to achieve high TEM$_{00}$ mode output in a linearly polarized beam. The 90° quartz rotator eliminates bifocusing and the convex rear mirror of the resonator compensates for the positive thermal lensing of the Nd : YAG crystal. Instead of using a convex resonator mirror, a negative lens is sometimes inserted between the

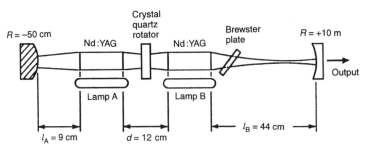

FIGURE 7.10. Schematic diagram of a birefringence-compensated laser [16].

two rods together with the quartz rotator. The resonator depicted in Fig. 7.10 also contains a Brewster plate to linearly polarize the laser output. Without the quartz rotator, the insertion of the polarizer generates a large depolarization loss for the TEM_{00} mode. With this technique, linearly polarized outputs with negligible power loss and TEM_{00} output of 50 to 70% of the multimode power have been achieved.

Other measures for achieving high TEM_{00} mode output are more fundamental and address the origin of thermal distortions. Changing from an arc lamp-pumped system to diode pumping drastically reduces the thermal distortions in the laser crystal. For the same output power, a diode pump source generates only about one-third the amount of heat in an Nd : YAG crystal compared to an arc lamp. Also, instead of using a cylindrical rod, a rectangular slab with an internal zigzag path minimizes wavefront distortions in the resonator. As explained in Section 7.2 such a structure provides a high degree of self-compensation of optical beam distortions. If the TEM_{00} output beam has to be linearly polarized, a naturally birefringent crystal such as Nd : YLF can be advantageous in particular situations. The natural birefringence in Nd : YLF is much larger than the thermally induced birefringence.

In an amplifier, wavefront distortions can be eliminated with an optical phase conjugated mirror as discussed in Section 10.4. This approach requires an injection seeded, narrow bandwidth oscillator with a diffraction limited output, and a double pass amplifier. The oscillator typically has a low power output because only aberrations in the amplifier are corrected. Compared to spherical diverging lenses, optical phase conjugation does provide complete correction of wavefront distortions over a large dynamic range.

7.2   Slab and Disk Geometries

Practical limitations arise in the operation of any solid-rod laser owing to the thermal gradients required to dissipate heat from the rod. As was discussed in Section 7.1, the thermally loaded cylindrical laser medium exhibits optical distortions that include thermal focusing, stress-induced biaxial focusing, and stress-induced birefringence. These thermally induced effects severely degrade the optical quality of the laser beam and eventually limit the laser output power, either because of an unacceptably poor beam pattern or because of thermal stress-induced breakage of the rod.

The limitations imposed by the rod geometry have long been recognized and a number of designs have emerged, such as slab and disk geometries, in an attempt to reduce the effect of thermal gradients on beam quality.

7.2.1   Rectangular-Slab Laser

The rectangular-slab laser provides a larger cooling surface and essentially a one-dimensional temperature gradient across the thickness of the slab. Figure 7.11

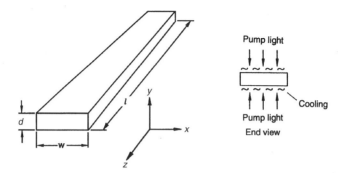

FIGURE 7.11. Geometry of a rectangular-slab laser.

shows the geometry of a rectangular slab laser. The z-axis coincides with the optical axis of the slab.

The slab has thickness d and width w. The upper and lower surfaces are maintained at a constant temperature by water-cooling, and the sides are uncooled. Provided the slab is uniformally pumped through the top and bottom surfaces, the thermal gradients are negligible in the x- and z-directions and the thermal analysis is reduced to a one-dimensional case, that is, temperature and stress are a function of y only. This, of course, is only true for an infinitely large plate in x and y, and uniform pumping and cooling. Under these conditions we find that the temperature assumes a parabolic profile (Fig. 7.12(a)).

The maximum temperature that occurs between the surface and the center of the slab ($y = d/2$) is given by

$$\Delta T = \frac{d^2}{8K} Q, \qquad (7.44)$$

where Q is the heat deposition, d is the thickness, and K is the thermal conductivity of the slab. For example, a heat deposition of 2 W/cm^3, in a 1 cm thick glass

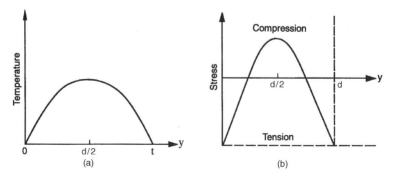

FIGURE 7.12. (a) Temperature profile and (b) stress in a rectangular slab.

slab ($K = 0.01$ W/cm^3) will create a temperature difference between the cooled surfaces and the center of $25°$ C.

The temperature rise causes stress in the slab according to

$$\sigma_{max} = \frac{2\alpha E}{3(1 - v)} \Delta T = \frac{\alpha E d^2}{12(1 - v)K} Q, \qquad (7.45)$$

where σ_{max} is the surface stress for the slab. The surfaces are in tension and the center is under compression as shown in Fig. 7.12(b).

If we introduce the "thermal-shock parameter" from Section 7.1 we can calculate the maximum temperature difference allowed between the surface and the center before thermal fracture occurs; this is

$$\Delta T_{max} = \frac{3R_s}{2K}. \qquad (7.46)$$

With $R_s = 1$ W/cm for Nd : glass, one obtains

$$\Delta T_{max} = 150° \text{ C}.$$

Stress fracture at the surface limits the total thermal power absorbed by the slab per unit of face area. For slabs of finite width w, the power per unit length at the stress fracture limit is given by

$$\frac{P_h}{l} = 12\sigma_{max} \frac{(1 - v)K}{\alpha E} \left(\frac{w}{d}\right), \qquad (7.47)$$

where w/d is the aspect ratio of a finite slab. It is interesting to compare the surface stress of a rod and slab for the same thermal power absorbed per unit length. From (7.14), (7.47) it follows that

$$\frac{(P_h/l)_{rod}}{(P_h/l)_{slab}} = \frac{2\pi}{3} \left(\frac{d}{w}\right). \qquad (7.48)$$

Thus, for a superior power handling capability relative to a rod, the aspect ratio of the slab must be greater than 2.

The temperature and stress profile leads to a birefringent cylindrical lens. The focal lengths of the birefringent lens are [17]

$$f_x = \frac{1}{2l(S_0 - S_1)}, \qquad f_y = \frac{1}{2l(S_0 - S_2)}, \qquad (7.49)$$

for x and y polarized light, respectively. The parameter S_0 is the contribution from thermal focusing, that is,

$$S_0 = \frac{dn}{dT} \left(\frac{Q}{2K}\right), \qquad (7.50)$$

and the parameters S_1 and S_2 are related to stress-induced focusing

$$S_1 = \frac{Q\alpha E}{2(1-\nu)K}(B_\perp + B_\parallel), \qquad S_2 = \frac{Q\alpha E}{(1-\nu)K}B_\perp, \qquad (7.51)$$

where $B_\parallel$ and $B_\perp$ are the stress optic coefficients for stress applied parallel and perpendicular to the polarization axis. Comparing the focal length of a rod (7.29) with the focal length of a slab (7.49) and considering only thermal focusing, then the slab has twice the focal power of a rod. In [17] it has been shown that if the stress-induced focusing is included, the slab still shows about twice the focal power compared to a rod.

In a slab, however, for incident radiation polarized along either the x- or y-directions the stress-induced depolarization is zero. In a rectangular rod, the main axes of the resulting index ellipse are oriented parallel to the x-and y-axes of the rod, and a beam polarized along either one of these axes can propagate in a direction parallel to the z-axis without being depolarized. This is, as mentioned earlier, in contrast to the situation in a cylindrical rod, where a plane-polarized beam does suffer depolarization because the direction of the main axes of the index ellipsoid vary from point to point within the cross section of the rod. The advantage of the rectangular geometry is obtained at some cost: The focal lengths are shorter than for the cylindrical rod, and cylindrical rather than spherical compensating optics are needed. The idea of the zigzag path in a slab, which will be discussed next, is to eliminate to a first-order focusing in a slab.

7.2.2   Slab Laser with Zigzag Optical Path

In the previous section we considered propagation straight through an infinite slab. Due to the rectilinear pumping and cooling geometry of the lasing medium, thermal gradients and thermally induced stresses are present only in the y-direction. Therefore, for light polarized in either the x- or y-directions, stress-induced biaxial focusing and depolarization losses are eliminated. However, the slab still behaves as a thin cylindrical lens with a focal length shorter by a factor of 2 relative to that of a rod.

The cylindrical focusing in the slab can be eliminated by choosing propagation along a zigzag optical path. In the zigzag geometry, the optical beam does not travel parallel to the z-axis, as was the case for the straight-through optical path. Instead the beam traverses the slab at an angle with respect to the x–z-plane using total internal reflection from the slab y faces. This geometry is depicted in Fig. 7.13.

The laser beam is introduced into the slab with suitable entrance/exit optics, usually through surfaces at Brewster's angle, as shown in Fig. 7.13(b). In order to maintain total internal reflection for the desired path length the two opposing faces have to be highly polished and have to be fabricated with a high degree of parallelism. The same two optical faces are also employed for pumping and cooling of the slab.

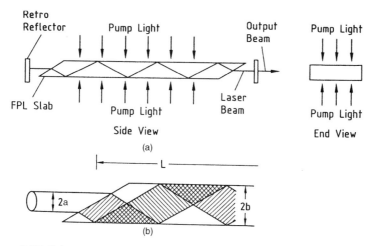

FIGURE 7.13. Schematic of a zigzag laser with Brewster angle faces; (a) schematic and (b) detail of Brewster angle entrance. The coordinate system is the same, as chosen in Fig. 7.12.

In the ideal case, this geometry results in a one-dimensional temperature gradient perpendicular to the faces and a thermal stress parallel to the faces. Since the thermal profile is symmetrical relative to the center plane of the slab, the thermal stress averaged from one slab surface to the other is zero. Thus, for a beam traversing from one slab surface to the other, the stress-optic distortion is compensated for to a high degree. Also, since all parts of a beam wavefront pass through the same temperature gradients in a surface-to-surface transit, no distortion results from the variation of refractive index with temperature. Thus, in this geometry, thermal distortion effects are fully compensated for within the host material in the ideal case.

The advantage of this configuration is the combination of two ideas: the elimination of stress induced birefringence by virtue of the rectangular geometry and the elimination of thermal and stress-induced focusing by optical propagation along a zigzag path.

A detailed model for calculating temperature and stress distributions in an infinitely long slab (plain strain approximation) has been described in [17]. Despite the inherent advantages, at least on a conceptual level, of a zigzag slab to overcome the thermal limitations of rod lasers, a number of practical engineering issues have to be solved before the potential advantages of the zigzag slab laser over the rod geometry can be realized. Distortions can only be eliminated in a slab of infinite extent, which is uniformly pumped and cooled. In practice, the slab laser approximates only an ideal infinite slab. For a slab of finite width and length, pump- or cooling-induced gradients across the slab width (normal to the plane of reflection), as well as end effects, give rise to distortions, and unavoidable temperature nonuniformities. Residual thermal gradients and thermally induced strain have in some cases resulted in beam qualities considerably below expectations.

Earlier work on slab lasers concentrated on Nd : glass slabs pumped with flash-lamps. A typical example of a flashlamp pumped zigzag Nd : glass slab laser is shown in Fig. 6.26.

In recent years, design efforts have concentrated mainly on systems utilizing Nd : YAG slabs. Flashlamp systems have been built in the multi-kilowatt regime, and diode-pumped slab lasers range from small conductively cooled systems to lasers with kilowatt output. Improvements in the growth technology of Nd : YAG boules has made it possible to fabricate Nd : YAG slabs up to 20 cm in length and 2.5 cm wide. The superior thermal properties of this crystal, compared to glass, has led to dramatic performance improvements and also opened the door for conductively cooled slabs in combination with diode pumping.

A large diode-pumped, liquid-cooled zigzag Nd : YAG slab laser is illustrated in Fig. 6.34. A photograph and a drawing of a conductively cooled Nd : YAG zigzag slab laser is shown in Fig. 6.33. The slab is designed for 13 bounces and near normal incidence AR-coated input faces. This type of slab geometry allows both S and P polarizations to be transmitted without loss. The slab dimensions are 138 mm × 9.5 mm × 8 mm. The side of the slab, mounted to the heat sink, was coated with a high reflection coating at 808 nm to allow two passes of the pump light through the slab.

7.2.3   Disk Amplifiers

Solid-state glass-disk amplifiers were considered very early in the development of high-brightness pulsed laser systems in order to solve problems of cooling, aperture size, gain uniformity, and pumping efficiency. In a disk laser amplifier (see Fig. 6.42) pump radiation is distributed uniformly over the faces of the slab. Pump-induced slab heating is uniform laterally, thus allowing transverse tempera-ture gradients to be minimized. Such temperature gradients give rise to distortion in the slab just as in the rod geometry. In contrast to the rod geometry, however, the transverse gradients in the slab geometry arise only from spurious effects in pumping and cooling, whereas in the rod geometry, transverse temperature gradi-ents arise from the principal cooling mechanism.

The Brewster angle disk lasers have been primarily developed for very high peak-power and relatively low average-power operation, with the slabs and coolant all approaching thermal equilibrium in the interval between pulses. To operate the Brewster angle slab laser at high average power requires adequate cooling as, for example, with forced convection over the faces. In addition, since the laser beam must pass through the coolant, the problem of thermal distortion in the coolant must be dealt with. This operating mode requires that the slabs be sufficiently thin so that the cooling time is several times less than the laser repeti-tion period. Figure 7.14 shows a configuration of a gas-cooled disk module which could potentially be developed for high average power operation, particularly if the Nd : glass slabs are replaced with crystalline laser host materials.

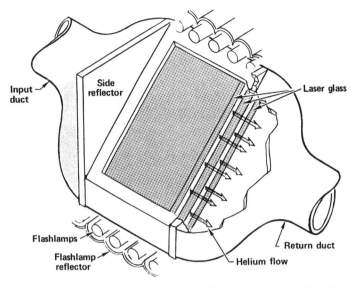

FIGURE 7.14. Concept of a gas-cooled disk amplifier module designed for high average power operation [18].

7.3   End-Pumped Configurations

So far, we have discussed cylindrical rods and rectangular slab configurations which are pumped from the side. As we have seen, analytical solutions generally assume uniform heating in an infinitely long rod or slab structure. End effects are treated separately as a distortion of the two-dimensional temperature distribution.

In contrast to transversally pumped systems, heat deposition in end-pumped lasers is very inhomogeneous. The very localized heat deposition leads to highly nonuniform and complex temperature and stress profiles. Besides the temperature and stress-dependent variations of the refractive index, the contribution of end bulging to the formation of a thermal lens can be substantial in end-pumped lasers. Inhomogeneous local heating and nonuniform temperature distribution in the laser crystal lead to a degradation of the beam quality due to the highly aberrated nature of the thermal lens.

With the availability of high-power diode lasers, end-pumping, or longitudinal-pumping, of laser crystals has become a very important technology. End-pumped lasers are described in Chapters 3, 6, and 9. Here, we are concerned with the thermal aspects of end-pumping.

The most common end-pumped configuration is a cylindrical rod mounted in a temperature-controlled heat sink with its cylindrical surface maintained at a constant temperature. The attraction of this arrangement is the relative simplicity of its implementation, but owing to the predominantly radial heat flow, it has the disadvantage of strong thermal lensing. Figure 7.15 shows the calculated temperature distribution in a conductively cooled end-pumped laser rod. An end-pumped

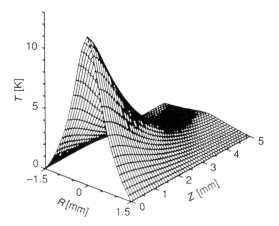

FIGURE 7.15. Temperature distribution in an end-pumped laser rod [19].

laser rod has a temperature profile across the pumped region which is a function of the distribution of pump radiation. From the edge of the pumped region, the temperature decays logarithmically to the cooled cylindrical surface of the rod. Along the axis of the rod, the temperature profile will decay exponentially due to the exponential absorption of pump radiation.

The large thermal gradients and associated stresses in an end-pumped laser are illustrated in Figs. 7.16(a) and 7.17(a). Shown are the temperature and stress distributions calculated by finite-element analysis for an Nd : YAG crystal. In this analysis, a 15 W pump beam from a diode array was assumed to be focused onto an Nd : YAG rod of 4.75 mm radius. The pump beam, which enters the laser crystal from the left along the z-axis in Figs. 7.16 and 7.17, has a Gaussian intensity distribution and a spot-size radius of 0.5 mm in the x-direction. It was assumed that 32% of the incident pump radiation is converted to heat. In the case considered here, the absorption length is much larger than the pump spot size. For the 1.3 at.%- doped rod an absorption coefficent of 4.5 cm^{-1} was assumed. Therefore, 90% of the pump radiation is absorbed in a 5 mm long path length.

Figure 7.16(a) illustrates the temperature distribution in one-half of the cross section of the rod. The isotherms represent the temperature difference from the heat sink surrounding the cylindrical surface of the rod. The temperature reaches a maximum of 59.5 K at the center of the pumped surface. The temperature gradients lead to thermal stress. The corresponding stresses in the Nd : YAG rod are indicated in Fig. 7.17(a). Positive values of the isobars indicate tensile stresses, and negative values show areas under compression. The core of the crystal is in compression, while the outer regions are in tension. The highest stress occurs on the face of the end-pumped rod, but in contrast to the temperature profile, not at the center, but at the edge of the pump beam (about $2w_p$ from the center). The heat generated by the pump radiation causes an axial expansion of the rod, that is, the front face begins to bulge as the crystal heats up. In the example above, the

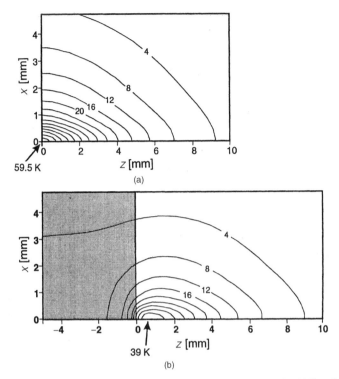

FIGURE 7.16. Modeled temperature profiles of an end-pumped Nd : YAG rod: (a) with uniform doping and (b) with a 5 mm long undoped end cap. The maximum values and their locations are indicated by arrows. The temperature difference beween two isotherms is 4 K [20].

bending of the pumped surface resulted in a dioptic power of $1.4\,m^{-1}$. The axial expansion of the rod leads to the high tensile stress at the pump face, as indicated in Fig. 7.17(a).

The above-mentioned example illustrates a number of characteristic features of end-pumped systems. In order to maximize gain, and to match the TEM_{00} resonator mode, the pump beam is tightly focused, which leads to high-pump-power irradiance incident on the end of the laser rod. As a result, the input face is under a high thermal load. The thermal stress, which leads to strong thermal lensing, is often high enough to cause fracture of the end face of the laser rod.

The thermal management of end-pumped lasers can be greatly improved by the use of composite rods. Laser crystals, such as Nd : YAG, Nd : YVO_4, and Yb : YAG, have become available with sections of undoped host material on one or both ends. These end caps are diffusion bonded to the doped laser crystal. Composite rods are proven to provide a very effective way to reduce temperature and stresses at the face of end-pumped lasers. This is illustrated in Figs. 7.16(b) and 7.17(b) that show the result of an analysis carried out for an Nd : YAG rod with a

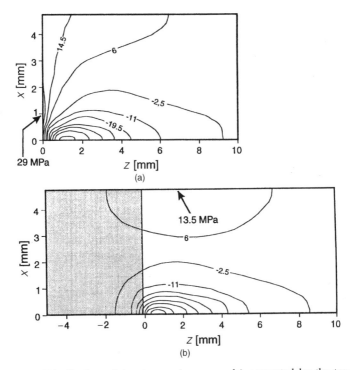

FIGURE 7.17. Distribution of the stresses (megapascals) generated by the temperature gradients illustrated in Fig. 7.16 (a) for the uniformly doped crystal and (b) the rod with an undoped end cap [20].

5 mm long undoped YAG section at the pumped front end. The other parameters are the same as for the uniformly doped crystal. Because the undoped region is transparent to the impinging pump radiation, there is no thermal load generated at the rod pump face.

The diffusion bond provides uninhibited heat flow from the doped to the undoped region. A large temperature reduction of about 35% is achieved for the composite rod, where a significant part of heat flow occurs into the undoped end cap. The undoped end cap prevents the doped pump face from expanding along its axis; therefore no surface deformation takes place at this interface. This leads to strong compressive stresses (as shown in Fig. 7.17(b)) instead of tensile stresses, as was the case for the unrestrained surface (Fig. 7.17(a)). Since the Nd:YAG crystal, like most materials, is much stronger under compression than under tension, the undoped end cap increases the stress fracture limit and thereby the maximum thermal loading that can be tolerated by the end-pumped laser.

An important parameter for the design of the optical resonator is the amount of thermal lensing caused by the temperature distribution in an end-pumped laser crystal. In a first approximation, the thermally induced lens can be described by considering only the temperature-dependent part of the refractive index. Also assuming only radial heat flow in a rod that is in contact with a thermal heat sink of fixed temperature, an analytical solution for the focal length of the thermal lens was derived in [21]. A Gaussian pump beam incident on the crystal has been assumed

$$I(r, z) = I_0 \exp(-2r^2/w_p^2) \exp(-\alpha_0 z), \qquad (7.52)$$

where α_0 is the absorption coefficient and w_p is the $(1/e^2)$ Gaussian radius of the pump beam. With P_h the fraction of the pump power that results in heating, the effective focal length for the entire rod can be expressed by

$$f = \frac{\pi K w_p^2}{P_h(dn/dT)} \left(\frac{1}{1 - \exp(-\alpha_0 l)} \right), \qquad (7.53)$$

where K is the thermal conductivity of the laser material and dn/dT is the change of refractive index with temperature. From (7.53) it follows that the effective focal length depends on the square of the pump-beam radius w_p. Therefore, it is desirable to use the largest pump-beam radius consistent with resonator-mode match.

The dependence of thermal lensing on pump power is illustrated in Fig. 7.18. An end-pumped Nd : YAG rod with a length of 20 mm and a radius of 4.8 mm was pumped with a fiber-coupled laser-diode array. The output from the fiber bundle was imaged onto the crystal surface into a pump spot with radius $w_p = 340\,\mu$m. Figure 7.18 displays the measured thermal lens and the calculated values according to (7.53). In using this equation, we assume that 32% of the pump power results in heating. The material parameters required for Nd : YAG are $dn/dT = 7.3 \times 10^{-6} K^{-1}$, $\alpha_0 = 4.1\,cm^{-1}$, $K = 0.13\,W/cm\,K$.

Equation (7.53) is an approximation for low and medium pump intensities. At high pump levels, the contribution of end bulging to thermal lensing is substantial and in Nd : YVO$_4$ can be as high as the thermal index variations [23]. As

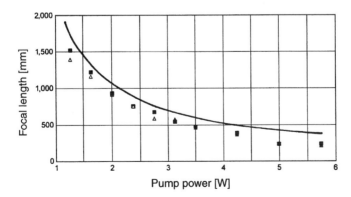

FIGURE 7.18. Thermal lensing in an end-pumped Nd : YAG laser. Measured values are from [22].

discussed above, end bulging due to thermal deformation can be prevented by diffusion bonding of an undoped end cap at the pump side of the laser. Another approach to reduce bulging is the application of mechanical stress to the crystal to compensate for thermally induced deformation. In this approach a thin crystal is compressed between a sapphire window and a copper heat sink. Pump and resonator beam enter through the sapphire window. At an applied pressure on the order of 700 MPa a marked reduction of thermal lensing was observed [24].

Summary

In solid-state lasers only a small fraction of electrical input power is converted to laser radiation; the remainder of the input power is converted to heat. Solid state lasers require cooling of the pump source, active medium, and in case of flash lamp pumping, cooling of the pump enclosure. In this chapter we are concerned with the thermally induced effects caused by internal heating and removal of waste heat from the active medium. Heating of the laser medium is caused by nonradiative transitions within the active ions and by absorption of pump radiation by the host material. The amount of heat generated in a Nd:YAG crystal is about equal to the laser output power for laser diode pumping; in flash lamp pumped systems the heat generated is about three times as high as the laser output.

The combination of volumetric heating of the laser medium by the absorbed pump radiation and surface cooling required for heat extraction leads to a nonuniform temperature distribution. The thermal gradients cause optical distortions that can severely degrade the optical quality of the laser beam and eventually limit the laser output power, either because of an unacceptable poor beam pattern or because of thermal stress-induced breakage of the laser material. The optical distortions are due to temperature and stress-dependent variations of the refractive index.

In long cylindrical laser rods that are transversely pumped by arc lamps or diode lasers, the rods develop a quadratic radial temperature, refractive index and thermal strain profile. This leads to thermal focusing of the laser rod, and in case of Nd:YAG, where the strain is different in radial and tangential directions, gives rise through the photoelastic effect, to thermally induced birefringence and stress-induced biaxial focusing.

The two-dimensional radial stress profile inherent in cylindrical geometrics is reduced to a one-dimensional plane stress distribution in a thin rectangular slab cooled only at the two opposing large faces. For incident radiation polarized along the major axis of the slab the stress-induced birefringence is zero. Cylindrical focusing in a slab can be eliminated by choosing a propagation direction along a zigzag path, which averages the distortions caused by the one-dimensional thermal gradient.

In end-pumped lasers a short cylindrical or rectangular laser crystal is mounted in a heat sink. Heat deposition in end-pumped lasers is very inhomogeneous. The very localized heat deposition leads to highly nonuniform and complex temperature and stress profiles, depending on whether cooling is through the lateral or front surface of the crystal. Also in end-pumped configurations the pumped front surface tends to bend owing to the locally varying expansion of the material.

In disk-type geometries the pump and laser radiation propagate in the same direction just as in end-pumped systems. However, heat extraction is also in the longitudinal direction. The idea behind the design of disk-shaped lasers is a reduction or elimination of thermally induced index of refraction gradients across the beam. Thermal gradients are predominantly in the direction of beam propagation. In very thin disks, such as utilized in Yb : YAG lasers, heat is extracted from the back side, whereas the pump and laser beam enter the disk through the front face. Very large disks employed in Nd : glass lasers are air or gas cooled on both sides.

References

[1] B. Comaskey, B.D. Moran, G.F. Albrecht, and R.J. Beach: *IEEE J. Quantum Electron.* **QE-31**, 1261 (1995).

[2] T.Y. Fan: *IEEE J. Quantum Electron.* **QE-29**, 1457 (1993).

[3] T.S. Chen, V.L. Anderson, and O. Kahan: *IEEE J. Quantum Electron.* **QE-26**, 6 (1990).

[4] M.S. Mangir and D.A. Rockwell: *IEEE J. Quantum Electron.* **QE-22**, 574 (1986).

[5] D.S. Sumida, D.A. Rockwell, and M.S. Mangir: *IEEE J. Quantum Electron.* **QE-24**, 985 (1988).

[6] H.S. Carslaw and J.C. Jaeger: *Conduction of Heat in Solids* (Oxford University Press, London), 1948, p. 191.

[7] W. Koechner: *Appl. Opt.* **9**, 1429 (1970).

[8] S. Timoshenko and J.N. Goodier: *Theory of Elasticity*, 3rd ed. (McGraw-Hill, Singapore), 1982.

[9] M. Born and E. Wolf: *Principles of Optics* (Pergamon, London), 1965.

[10] J.F. Nye: *Physical Properties of Crystals* (Clarendon, Oxford, UK), 1985, reprinted 1993.

[11] J.D. Foster and L.M. Osterink: *J. Appl. Phys.* **41**, 3656 (1970).

[12] W. Koechner and D.K. Rice: *IEEE J. Quantum Electron.* **QE-6**, 557 (1970).

[13] W. Koechner: *Appl. Opt.* **9**, 2548 (1970).

[14] G.D. Baldwin: Q-switched evaluation of Nd: CaLaSOAP. Final Report AFAL-TR-72-334, Air Force Avionics Laboratory, Wright-Patterson AFB, Ohio (September 1972).

[15] S.Z. Kurtev, O.E. Denchev, and S.D. Savov: *Appl. Opt.* **32**, 278 (1993).

[16] W.C. Scott and M. de Wit: *Appl. Phys. Lett.* **18**, 3 (1971).

[17] J.M. Eggleston, T.J. Kane, K. Kuhn, J. Unternahrer, and R.L. Byer: *IEEE J. Quantum Electron.* **QE-20**, 289 (1984).

[18] M.A. Summers, J.B. Trenholme, W.L. Gagnon, R.J. Gelinas, S.E. Stokowski, J.E. Marion, H.L. Julien, J.A. Blink, D.A. Bender, M.O. Riley, and R.F. Steinkraus: *High Power and Solid State Lasers* (edited by W. Simmons), *SPIE Proc.* **622**, 2 (1986).

[19] J. Frauchiger, P. Albers, and H.P. Weber: *IEEE J. Quantum Electron.* **QE-28**, 1046 (1992).

[20] R. Weber, B. Neuenschwander, M. MacDonald, M.B. Roos, and H.P. Weber: *IEEE J. Quantum Electron.* **QE-34**, 1046 (1998).

[21] M.E. Innocenzi, H.T. Yura, C.L. Fincher, and R.A. Fields: *Appl. Phys. Lett.* **56**, 1831 (1990).

[22] B. Neuenschwander, R. Weber, and H.P. Weber: *IEEE J. Quantum Electron.* **QE-31**, 1082 (1995).

[23] X. Peng, A. Asundi, Y. Chen, and Z. Xiong: *Appl. Opt.* **40**, 1396 (2001).

[24] Y. Liao, R.J.D. Miller, and M.R. Armstrong: *Opt. Lett.* **24**, 1343 (1999).

Exercises

1. Evaluate (7.34) for Nd : YAG and, ignoring end effects, find the material constant that determines the thermal lensing-induced focal length of a laser rod made of this material that is uniformly pumped.

2. Is the assumption of uniform pumping in the previous problem strictly valid? If you think it is, explain in words and sketches why this is so. If you think it is not, explain in words and sketches why it is not so.

3. If an Nd : YAG laser rod with flat and parallel faces is placed centered between two flat resonator mirrors spaced by 1 m, is the resonator stable or unstable? When the pump lamp (assume an ideal elliptical pump cavity and an ideal cw discharge lamp) power reaches the threshold value for oscillation that you would calculate using the formalism of Chapter 3, you find that lasing does not begin. Instead, you have to pump harder. At what pump power would you exptect it to start lasing and why?

4. When the laser in Problem 3 is lasing you try to get more output by pumping still harder. You do up to a point and then the lasing stops. At what pump power would you expect the lasing to stop and why?

8

Q-Switching

8.1 Q-Switch Theory

8.2 Mechanical Devices

8.3 Electro-Optical Q-Switches

8.4 Acousto-Optic Q-Switches

8.5 Passive Q-Switch

    References

    Exercises

A mode of laser operation extensively employed for the generation of high pulse power is known as Q-switching. It has been so designated because the optical Q of the resonant cavity is altered when this technique is used. As was discussed in Chapter 3, the quality factor Q is defined as the ratio of the energy stored in the cavity to the energy loss per cycle. Consequently, the higher the quality factor, the lower the losses.

In the technique of Q-switching, energy is stored in the amplifying medium by optical pumping while the cavity Q is lowered to prevent the onset of laser emission. Although the energy stored and the gain in the active medium are high, the cavity losses are also high, lasing action is prohibited, and the population inversion reaches a level far above the threshold for normal lasing action. The time for which the energy may be stored is on the order of τ_f, the lifetime of the upper level of the laser transition. When a high cavity Q is restored, the stored energy is suddenly released in the form of a very short pulse of light. Because of the high gain created by the stored energy in the active material, the excess excitation is discharged in an extremely short time. The peak power of the resulting pulse exceeds that obtainable from an ordinary long pulse by several orders of magnitude.

Figure 8.1 shows a typical time sequence of the generation of a Q-switched pulse. Lasing action is disabled in the cavity by a low Q of the cavity. Toward the end of the flashlamp pulse, when the inversion has reached its peak value, the Q of the resonator is switched to some high value. At this point a photon flux starts to build up in the cavity, and a Q-switch pulse is emitted. As illustrated in

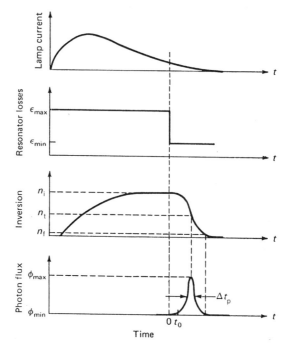

FIGURE 8.1. Development of a Q-switched laser pulse. Shown is the flashlamp output, resonator loss, population inversion, and photon flux as a function of time.

Fig. 8.1, the emission of the Q-switched laser pulse does not occur until after an appreciable delay, during which time the radiation density builds up exponentially from noise.

8.1   Q-Switch Theory

A number of important features of a Q-switched pulse, such as energy content, peak power, pulsewidth, rise and fall times, and pulse formation time, can be obtained from the rate equations discussed in Chapter 1. In all cases of interest the Q-switched pulse duration is so short that we can neglect both spontaneous emission and optical pumping in writing the rate equations.

From (1.61, 1.58), it follows that

$$\frac{\partial \phi}{\partial t} = \phi \left(c\sigma n \frac{l}{L} - \frac{\varepsilon}{t_r} \right) \tag{8.1}$$

and

$$\frac{\partial n}{\partial t} = -\gamma n \phi \sigma c. \tag{8.2}$$

In (8.1) we expressed the photon lifetime τ_c by the round-trip time t_r and the fractional loss ε per round trip according to (3.8). Also, a distinction is made between the length of the active material l and the length of the resonator L. Q-switching is accomplished by making ε an explicit function of time (e.g., rotating mirror or Pockels cell Q-switch) or a function of the photon density (e.g., saturable absorber Q-switch). The losses in a cavity can be represented by

$$\varepsilon = -\ln R + \delta + \zeta(t), \tag{8.3}$$

where the first term represents the output coupling losses determined by the mirror reflectivity R, the second term contains all the incidental losses such as scattering, diffraction, and absorption, and $\zeta(t)$ represents the cavity loss introduced by the Q-switch. For a particular explicit form of $\zeta(t, \phi)$, the coupled rate equations can be solved numerically with the boundary condition $\zeta(t < 0) = \zeta_{max}$; $\zeta(t \geq 0) = 0$. In many instances Q-switches are so fast that no significant change of population inversion takes place during the switching process; in these cases ζ can be approximated by a step function.

In the ideal case, where the transition from low Q to high Q is made instantaneously, the solution to the rate equations is particularly simple. In this case we assume that at $t = 0$ the laser has an initial population inversion n_i, and the radiation in the cavity has some small but finite photon density ϕ_i. Initially, the photon density is low while the laser is being pumped and the cavity losses are $\varepsilon_{max} = -\ln R + \delta + \zeta_{max}$ as illustrated in Fig. 8.1. The losses are suddenly reduced to $\varepsilon_{min} = -\ln R + L$. The photon density rises from ϕ_i, reaches a peak ϕ_{max} many orders of magnitude higher than ϕ_i, and then declines to zero. The population inversion is a monotone decreasing function of time starting at the initial inversion n_i and ending at the final inversion n_f. We note that the value for n_f is below the threshold inversion n_t for normal lasing operation. At n_t the photon flux is maximum and the rate of change of the inversion dn/dt is still large and negative, so that n falls below the threshold value n_t and finally reaches the value n_f. If n_i is not too far above n_t, that is, initial gain is close to threshold, then the final inversion n_f is about the same amount below threshold as n_i is above and the output pulse is symmetric. On the other hand, if the active material is pumped considerably above threshold, the gain drops quickly in a few cavity transit times t_r to where it equalizes the losses. After the maximum peak power is reached at n_t, there are enough photons left inside the laser cavity to erase the remaining population excess and drive it quickly to zero. In this case the major portion of the decay proceeds with a characteristic time constant τ_c, which is the cavity time constant.

The equations describing the operation of rapidly Q-switched lasers involves the simultaneous solution of two coupled differential equations for the time rate of change of the internal photon density in the resonator, (8.1), and the population inversion density in the active medium, (8.2). We can express the output energy of the Q-switched laser as follows [1]:

$$E_{out} = \frac{h\nu A}{2\sigma\gamma} \ln\left(\frac{1}{R}\right) \ln\left(\frac{n_i}{n_f}\right), \tag{8.4}$$

where $h\nu$ is the laser photon energy and A is the effective beam cross sectional area. The initial and final population inversion densities, n_i and n_f, are related by the transcendental equation

$$n_i - n_f = n_t \ln\left(\frac{n_i}{n_f}\right), \tag{8.5}$$

where n_t is the population inversion density at threshold, that is,

$$n_t = \frac{1}{2\sigma\ell}\left(\ln\frac{1}{R} + \delta\right). \tag{8.6}$$

The pulse width of the Q-switch pulse can also be expressed as a function of the inversion levels, n_i, n_f, n_t,

$$\Delta t_p = \tau_c \frac{n_i - n_f}{n_i - n_t[1 + \ln(n_i/n_t)]}. \tag{8.7}$$

The equations for pulse energy, pulse width, and therefore peak power, are expressed in terms of the initial and final population inversion densities which depend not only on the particular choice of output coupler, but which are also related via a transcendental equation.

An analytical solution has been reported in [2], which reveals that key parameters such as optimum reflectivity, output energy, extraction efficiency, pulse width, peak power, etcetera, can all be expressed as a function of a single dimensionless variable $z = 2g_0\ell/\delta$, where $2g_0\ell$ is the logarithmic small-signal gain and δ is the round-trip loss.

It is interesting to note that in Chapter 3 we derived expressions for T_{opt}, P_{opt}, and η_E for normal mode operation of the oscillator which were also functions of the parameter z. We will summarize the results of [2] because the equations presented in this work are particularly useful to the laser designer. The following expression for the optimum reflectivity was derived

$$R_{opt} = \exp\left[-\delta\left(\frac{z - 1 - \ln z}{\ln z}\right)\right]. \tag{8.8}$$

For values of R_{opt} close to one, this expression simplifies to

$$T_{opt} = \delta\left(\frac{z - 1}{\ln^z} - 1\right). \tag{8.9}$$

This equation has a similar functional form as (3.70), which was derived for the long pulse operation. In both cases the optimum transmission increases for larger values of z and δ.

The energy output for an optimized system is

$$E_{\text{out}} = E_{\text{sc}}(z - 1 - \ln z), \tag{8.10}$$

where E_{sc} is a scale factor with the dimension of energy which contains a number of constants

$$E_{\text{sc}} = Ah\nu\delta/2\sigma\gamma,$$

where A is the beam cross section, $h\nu$ is the photon energy, σ is the stimulated emission cross section, δ is the round-trip loss, and γ is one for a four-level laser. The FWHM pulse width versus z is obtained from

$$t_{\text{p}} = \frac{t_{\text{r}}}{\delta} \left(\frac{\ln z}{z[1 - a(1 - \ln a)]} \right), \tag{8.11}$$

where t_{r} is the cavity round-trip time and $a = (z - 1)/(z \ln z)$.

In the limit of large z, the output energy approaches the total useful stored energy in the gain medium

$$E_{\text{st}} = \frac{Ah\nu\delta}{2\sigma\gamma} z = \frac{V h\nu n_{\text{i}}}{\gamma}. \tag{8.12}$$

With (8.10), (8.12) one can define an energy extraction efficiency

$$\eta_{\text{E}} = 1 - \left(\frac{1 + \ln z}{z} \right) \tag{8.13}$$

which is plotted in Fig. 8.2. As one would expect, a high gain-to-loss ratio leads to a high Q-switch extraction efficiency. For a ratio of logarithmic gain to loss of

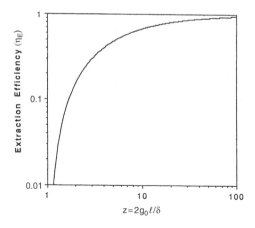

FIGURE 8.2. Q-switch extraction efficiency as a function of z.

about 10, an extraction efficiency of 70% is achieved. For higher factors of z, the extraction efficiency increases only very slowly.

The reader is reminded that the shape of the curve for η_E is similar to the results obtained for the free-running laser discussed in Chapter 3, see Fig. 3.7, which also depends only on the ratio of $2g_0\ell/\delta$. It is also important to remember that besides the Q-switch extraction efficiency expressed by (8.13), the total energy extraction from a Q-switched laser also depends on the fluorescence losses and ASE depopulation losses prior to opening of the Q-switch. The overall efficiency of the Q-switch process was defined in Section 3.4.1 as the product of the Q-switch extraction efficiency, storage efficiency, and depopulation efficiency.

Laser-design trade-offs and performance projections and system optimization can be accomplished quickly with the help of (8.8)–(8.13). For example, we consider the design parameters for a Q-switched Nd:YAG laser with a desired multimode output of 100 mJ. The laser crystal has a diameter of 5 mm and the laser resonator is 30 cm long. Assuming a 5% round-trip cavity loss ($\delta = 0.05$), and with $h\nu = 1.86 \times 10^{-19}$ J and $\sigma = 2.8 \times 10^{-19}$, cm^2 we calculate $E_{sc} = 3.2 \times 10^{-3}$ J. This requires a ratio of $E_{out}/E_{sc} = 31$ in order to achieve the desired output energy. From (8.10), one therefore obtains a value $2g_0\ell/\delta = 35$, or a single-pass power gain of the rod of $G = \exp(g_0\ell) = 2.4$. Extraction efficiency follows from (8.12) to be around 87% for $z = 35$ and the optimum output coupler has a reflectivity of $R = 0.65$ according to (8.9). Since the cavity transit round-trip time for the given resonator length is about 2 ns, the expected pulse width from the laser is $t_p = 11$ ns according to (8.11). The peak power of the Q-switch pulse follows from the parameters already calculated and is $P_p = E_{out}/t_p = 9$ MW.

8.1.1 Continuously Pumped, Repetitively Q-Switched Systems

A very important class of laser systems, employed extensively in micromachining applications, is the cw-pumped, repetitively Q-switched Nd:YAG laser. In these laser systems the population inversion undergoes a cyclic variation, as shown in Fig. 8.3. Between Q-switches the population inversion rises from a value n_f to a value n_i. The buildup of the inversion under the influence of a continuous pumping rate and spontaneous decay is described as a function of time by

$$n(t) = n_\infty - (n_\infty - n_f) \exp\left(-\frac{t}{\tau_f}\right), \tag{8.14}$$

where n_∞ is the asymptotic value of the population inversion which is approached as t becomes large compared to the spontaneous decay time τ_f. The value n_∞, which depends on the pump input power, is reached only at repetition rates small compared to $1/\tau_f$. For repetition rates larger than $1/\tau_f$, the curves representing the buildup of the population are shorter segments of the same exponential curve followed for the lower repetition rates.

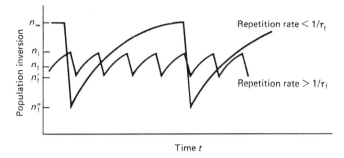

FIGURE 8.3. Population inversion versus time in a continuously pumped Q-switched laser. Shown is the inversion for two different repetition rates. At repetition rates less than $1/\tau_f$, the inversion approaches the asymptotic value n_∞.

During the emission of a Q-switch pulse, the inversion changes from n_i to n_f. Figure 8.4 shows the development of the Q-switched pulse on an expanded time scale. At $t = 0$ the cavity Q factor starts to increase until it reaches its maximum value $Q_{\max}$ at $t = t_1$. Pulse formation ensues until the full pulse output is achieved at $t = t_2$. Stimulated emission ceases at $t = t_3$; at this time continued pumping causes the inversion to start to increase. At the point where the inversion begins to increase, $t = t_3$, the cavity Q begins to decrease, reaching its minimum value at $t = t_4$. During the time period t_3–t_5, the inversion is allowed to build up to its initial value n_i.

The theory of Q-switch operation [1], summarized at the beginning of this chapter for the case of single Q-switched pulses, can also be applied to the case of repetitive Q-switches, with some modifications that take into account the effects of continuous pumping. Equations (8.1), (8.2) are applicable, however we will set $\gamma = 1$ and $n = n_2$ because all repetitively Q-switched lasers of practical use are

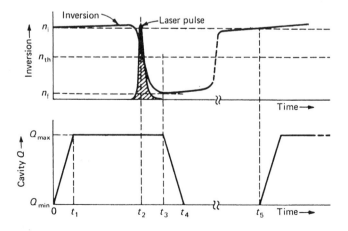

FIGURE 8.4. Development of a Q-switched pulse in a cw-pumped system [3].

four-level systems. The inversion levels n_f and n_i are connected by (8.7). During the low-Q portion of the cycle, the inversion n_2 is described by the differential equation

$$\frac{dn_2}{dt} = W_p(n_{tot} - n_2) - \frac{n_2}{\tau_f}. \tag{8.15}$$

With the assumption that $n_2 \ll n_{tot}$ and with

$$n_\infty = W_p \tau_f n_{tot} \tag{8.16}$$

we obtain (8.14) as a solution. For repetitive Q-switching at a repetition rate f, the maximum time available for the inversion to build up between pulses is $1/f$. Therefore,

$$n_i = n_\infty - (n_\infty - n_f) \exp\left(-\frac{1}{\tau_f f}\right) \tag{8.17}$$

in order for the inversion to return to its original value after each Q-switch cycle.

During each cycle a total energy $(n_i - n_f)$ enters the coherent electromagnetic field. Of this a fraction, $T/(T + \delta)$, appears as laser output. Therefore, the Q-switched average power P_{av} at a repetition rate f is given by

$$P_{av} = \frac{Tf}{T + \delta}(n_i - n_f)h\nu V. \tag{8.18}$$

The peak power P_p of the Q-switched pulses is obtained from (8.6). The effective pulse width t_p can be calculated from the peak and average powers according to $t_p = P_{av}/P_p f$. It is convenient to calculate the ratios P_p/P_{cw} and P_{av}/P_{cw}, where P_p is the Q-switched peak power and P_{cw} is the cw power from the laser at the same pumping level. For cw operation the time derivatives in (8.1) and (8.2) are zero, and we have

$$P_{cw} = \frac{T}{T + \delta}\left(\frac{n_\infty - n_t}{\tau_f}\right)h\nu V. \tag{8.19}$$

Because of the transcendental functions expressed by (8.6), (8.7), and (8.17), the ratios P_p/P_{cw} and P_{av}/P_{cw} cannot be expressed in closed form. The result of numerical calculations is shown in Figs. 8.5 and 8.6.

In Fig. 8.5 the ratio of Q-switched peak power output to the maximum cw output is plotted versus repetition rate for an Nd:YAG laser. For repetition rates below approximately 800 Hz ($\tau_f f \approx 0.20$) the peak power is independent of the repetition rate. At these low repetition rates there is sufficient time between pulses for the inversion to reach the maximum value n_∞. In the transition region between 0.8 and 3 kHz, peak power starts to decrease as the repetition rate is increased. Above 3 kHz, the peak power decreases very rapidly for higher repetition rates.

Figure 8.6 shows the ratio of Q-switched average power to cw power as a function of repetition rate. Above a repetition rate of approximately 10 kHz, the

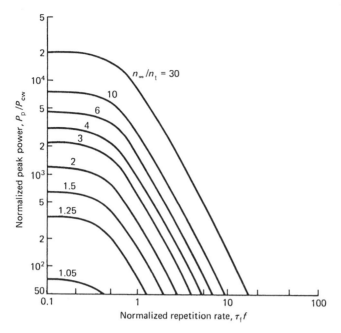

FIGURE 8.5. Ratio of peak power to cw power as a function of the repetition rate for a typical cw-pumped Nd : YAG laser. The parameter n_∞/n_t expresses the fractional inversion above threshold [4].

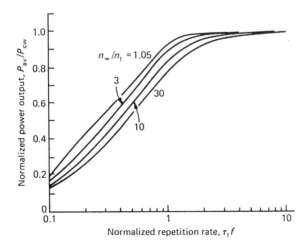

FIGURE 8.6. Ratio of Q-switched average power to maximum cw power as a function of normalized repetition rate f [4].

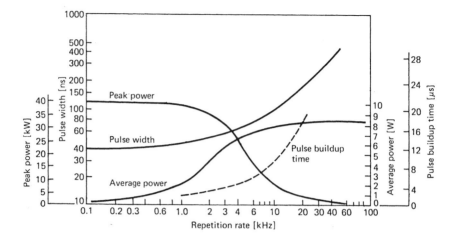

FIGURE 8.7. Performance of a cw-pumped Nd : YAG laser system. Plotted is the peak power, average power, pulse buildup time, and pulse width as a function of repetition rate.

Q-switch average power approaches the cw power. At low repetition rates the average power is proportional to the repetition rate. In Fig. 8.7 the experimentally determined peak power, average power pulse width, and pulse buildup time is plotted as a function of repetition rate for an Nd : YAG laser. In accordance with theory, for higher repetition rates the pulse width and the pulse buildup time increase as a result of the reduction of gain.

In the following sections we will describe and compare different Q-switch techniques.

8.2   Mechanical Devices

Q-switches have been designed based upon rotational, oscillatory, or translational motion of optical components. These techniques have in common that they inhibit laser action during the pump cycle by either blocking the light path, causing a mirror misalignment, or reducing the reflectivity of one of the resonator mirrors. Near the end of the flashlamp pulse, when maximum energy has been stored in the laser rod, a high Q-condition is established and a Q-switch pulse is emitted from the laser.

Mechanical Q-switches are relatively slow and tend to emit multiple pulses; also, the mechanical wear requires frequent maintenance. Owing to these disadvantages mechanical Q-switches have been replaced by electro-optic or acousto-optic devices.

8.3   Electro-Optical Q-Switches

Very fast electronically controlled optical shutters can be designed by exploiting the electro-optic effect in crystals or liquids. The key element in such a shutter is an electro-optic element that becomes birefringent under the influence of an external field. Birefringence in a medium is characterized by two orthogonal directions, called the "fast" and "slow" axes, which have different indices of refraction. An optical beam, initially plane-polarized at 45° to these axes and directed normal to their plane, will split into two orthogonal components, traveling along the same path but at different velocities. Hence, the electro-optic effect causes a phase difference between the two beams. After traversing the medium, the combination of the two components results, depending on the voltage applied, in either an elliptical, circular, or linearly polarized beam. For Q-switch operation only two particular voltages leading to a quarter-wave and half-wave retardation are of interest. In the first case, the incident linearly polarized light is circular polarized after passing the cell, and in the second case the output beam is linearly polarized; however, the plane of polarization has been rotated 90°.

The two most common arrangements for Q-switching are shown in Fig. 8.8. In Fig. 8.8(a) the electro-optic cell is located between a polarizer and the rear mirror.

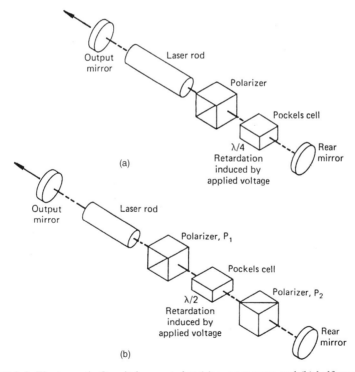

FIGURE 8.8. Electro-optic Q-switch operated at (a) quarter-wave and (b) half-wave retardation voltage.

The inclusion of the polarizer is not essential if the laser radiation is polarized, such as, for example, in Nd : YLF. The sequence of operation is as follows: During the flashlamp pulse, a voltage $V_{1/4}$ is applied to the electro-optic cell such that the linearly polarized light passed through the polarizer is circularly polarized. After being reflected at the mirror, the radiation again passes through the electro-optic cell and undergoes another $\lambda/4$ retardation, becoming linearly polarized but at 90° to its original direction. This radiation is ejected from the laser cavity by the polarizer, thus preventing optical feedback. Toward the end of the flashlamp pulse the voltage on the cell is switched off, permitting the polarizer–cell combination to pass a linearly polarized beam without loss. Oscillation within the cavity will build up, and after a short delay a Q-switch pulse will be emitted from the cavity.

In the arrangement of Fig. 8.8(b) an electric voltage must first be applied to the cell to transmit the beam. In this so-called pulse-on Q-switch, the cell is located between two crossed polarizers. As before, polarizer P_1, located between the laser rod and the cell, is not required if the active medium emits a polarized beam. During the flashlamp pulse, with no voltage applied to the cell, the cavity Q is at a minimum due to the crossed polarizers. At the end of the pump pulse a voltage $V_{1/2}$ is applied to the cell, which causes a 90° rotation of the incoming beam. The light is therefore transmitted by the second polarizer P_2. Upon reflection at the mirror the light passes again through polarizer P_2 and the cell, where it experiences another 90° rotation. Light traveling toward the polarizer P_1 has experienced a 180° rotation and is therefore transmitted through P_1.

Two types of electro-optic effects have been utilized in laser Q-switches: the Pockels effect, which occurs in crystals that lack a center of point symmetry, and the Kerr effect, which occurs in certain liquids. Pockels cells, which require a factor of 5 to 10 lower voltage than Kerr cells, are the most widely used active devices for Q-switching pulsed lasers.

Pockels Cell Q-Switch

The Pockels cell contains an electro-optic crystal in which a refractive index change is produced by an externally applied electric field. Crystals are classified into 32-point groups according to their structure. Only 20-point groups, namely those that lack a center on symmetry, exhibit a nonvanishing electro-optic effect. The index change produced by an externally applied field is described in each case by a 6×3 matrix of electro-optic coefficients. The number of coefficients is greatly reduced by the structural characteristics of crystals. The most commonly used electro-optic materials have only a few distinct coefficients. The location of a coefficient in the matrix array interrelates the crystal orientation, the applied field direction, and the polarization and direction of the optical beam. The magnitude of each coefficient determines the strength of the electro-optic effect for each geometry.

The basic requirements for a crystal to be useful as an electro-optic Q-switch are: a good optical quality combined with a high laser damage threshold and a large electro-optic coefficient for light propagating parallel to the optic axis. The

latter requirement is important because the two-phase-shifted orthogonal components of the beam travel along the same path only if the direction of the light
beam is either parallel or normal to the optical axis of the crystal. For other directions the fast and slow axes of the beam include a small angle. Two crystals
that meet these criteria and are widely employed in electro-optical Q-switches are
potassium dihydrogen phosphate (KDP) and lithium niobate ($LiNbO_3$).

KDP (KH_2PO_H) and its isomorph KD*P (KD_2PO_H) are grown at room temperature from a water solution that yields large distortion-free single crystals. The
attributes of this family of crystals are their high damage threshold and excellent optical quality combined with a large electro-optic coefficient. Crystals with
cross sections up to 100 mm have been produced. A disadvantage is the fact that
the crystals are fairly soft and hygroscopic and must be protected from the environment by enclosing them in cells that are hermetically sealed or filled with an
index matching fluid. In order to avoid walk-off between the fast and slow beam
axes, the electric field has to be applied longitudinally in the same direction as
the beam propagation axis and the optical axis of the crystal. The electric field is
applied to the crystal by means of a pair of electrodes that contain openings for
passage of the laser beam.

Lithium niobate is grown from the melt; it is hard and nonhygroscopic. The
crystals can be antireflection coated and do not need a special protection from
the environment. Also, the quarter and half-wave voltages are about a factor of 2
lower compared to KD*P. In $LiNbO_3$, the electric field is applied perpendicularly
to the beam propagation and optical axis of the crystal. As we will see, this has
the advantage of reducing the required voltage by the width to length ratio of the
crystal. A drawback of $LiNbO_3$ is that the crystal is only available in relatively
small sizes, with cross sections of about 10 mm, and the laser damage threshold
is considerably below that of KD*P.

KDP and KD*P

Potassium dihydrogen phosphate (KH_2PO_4) or KDP and the deuterated form,
which is (KD_2PO_4) or KD*P, are widely used crystals for Pockels cell Q-switches.
The latter is usually preferred because of its larger electro-optic coefficient. The
dependence of the index of refraction on the electric field can be described in
terms of a change in orientation and dimensions of the index ellipsoid. The crystals are uniaxial, that is, in the absence of an electric field the index ellipsoid is
an ellipse of revolution about the optic (z) axis. As indicated in Fig. 8.9 the index ellipsoid projects as a circle on a plane perpendicular to the optic axis. The
circle indicates that the crystal is not birefringent in the direction of the optic axis.
When an electric field is applied parallel to the crystal optic axis, the cross section
of the ellipsoid becomes an ellipse with axes x' and y', making a 45° angle with
the x and y crystallographic axes. This angle is independent of the magnitude of
the electric field. The length of the ellipse axes in the x'- and y'-directions are
proportional to the reciprocals of the indices of refraction in these two directions.

We will now express the phase shift between the orthogonal components corresponding to a wave polarized in the x'- and y'-directions. Changes of the refractive

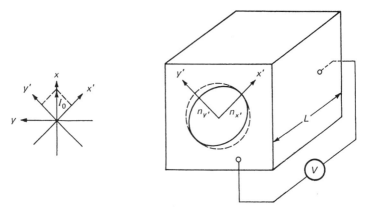

FIGURE 8.9. Change of the index ellipsoid in a KDP crystal when an electric field is applied parallel to the z-axis, I_0 is an incident wave polarized in the x-direction, x and y are the crystallographic axes, and x' and y' are the electrically induced axes.

index Δn are related by the electro-optic tensor r_{ij} of rank 3 to the applied field

$$\Delta \left(\frac{1}{n_i^2} \right) = \sum_{j=1}^{3} r_{ij} E_j, \qquad (8.20)$$

where $i = 1, \ldots, 6$ and $j = 1, \ldots, 3$.

Generally there exist 18 linear electro-optic coefficients r_{ij}. However, in crystals of high symmetry, many of these vanish. For phosphates of the KDP family, r_{63} is the only independent electro-optic coefficient that describes the changes in the ellipsoid when a longitudinal field is applied to the crystal. The change of refractive index in the x'- and y'-directions is

$$n_{x'} = n_0 + \tfrac{1}{2} n_0^3 r_{63} E_z, \qquad n_{y'} = n_0 - \tfrac{1}{2} n_0^3 r_{63} E_z, \qquad (8.21)$$

where n_0 is the ordinary index of refraction and E_z is the electric field in the z-direction. The difference in the index of refraction for the two orthogonal components is then

$$\Delta n_0 = n_0^3 r_{63} E_z. \qquad (8.22)$$

For a crystal of length l, this leads to a path-length difference $\Delta n_0 l$ and a phase difference of $\Delta \varphi = (2\pi/\lambda) \Delta n_0 l$.

The phase difference δ in a crystal of length l is related to the voltage $V_z = E_z l$ applied across the faces by

$$\Delta \varphi = \frac{2\pi}{\lambda} n_0^3 r_{63} V_z. \qquad (8.23)$$

It should be noted that δ is a linear function of voltage and is independent of the crystal dimensions. If linearly polarized light is propagated through the crystal with the direction of polarization parallel to the x- or y-axis, as shown in Fig. 8.9, the components of this vector parallel to the electrically induced axes x' and y' will suffer a relative phase shift δ. In general, orthogonal components undergoing a relative phase shift produce elliptically polarized waves. Thus, the application of voltage in this configuration changes linearly polarized light to elliptically polarized light. If the light then passes through a polarizer, the resulting light intensity will be a function of the ellipticity and therefore the voltage applied to the crystal. A simple derivation shows that, with the analyzer axis oriented at right angles to the input polarization direction, the voltage and the transmitted light intensity I are related by [5]

$$I = I_0 \sin^2 \frac{\Delta\varphi}{2}, \tag{8.24}$$

where I_0 is the input light intensity.

For Q-switch operation, two particular values of phase shift are of interest: these are the $\lambda/4$ and $\lambda/2$ wave retardations that correspond to a phase shift of $\pi/2$ and π. With linearly polarized light being applied, for example, in the x-direction, as shown in Fig. 8.9, the output from the crystal is circularly polarized if $\Delta\varphi = \pi/2$. For $\Delta\varphi = \pi$ the output beam is linearly polarized, but the plane of polarization has been rotated $90°$.

From (8.23) it follows that the voltage required to produce a retardation of π is

$$V_{1/2} = \frac{\lambda}{2n_0^3 r_{63}}. \tag{8.25}$$

KD*P has an index of refraction of $n = 1.51$ and an electro-optic constant $r_{63} = 26.4 \times 10^{-6} \, \mu\text{m/V}$. With these values we calculate a voltage of $V_{\lambda/2} = 5.8 \, \text{kV}$ at $1.06 \, \mu\text{m}$ in order to produce a half-wave retardation.

LiNbO$_3$

The crystal lithium niobate is employed in transverse field electro-optic Q-switches. In an arrangement as shown in Fig. 8.10 light propagates along the c-axis of the crystal. If the light is polarized parallel to the a-axis and an electric field is applied parallel to the a axis, the half-wave retardation is

$$V_{1/2} = \frac{\lambda d}{2r_{22}n_0^3 l}, \tag{8.26}$$

where l is the length of the crystal in the c-direction, d is the distance between the electrodes along the a-axis, and r_{22} is the electro-optic coefficient.

The half-wave voltage is directly proportional to the distance between the electrodes and inversely proportional to the path length. At a wavelength of $\lambda = 1.064 \, \mu\text{m}$, the linear electro-optic coefficient is $r_{22} = 5.61 \times 10^{-6} \, \mu\text{m/V}$ and the refractive index of the ordinary ray is $n_0 = 2.237$ in LiNbO$_3$. The theoretical half-wave voltage obtained from (8.26) for a typical crystal size of $9 \, \text{mm} \times 9 \, \text{mm} \times 25 \, \text{mm}$

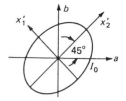

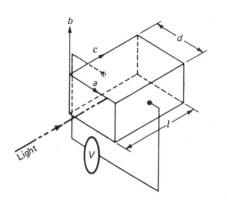

I_0 orientation of linearly polarized light; x'_1-, x'_2- 2 axes of induced birefringence

FIGURE 8.10. Electro-optic Q-switch employing an $LiNbO_3$ crystal.

is 3.0 kV. The high refractive index of $LiNbO_3$ and the geometrical factor d/ℓ are responsible for the lower voltage requirements of $LiNbO_3$ as compared to KD^*P.

Examples of Pockels Cell Q-Switched Lasers

Output energies between 100 and 250 mJ are obtained from Nd : YAG oscillators. The energy extraction is limited by the high gain of this material, which leads to prelasing and subsequently to a depopulation of the inversion. The width of the Q-switched pulses is usually between 10 and 25 ns, and the pulse buildup time is between 50 and 100 ns in typical systems.

For typical KD^*P Pockels cell Q-switches the maximum power density is around $250 \, MW/cm^2$ and for $LiNbO_3$ Q-switches about $50 \, MW/cm^2$. Switching a Pockels cell to obtain a quarter-wave retardation has the advantage that it requires only half the voltage of a half-wave retardation device. But the voltage has to be applied to the crystal for the duration of the pump pulse (i.e., about 240 μs for Nd : YAG). In the case of a half-wave retardation, the voltage is only applied to the crystal during the pulse build-up time at the end of the pump cycle (typically about 100 ns). As far as the electronics is concerned, for $\lambda/4$, switching a fast turn-off of the high voltage is critical and, for $\lambda/2$, switching a fast rise-time of the high voltage pulse is important to minimize resonator losses.

Operation of an electro-optically Q-switched laser requires fast switching of voltages in the multi-kilovolt regime. The driver for the Pockels cell must be a high-speed, high-voltage switch that also must deliver a sizeable current. The cell has a few tens of picofarads capacitance which is charged (or discharged) to several kilovolts in a few nanoseconds. The resulting current is of the order of 10 to 20 A. Common switching techniques include the use of MOSFET's, SCR's, and avalanche transistors.

A typical circuit is based on a Marx bank, in which a number of capacitors are charged in parallel and then connected in series by means of semiconductor switches. The advantage of the Marx bank circuit is the fact that the voltage

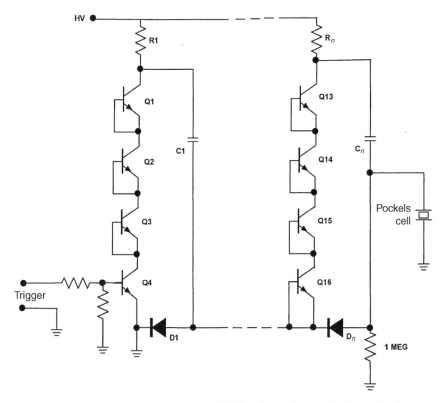

FIGURE 8.11. Circuit diagram of Pockels cell drive electronics employing avalanche transistors.

requirement of the power supply is only a fraction of the voltage generated at the crystal. Figure 8.11 illustrates such a design. The capacitors C_1 to C_n are all charged through resistors R_1 and R_n to about 1 kV. A string of transistors is connected in parallel to each capacitor and associated diode D. The transistors are operated close to their avalanche breakdown voltage. As soon as one transistor is triggered, all the transistors are switched on, which connects all capacitors in series and a large negative voltage appears on the Pockels cell. Typically four to six strings of avalanche transistors are employed to generate voltages in the (2.5–6) kV range starting with 0.8 to 1 kV at the power supply.

8.4 Acousto-Optic Q-Switches

In acousto-optic Q-switches, an ultrasonic wave is launched into a block of transparent optical material, usually fused silica. By switching on the acoustic field, a fraction of the energy of the main beam is diffracted out of the resonator, thus introducing a loss mechanism that prevents laser action. When the acoustic field is switched off, full transmission through the Q-switch cell is restored and a laser pulse is created.

The acousto-optic Q-switch is the device of choice for repetitively Q-switching cw lasers. The low gain characteristics of cw-pumped solid-state lasers do not require very high extinction ratios but do demand an exceptionally low insertion loss. Since high optical-quality fused silica with antireflection coatings can be used as the active medium in the acousto-optical Q-switch, the overall insertion loss of the inactive Q-switch can be reduced to less than 0.5% per pass. The low-insertion loss of the acousto-optic Q-switch offers the convenience of converting from Q-switched to cw operation simply by removing the rf drive power.

An acousto-optic Q-switch is activated by the application of rf power to a transducer that is attached to the transparent medium. The resulting acoustic wave gives rise to a sinusoidal modulation of the density of the medium. The index to refraction is coupled to these periodic variations of the density and strain via the photoelastic effect. Considering only the one-dimensional case for an acoustic wave traveling in the y-direction, we can write

$$n(y, t) = n_0 + \Delta n_0 \sin(w_s t - k_s y), \tag{8.27}$$

where n_0 is the average refractive index of the medium, Δn_0 is the amplitude of the index change, and $w_s = 2\pi v_s$ and $k_s = 2\pi/\lambda_s$ are the angular frequency and the wave vector of the sound wave which travels with the sound velocity $v_s = v_s \lambda_s$ through the medium.

The refractive index change described by (8.27) forms a traveling-wave phase grating across the width of the optical beam. The grating has a period equal to the acoustic wave length and an amplitude proportional to the sound amplitude. A portion of the optical beam which passes through the region occupied by the acoustic wave is diffracted by this phase grating. Acousto-optic Q-switches are operated at ultrasonic frequencies in the tens of megahertz, and the beam interaction length is on the order of several centimeters. In this so-called Bragg regime, the grating acts as thick phase grating. The diffracted beam is primarily confined to a single direction and has a maximum at the Bragg angle. The condition for Bragg scattering to occur is $l\lambda \gg \lambda_s^2$, where l is the interaction length and λ and λ_s are the optical and acoustical wavelengths in the medium, respectively.

In the case of a short interaction length or long acoustical wavelength the acousto-optic cell becomes a thin phase grating and the light beam is diffracted into many orders. This is the regime of Raman–Nath scattering. Because of the higher rf power requirements acousto-optic Q-switches are not used in this regime.

Figure 8.12 displays the geometry for Bragg reflection. The incident beam appears to be reflected from the acoustic wavefronts as if they were mirrors. This phenomenon is called acoustic Bragg reflection in analogy to the selective reflection of X-rays by the lattice of crystals first observed by Bragg. Conservation of energy and momentum of the photon–phonon interaction leads to the condition

$$\sin \Theta = \lambda/2\lambda_s, \tag{8.28}$$

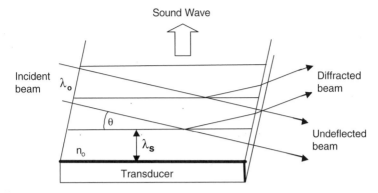

FIGURE 8.12. Geometry for Bragg reflection.

where Θ is the angle of incidence between the optical beam and the acoustic wave, and $\lambda = \lambda_0/n_0$ is the optical wavelength in the diffracting medium. We can derive the Bragg condition (8.28) with the aid of Fig. 8.13 if we assume for the moment that there is no motion of the sound wave. Rays A and B are reflected by two successive acoustic wavefronts. The rays differ in pathlength by the distance $2s$. In order for the rays to be in phase and reinforce we require $2s = \lambda$. From Fig. 8.13 we also obtain the relationship $\sin\Theta = s/\lambda_s$. Elimination of s in these two equations yields the Bragg condition. The beam observed at the Bragg angle arrives from reflections at successive acoustic wavefronts. The rays from each wavefront add in phase and the total intensity is, therefore, N times the intensity of a single reflection.

The difference in direction of the diffracted and incident beam inside the medium is twice the Bragg angle, that is, $2\sin\theta$ or 2θ, since the angle is very small. Reflection at the external boundary of the device will increase the angle by n, accordingly we obtain

$$2\Theta' = \lambda_0/\lambda_s. \tag{8.29}$$

The external angular separation of the two beams is equal in radians to the ratio of the optical to acoustic wavelengths.

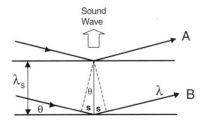

FIGURE 8.13. Reflection of rays A and B by two successive acoustic wavefronts.

A typical acousto-optic Q-switch uses fused silica as the ultrasonic medium and an rf frequency to the transducer for launching an acoustic wave of $v_s = 50$ MHz. The velocity of sound in fused silica is $v_s = 5.95$ km/s. With these numbers and $\lambda_0 = 1.06\,\mu$m, we obtain $\lambda_s = 119\,\mu$m and $2\theta' = 8.9$ mr.

Although of no particular consequence for Q-switch operation, it is worth mentioning that the diffracted beam is frequency shifted. The diffracted beam may be though of as having been Doppler-shifted during diffraction from the acoustic traveling wave. It has the frequency $(v \pm v_s)$ depending on whether the optical beam is incident in front of the oncoming acoustic wave (see Fig. 8.12) or from behind the receding wave.

We will now calculate the partitioning of the power between the incoming and diffracted beam as a function of acoustic power and material parameters. A rigorous treatment of the subject that requires the solution of two coupled traveling wave equations can be found in [6], [7].

The total electric field $E(x, t)$ is taken as the sum of two waves along the distance x. The diffracted wave grows at a rate [8]

$$\frac{dE_1(x)}{dx} = \frac{1}{2}E_2\frac{d\varphi}{dx} \tag{8.30}$$

and the undeflected beam decreases at a corresponding rate

$$\frac{dE_2(x)}{dx} = -\frac{1}{2}E_1\frac{d\varphi}{dx}, \tag{8.31}$$

where E_1, E_2 are the electric field strength of the diffracted and undeflected beam, respectively, and $d\varphi$ is the phase excursion of the beam over an incremental length x given by

$$d\varphi = \beta_0\frac{\Delta n_0}{n_0}\,dx, \tag{8.32}$$

where β_0 is the propagation constant $\beta_0 = 2\pi/\lambda_0$ in free space. Equations (8.30) and (8.31) have the solution

$$E_1 = E_3\sin(\Delta\varphi/2) \tag{8.33}$$

and

$$E_2 = E_3\cos(\Delta\varphi/2), \tag{8.34}$$

where E_3 is the amplitude of the wave entering the crystal.

The diffracted power P_1 as a function of the incident power P_3 follows from (8.33)

$$P_1/P_3 = \sin^2(\Delta\varphi/2) \tag{8.35}$$

and the reduction of power of the undeflected beam is obtained from (8.34)

$$P_2/P_3 = \cos^2(\Delta\varphi/2), \tag{8.36}$$

where $\Delta\varphi$ is the total phase shift the beam experienced over the interaction length ℓ,

$$\Delta\varphi = \ell\beta_0\Delta n_0. \tag{8.37}$$

The sum of the powers of the undeflected and diffracted beams remains constant $P_1 + P_2 = P_3$. Now we have to connect a change of Δn_0 with the acoustic power P_{ac} in the medium.

The change of refractive index as a function of strain s is given by [5]

$$\Delta n_0 = \frac{n_0^3 p_{ij}}{2}s, \tag{8.38}$$

where p_{ij} is the elasto-optic coefficient in the medium, and the strain s is related to the acoustic power P_{ac} by [6]

$$s = \left[\frac{2P_{ac}}{\rho_0 v_s^3 \ell w}\right]^{1/2}. \tag{8.39}$$

In the above equation v_s is the velocity of sound in the medium, ρ_0 is the mass density, and ℓ and w are the length and width of the transducer that launches the acoustic wave into the medium. Introducing (8.37)–(8.39) into (8.35) yields

$$\frac{P_1}{P_3} = \sin^2\left[\frac{\pi}{\lambda}\left(\frac{\ell P_{ac} M_{ac}}{2w}\right)^{1/2}\right]. \tag{8.40}$$

The quantity M_{ac} is a material constant that determines the inherent efficiency of diffraction and is called the figure of merit

$$M_{ac} = n_0^6 p_{ij}^2 / \rho_0 v_s^3. \tag{8.41}$$

It is apparent from (8.40) that the amount of diffracted power depends on the material parameters expressed by M_{ac}, the ratio of length to width of the interaction path and the acoustical power P_{ac}. In a given material such as, for example, fused silica, the value of the photoelastic coefficient p_{ij} in (8.41) depends on the plane of polarization of the light beam with respect to the ultrasonic propagation direction and on the type of ultrasonic wave, that is, longitudinal or shear wave. With shear wave generation the particle motion is transverse to the direction of the acoustic wave propagation direction. In this case the dynamic optical loss is independent of polarization in isotropic materials such as fused quartz.

Table 8.1 lists the pertinent material parameters for an acousto-optic Q-switch fabricated from fused silica. The p coefficients and velocity of sound in fused silica are from [10]. The figure of merit M_{ac} follows from (8.41) with $n_0 = 1.45$ at $1.06\,\mu$m and $\rho_0 = 2.2\,\text{g/cm}^3$. The acoustic power requirement for 1% diffraction has been calculated from (8.40) for $l/w = 10$ and $\lambda = 1.06\,\mu$m. The dimension of P_{ac} in (8.40) is erg/s ($1\,\text{erg/s} = 10^{-7}$ W.)

TABLE 8.1. Material parameters of acousto-optic Q-switches employing fused silica.

Acoustic wave	p coefficient	Polarization of optical beam with respect to acoustic wave vector	Velocity of sound $\times 10^5$ [cm/s]	Figure of merit M_{ac} $\times 10^{-18}$ [s³/g]	Acoustical power P_{ac} [W] for 1% diffraction $(l/w = 10)$
Shear wave	$p_{44} = 0.075$	Independent	3.76	0.45	0.51
Longitudinal	$p_{11} = 0.121$	Parallel	5.95	0.30	0.77
Longitudinal	$p_{12} = 0.270$	Perpendicular	5.95	1.46	0.16

8.4.1 Device Characteristics

Figure 8.14 shows an optical schematic of a cw-pumped laser that contains an acousto-optic Q-switch: The Q-switch consists of a fused silica block to which a crystalline quartz or an $LiNbO_3$ transducer is bonded. Both the transducer and the fused silica interface contain vacuum-deposited electrodes to allow for electrical connections. An inductive impedance-matching network couples the signal of the rf generator to the quartz transducer. The ultrasonic wave is launched into the Q-switch block by the piezoelectric transducer that converts electrical energy into ultrasonic energy. The laser is returned to the high Q-state by switching off the driving voltage to the transducer. With no ultrasonic wave propagating through it, the fused silica block returns to its usual state of high optical transmission and a Q-switch pulse is emitted. The width of the transducer is typically 3 mm, which is about twice the beam diameter for most TEM_{00}-mode Nd : YAG lasers. The length of the Bragg cell and the transducer is around 50 mm. From (8.40) it follows that a large length-to-width ratio reduces the acoustic power requirement for a given diffraction ratio P_1/P_3.

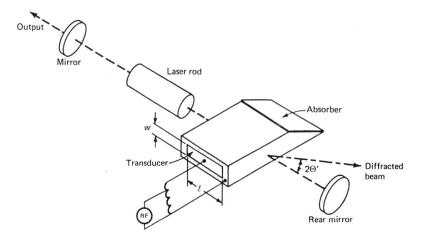

FIGURE 8.14. Acousto-optic Q-switch employed in a cw-pumped Nd : YAG laser.

Virtually all acousto-optic Q-switches are single-pass devices; i.e., the acoustic wave generated by the transducer is absorbed after traveling across the interaction region. The absorber, consisting of a piece of lead attached to the tapered end of the quartz block, prevents reflected acoustical waves from interfering with the incident light beam.

Although the figure of merit M_{ac} of fused quartz is quite low, its optical high-quality, low optical absorption, and high damage threshold make it superior to other, more efficient acousto-optic materials, such as lithium niobate ($LiNbO_3$), lead molybdate ($PbMoO_4$), tellurium dioxide (TeO_2), and dense flint glass. These materials are usually employed in low-power light modulators and optical scanners.

By properly choosing the parameters of the acousto-optic device, a large enough fraction of the laser beam can be deflected out of the resonator to provide an energy loss that inhibits laser action. The frequency of the rf generator determines the diffraction angle according to (8.29), whereas the magnitude of the diffracted power is controlled by the rf power according to (8.40). The frequency of the rf signal driving the transducer is typically in the 40 to 50 MHz range. For these frequencies we obtain, with the values from Table 8.1 for both shear wave and longitudinal wave Q-switches, scattering angles in fused quartz between 14.1 and 7.1 mr. These angles are large enough to deflect the diffracted beam out of the resonator.

The amount of acoustical power required to achieve a certain diffraction efficiency can be calculated from (8.40). Unpolarized lasers are usually Q-switched with shear-wave devices, whereas for polarized laser radiation longitudinal modulators are employed. A longitudinal Q-switch in which the large p_{12} coefficient is utilized requires substantially lower rf powers compared to a shear wave device.

For the practical case of a shear wave Bragg cell having a length of $\ell = 50$ mm and a width of $w = 3$ mm, we find from (8.40) that in order to deflect 35% of the laser beam out of the resonator an acoustic power of 11.7 W is required. Since the Q-switch diffracts radiation in both directions the total insertion loss per round trip is $\delta = (P_1/P_3)(2 - P_1/P_3)$, where (P_1/P_3) is the fraction of radiation diffracted out of the resonator in one pass. For the example above one obtains a round trip insertion loss of 58%, which is large enough to prevent laser action in most cw lasers. The conversion efficiency of rf power delivered to the transducer to acoustic power is on the order of 25 to 30%; therefore, at least 40 W of rf power have to be delivered to the transducer. If the laser output is polarized a Bragg cell with a longitudinal wave can be utilized, in this case only 12 W of rf power is required to achieve the same insertion loss.

The Q-switch must be able to switch from the high-loss to the low-loss state in less than the time required for the laser pulse to build up if maximum output energy is to be achieved. The turn-off time of an acousto-optical Q-switch is determined by the transit time of the sound wave across the beam diameter. Depending on the type of interaction, the transit time of the acoustic wave across a 1-mm diameter beam is between 266 and 168 ns. The relatively slow transit-time is not a serious drawback since the Q-switch pulse evolution time in most cw-pumped systems is on the order of several hundred nanoseconds.

8.5   Passive Q-Switch

A passive Q-switch consists of an optical element, such as a cell filled with organic dye or a doped crystal, which has a transmission characteristic as shown in Fig. 8.15. The material becomes more transparent as the fluence increases, and at high fluence levels the material "saturates" or "bleaches," resulting in a high transmission. The bleaching process in a saturable absorber is based on saturation of a spectral transition. If such a material with high absorption at the laser wavelength is placed inside the laser resonator, it will initially prevent laser oscillation. As the gain increases during a pump pulse and exceeds the round-trip losses, the intracavity flux increases dramatically causing the passive Q-switch to saturate. Under this condition the losses are low, and a Q-switch pulse builds up.

Since the passive Q-switch is switched by the laser radiation itself, it requires no high voltage, fast electro-optic driver, or rf generator. As an alternative to active methods, the passive Q-switch offers the advantage of an exceptional simple design, which leads to very small, robust, and low-cost systems. The major drawbacks of a passive Q-switch are the lack of a precision external trigger capability and a lower output compared to electro-optic or acousto-optic Q-switched lasers. The latter is due to the residual absorption of the saturated passive Q-switch which represents a rather high insertion loss.

Originally, saturable absorbers were based on different organic dyes, either dissolved in an organic solution or impregnated in thin films of cellulose acetate. The poor durability of dye-cell Q-switches, caused by the degradation of the light sensitive organic dye, and the low thermal limits of plastic materials severely re-

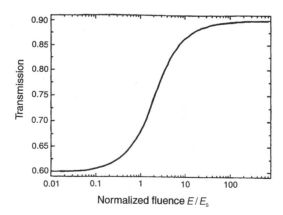

FIGURE 8.15. Nonlinear transmission of a saturable absorber versus fluence normalized to the saturation fluence E_s of the absorber.

stricted the applications of passive Q-switches in the past. The emergence of crystals doped with absorbing ions or containing color centers have greatly improved the durability and reliability of passive Q-switches.

The first new material to appear was the F_2^- : LiF color center crystal. The color centers are induced in the crystal by irradiation with gamma, electron, or neutron sources. Today, the most common material employed as a passive Q-switch is Cr^{4+} : YAG. The Cr^{4+} ions provide the high absorption cross section of the laser wavelength and the YAG crystal provides the desirable chemical, thermal, and mechanical properties required for long life.

A material exhibiting saturable absorption can be represented by a simple energy-level scheme such as that shown in Fig. 8.16. For the moment we will only consider levels 1–3. Absorption at the wavelength of interest occurs at the 1–3 transitions. We assume that the 3–2 transition is fast. For a material to be suitable as a passive Q-switch, the ground state absorption cross section has to be large and, simultaneously, the upper state lifetime (level 2) has to be long enough to enable considerable depletion of the ground state by the laser radiation. When the absorber is inserted into the laser cavity, it will look opaque to the laser radiation until the photon flux is large enough to depopulate the ground level. If the upper state is sufficiently populated the absorber becomes transparent to the laser radiation, a situation that is similar to a three-level laser material pumped to a zero inversion level.

Solutions of the rate equation lead to an absorption coefficient which is intensity dependent [11]

$$\alpha_0(E) = \frac{\alpha_0}{1 + E_i/E_s}, \tag{8.42}$$

where α_0 is the small-signal absorption coefficient and E_s is a saturation fluence

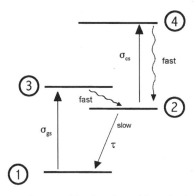

FIGURE 8.16. Energy levels of a saturable absorber with excited state absorption (σ_{gs} and σ_{es} is the ground state and excited state absorption, respectively, and τ is the excited state lifetime).

$$E_s = h\nu/\sigma_{gs}, \tag{8.43}$$

where σ_{gs} is the absorption cross section for the 1–3 transition.

Important characteristics of a saturable absorber are the initial transmission T_0, the fluence E_s at which saturation becomes appreciable, and the residual absorption which results in a T_{max} of the fully bleached absorber.

The small signal transmission of the absorber is

$$T_0 = \exp(-\alpha_0 \ell_s) = \exp(-n_g \sigma_{gs} \ell_s), \tag{8.44}$$

where ℓ_s is the thickness of the bleachable crystal and n_g is the ground state density. In order to calculate the transmission as a function of fluence, the photon flux and population density must be considered as a function of position within the absorbing medium.

Identical to the situation which occurs in pulse amplifiers (Section 4.1), differential equations for the population density and photon flux have to be solved. The solution is the Frantz–Nodvik equation, which is identical to (4.11) except that gain G, G_0 is replaced by transmission T_i, T_0. Therefore the energy transmission T_i of an ideal saturable absorber as a function of input fluence E_i is given by

$$T_i = \frac{E_s}{E_i} \ln[1 + (e^{E_i/E_s} - 1)T_0] \tag{8.45}$$

Equation (8.45) reduces to $T_i = T_0$ for $E_i < E_s$ and $T_i = 1$ for $E_i > E_s$.

In practical saturable absorbers, the transmission never reaches 100%. The reason for that is photon absorption by the excited atoms. A passive Q-switch requires a material which exhibits saturation of the ground state absorption. However, most materials also exhibit absorption from an excited state. This is illustrated in Fig. 8.16 by the transition from the excited state (level 2) to some higher level 4 which has an energy level corresponding to the laser transition. As the ground state is depleted, absorption takes place increasingly between levels 2 and 4. Excited State Absorption (ESA) results in a residual loss in the resonator when the ground state absorption has been saturated. The 2–4 transition does not saturate because of the fast relaxation of level 4. A saturable absorber is useful for Q-switching only as long as $\sigma_{gs} > \sigma_{es}$, where σ_{es} is the cross section for excited state absorption.

A saturable absorber with ESA can be described by a four-level model [11], [12]. In this case, maximum transmission T_{max} is given by

$$T_{max} = \exp(-n_g \sigma_{es} \ell_s). \tag{8.46}$$

In [12] an approximation in closed form that gives the shape of the transmission versus fluence curve in the presence of ESA, has been derived. For a nonideal absorber the transmission T_n can be approximated by

$$T_n = T_0 + \frac{T_i - T_0}{1 - T_0}(T_{max} - T_0), \tag{8.47}$$

where T_i is the transmission of an ideal absorber given by (8.45) and T_0 and T_{max} are the lower and upper limits of the transmission.

In Fig. 8.15, T_n is plotted for the case of $T_0 = 60\%$ and $T_{max} = 90\%$. For a given saturable absorber one needs to know the cross sections σ_{gs} and σ_{es}, the ground state concentration n_0, and the thickness l_s of the material. With these quantities known one can calculate T_0, T_{max}, and E_s and plot a transmission versus energy density curve as determined by (8.47).

Because of the importance of Cr^{4+} : YAG as a passive Q-switch material, we will briefly review the properties of this material. In order to produce Cr^{4+} : YAG, a small fraction of chromium ions in YAG are induced to change valence from the normal Cr^{3+} to Cr^{4+} with the addition of charge compensating impurities such as Mg^{2+} or Ca^{2+}. The crystal Cr^{4+} : YAG has broad absorption bands centered at 410, 480, 640, and 1050 nm. Published values for the cross section of the ground state vary greatly [12], [13]. The most recent measurements indicate $\sigma_{gs} = 7 \times 10^{-18}$ cm^2 and $\sigma_{es} = 2 \times 10^{-18}$ cm^2 for the excited-state absorption at the Nd : YAG wavelength [12]. The excited state lifetime (level 2 in Fig. 8.16) is 4.1 μs and the lifetime of the higher excited state (level 4) is 0.5 ns. With $h\nu = 1.87 \times 10^{-19}$ J at 1.06 μm and the above value for σ_{gs} one obtains a saturation fluence of $E_s = 27$ mJ/cm^2 for Cr^{4+} : YAG.

Commercially available Cr^{4+} : YAG passive Q-switches are specified by the low power transmission at the laser wavelength. Typical transmission values range from 30 to 50%, and the crystal thickness is usually between 1–5 mm. Values of the small signal absorption coefficient α_0 vary from 3–6 cm^{-1}. For example, for $\alpha_0 = 4$ cm^{-1} and $\ell_s = 2$ mm the low power transmission is $T_0 = 45\%$.

For a given pump power, that is, gain in the laser medium, there is an optimal choice of output coupler reflectivity and unsaturated absorber transmission. Design procedures which permit optimization of passively Q-switched lasers have been reported in [14], [15].

Summary

When the lifetime of the upper laser level is much longer than the desired pulse length of the laser output pulse, the laser medium can act as an energy storage medium. In this situation the upper laser level is able to integrate the power supplied by the pump source. The stored energy can be released in a short output pulse using the method of Q-switching. This chapter describes the theory and techniques of Q-switching solid-state lasers. Q-switching is a method to generate pulses in the nanosecond regime from a solid-state laser.

In analogy to electrical terminology the losses in an optical resonator can be characterized by a Q-factor that is defined as the ratio of energy stored in the resonator to the energy loss per cycle. The higher the quality factor Q, the lower the losses. In the Q-switch mode the loss of the optical resonator is made so large that a photon flux cannot build up and stimulated emission cannot occur. During this time the Q of the resonator is kept low, although the energy stored in the laser material is high. After there is considerable energy stored in the laser material, the loss in the resonator is suddenly removed, that is, the Q-factor is increased. This results in a sudden increase of photon flux and the energy stored in the gain medium is converted to optical radiation and a short powerful pulse is emitted.

A number of different methods have been devised to introduce a controllable dynamic loss into the resonator. A simple technique of Q-switching consists of the rotation of one of the resonator mirrors by a high-speed motor. In an acoustic-optic Q-switch a fraction of the optical beam is diffracted out of the resonator as a result of the modulation of the refractive index by an acoustic wave launched into a transparent solid. In a passive Q-switch the absorption of laser radiation is dependent on the intensity. The active element, either a liquid or a crystal, becomes more transparent as the fluence increases, and at very high fluence levels the material saturates and becomes transparent. In an electro-optic Q-switch, a high voltage applied to a suitable crystal rotates the plane of polarization. In conjunction with a polarizer in the resonator the device represents an electro-optical shutter.

References

[1] W.G. Wagner and B.A. Lengyel: *J. Appl. Phys.* **34**, 2040 (1963).

[2] J.J. Degnan: *IEEE J. Quantum Electron.* **QE-25**, 214 (1989).

[3] G.D. Baldwin: *IEEE J. Quantum Electron.* **QE-7**, 220 (1971).

[4] R.B. Chesler, M.A. Karr, and J.E. Geusic: *Proc. IEEE* **58**, 1899 (1970).

[5] M. Born and E. Wolf: *Principles of Optics,* 2nd ed. (Macmillan, New York), 1964.

[6] C.F. Quate, C.D. Wilkinson, and D.K. Winslow: *Proc. IEEE* **53**, 1604 (1965).

[7] N. Uchida and N. Niizeki: *Proc. IEEE* **61**, 1073 (1973).

[8] R. Adler: *IEEE Spectrum* **4**, 42 (1967).

[9] R.W. Dixon: *IEEE J. Quantum Electron.* **QE-3**, 85 (1967).

[10] R.W. Dixon: *J. Appl. Phys.* **38**, 5149 (1967).

[11] M. Hercher: *Appl. Opt* **6**, 947 (1967).

[12] Z. Burshtein, P. Blau, Y. Kalisky, Y. Shimony, and M.R. Kokta: *IEEE J. Quantum Electron.* **QE-34**, 292 (1998).

[13] Y. Shimony, Z. Burshtein, and Y. Kalisky: *IEEE J. Quantum Electron.* **QE-31**, 1738 (1995).

[14] G. Xiao and M. Bass: *IEEE J. Quantum Electron.* **QE-33**, 41 (1997) and **QE-34**, 1142 (1998).

[15] J.J. Degnan: *IEEE J. Quantum Electron.* **QE-31**, 1890 (1995).

Exercises

1. Explain why it is unwise to place the Q-switch in a Q-switched laser between the gain medium and the output mirror of the resonator.

2. An early mechanical Q-switch was a flat mirror rotated by a high-speed motor spinning with its axis perpendicular to the resonator axis. This is still used

today for infrared lasers where there are no acceptable electro-optic materials or polarizers to allow one to use electro-optic Q-switching. If the motor speed was 24,000 rpm and the necessary misalignment for hold-off were 1 mrad, how long would the rotating mirror require to switch from hold-off to lasing alignment? What would be the effects of using a 45–90–45° prism rotating so that its roof top or 90° ridge were parallel to the axis of the motor rotation? Would it be better to orient the roof top ridge to be perpendicular to the rotation axis of the prism? Explain your answers. Draw sketches to help both yourself and the reader of your solution to understand your reasoning.

3. Consider an ideal (that means lossless and very fast) Q-switch in a laser resonator where all the components have been designed or selected to have properties that are independent of temperature. On the other hand, there is nothing you can do about the temperature dependence of the stimulated emission cross section of the lasing medium, σ. Assume that you always pump this laser to the exact same initial inversion and find the temperature dependence of the output energy of the laser. If σ decreases with increasing temperature, as it does in Nd : YAG, what would you expect the output energy to do as temperature increases? What problems might this cause?

9

Mode-Locking

9.1 Pulse Formation

9.2 Passive Mode-Locking

9.3 Active Mode-Locking

9.4 Picosecond Lasers

9.5 Femtosecond Lasers

    References

    Exercises

As we have seen in the previous chapter, the minimum pulse width obtainable from a Q-switched laser is on the order of 10 ns because of the required pulse buildup time. With the cavity dumping technique, the pulse width can be reduced to a minimum of 1 to 2 ns. The limitation here is the length of the cavity, which determines the pulse length. Ultrashort pulses with pulse widths in the picosecond or femtosecond regime are obtained from solid-state lasers by mode-locking. Employing this technique, which phase-locks the longitudinal modes of the laser, the pulse width is inversely related to the bandwidth of the laser emission.

The output from laser oscillators is subject to strong fluctuations that originate from the interference of longitudinal resonator modes with random phase relations. These random fluctuations can be transformed into a powerful well-defined single pulse circulating in the laser resonator by the introduction of a suitable nonlinearity, or by an externally driven optical modulator. In the first case, the laser is referred to as passively mode-locked because the radiation itself, in combination with the passive nonlinear element, generates a periodic modulation that leads to a fixed phase relationship of the axial modes. In the second case, we speak of active mode-locking because an rf signal applied to a modulator provides a phase or frequency modulation, which leads to mode-locking.

9.1 Pulse Formation

In a free-running laser, both longitudinal and transverse modes oscillate simultaneously without fixed mode-to-mode amplitude and phase relationships. The resulting laser output is a time-averaged statistical mean value. Restricting oscillation of the laser to the TEM_{00} mode for the moment, we recall from Chapter 5 that in a typical laser cavity there are perhaps a few hundred axial modes that fall within the frequency region where the gain of the laser medium exceeds the losses of the resonator.

As an example for illustrating the formation of a mode-locked pulse, we consider a passively mode-locked laser containing as a nonlinear element a bleachable dye absorber. As mentioned in Chapter 8, a saturable absorber has a decreasing loss for increasing pulse intensities. In order for the passive mode-locking process to start spontaneously from the mode-beating fluctuations of a free-running laser, the nonlinear element must create an amplitude instability so that an intensive fluctuation experiences lower losses compared to less intensive parts of the radiation. A further requirement is that the reaction time of the nonlinear element be as short as the fluctuation itself in order to lock all the modes oscillating in the resonator.

A computer simulation of the evolution of a mode-locked pulse train from noise is shown in Fig. 9.1. This figure shows the transformation of irregular pulses into a single mode-locked pulse. The simultaneous Q-switched and mode-locked pulse evolution from noise can be divided into several stages.

In Figs. 9.1(a)–(c) the noiselike fluctuations are linearly amplified; however, a smoothening and broadening of the pulse structure can be seen. The chaotic sequence of fluctuations shown in Fig. 9.1(a) represents the laser radiation in the early stage of pulse generation. As the pump process continues, the gain increases above threshold and the noiselike signal is amplified. During the linear amplification a natural mode selection takes place because the frequency-dependent gain favors cavity modes in the center of the fluorescence line. As a result of the spectral narrowing caused by the amplification process, a smoothening and broadening of the amplitude fluctuations occurs, as shown in Figs. 9.1(b) and (c).

The following numbers are typical for the linear amplification process in Nd : glass. A large number of longitudinal modes is initially excited. For a typical cavity length of ~ 1 m, one calculates $\sim 4 \times 10^4$ cavity modes in Nd : glass with $\Delta \nu \approx 7500$ GHz. Assuming a typical pulse build-up time of $10\,\mu s$, the linear stage comprises ≈ 1500 round trips. The light intensity rises by many orders of magnitude to approximately 10^7 W/cm^2.

In Figs. 9.1(d)–(e) the peak-to-peak excursions of the fluctuations have increased and, in particular, the amplitude of the strongest pulse has been selectively emphasized. In this second phase of pulse evolution, the gain is still linear but the absorption of the dye cell becomes nonlinear because the intensity peaks in the laser cavity approach values of the saturation intensity I_s of the dye. In the nonlinear regime of the mode-locked laser we note two significant processes acting together.

First, there is a selection of one peak fluctuation or at least a small number. The most intense fluctuations at the end of the linear amplification phase preferentially bleach the dye and grow quickly in intensity. The large number of smaller fluctuations, on the contrary, encounter larger absorption in the dye cell and are effectively suppressed.

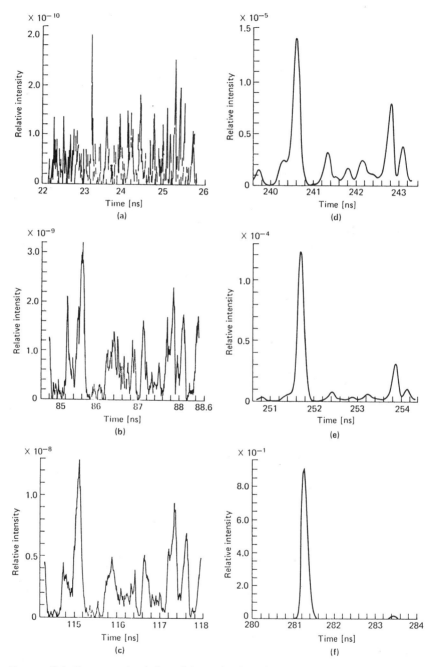

FIGURE 9.1. Computer simulation of the evaluation of a mode-locked pulse from noise: (a–c) regime of linear amplification and linear dye absorption, (d–e) nonlinear absorption in the dye cell, and (f) regime of nonlinear amplification, dye completely bleached [1].

The second effect is a narrowing of the existing pulses in time, which broadens the frequency spectrum. The shapes of the pulses are affected by the nonlinearity of the dye because the wings of the pulse are more strongly absorbed than the peak. The second phase ends when the absorbing transition in the dye cell is completely saturated. Under favorable conditions the final transmission is close to one; that is, the dye is transparent.

In Fig. 9.1(f) the background pulses have been completely suppressed. The final phase of the pulse evolution occurs when the intensity is sufficiently high for complete saturation of the absorber transition to take place and for the amplification to be nonlinear. This is the regime of high-peak power. During the nonlinear stage the pulse intensity quickly rises within ≈ 50 cavity round trips to a value of several gigawatts per square centimeter. Successive passages of the high-intensity radiation pulse through the resonator result in a pulse train appearing at the laser output. Finally, the population inversion is depleted and the pulse decays.

In Fig. 9.2 the spectral and temporal structure of the radiation inside a laser cavity are shown for a non-mode-locked laser. In the frequency domain, the radiation consists of a large number of discrete spectral lines spaced by the axial mode

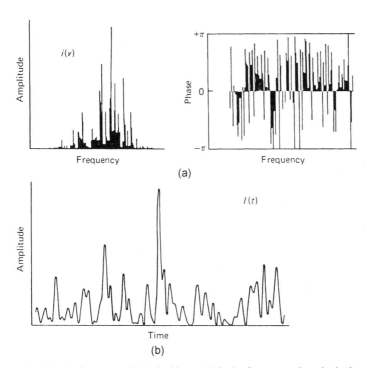

FIGURE 9.2. Signal of a non-mode-locked laser. (a) In the frequency domain the intensities $i(v)$ of the modes have a Rayleigh distribution about the Gaussian mean and the phases are randomly distributed. (b) In the time domain the intensity has the characteristic of thermal noise.

interval $c/2L$. Each mode oscillates independent of the others, and the phases are randomly distributed in the range $-\pi$ to $+\pi$. In the time domain, the field consists of an intensity distribution which has the characteristic of thermal noise.

If the oscillating modes are forced to maintain a fixed phase relationship to each other, the output as a function of time will vary in a well-defined manner. The laser is then said to be "mode-locked" or "phase-locked." Figure 9.3 shows the output signal of an ideally mode-locked laser. The spectral intensities have a Gaussian distribution, while the spectral phases are identically zero. In the time domain the signal is a single Gaussian pulse. As can be seen from this figure, mode-locking corresponds to correlating the spectral amplitudes and phases. When all the initial randomness has been removed, the correlation of the modes is complete and the radiation is localized in space in the form of a single pulse.

Because the radiation with intensity profiles $I(t)$ shown in Figs. 9.2(b) and 9.3(b) circulates around inside the cavity with a repetition rate determined by the round-trip transit time, these signals will repeat themselves and appear in the laser output at a rate of $c/2L$. Therefore, mode-locking results in a train of pulses whose repetition period is twice the cavity transit time, i.e.,

$$PRF = c/2L, \tag{9.1}$$

where PRF is the pulse repetition frequency and L is the resonator length.

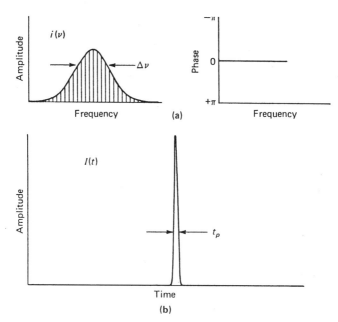

FIGURE 9.3. Signal structure of an ideally mode-locked laser. (a) The spectral intensities have a Gaussian distribution, while the spectral phases are identically zero. (b) In the time domain the signal is a transform-limited Gaussian pulse.

If we make some simplifying assumptions, we can obtain a general idea about the pulse width and peak power of the mode-locked output pulses. The structure of an optical pulse is completely defined by a phase and an intensity. Whether these refer to time or to frequency is immaterial since, if the description in one domain is complete, the profiles in the other are obtained from the Fourier transform. However, there is no one-to-one correspondence between the two intensity profiles $I(t)$ and $i(v)$ since each depends not only on the other but also on the associated phase function. The only general relationship between the two is

$$\Delta v t_p \geq \psi, \tag{9.2}$$

where t_p and Δv are the full width half-maximum of $I(t)$ and $i(v)$, respectively, and ψ is a constant of the order of unity. In particular, the shortest pulse obtainable for a given spectral bandwidth is said to be transform-limited; its duration is

$$t_p = \frac{\psi}{\Delta v}. \tag{9.3}$$

For mode-locking purposes Δv corresponds to the gain bandwidth Δv_L of the laser. The number of axial modes that are contained within the oscillating bandwidth is $N = \Delta v_L t_r$, where $t_r = 2L/c$ is the round-trip time in the optical resonator. The width of the individual mode-locked pulses is therefore

$$t_p \approx \frac{1}{\Delta v_L} \approx \frac{t_r}{N}. \tag{9.4}$$

Equation (9.4) expresses the well-known result from Fourier's theorem that the narrower the pulse width t_p, the larger the bandwidth required to generate the pulse. From (9.4) also follows the interesting fact that the pulse width of the mode-locked pulses roughly equals the cavity round-trip time divided by the number of phase-locked modes.

For simplicity, the spectral intensity in Fig. 9.3(a) was chosen to be a Gaussian function and hence the temporal profile has also a Gaussian distribution. The most general Gaussian optical pulse is given by

$$E(t) = \left(\frac{E_0}{2}\right) \exp(-\alpha t^2) \exp[j(\omega_p t + \beta t^2)]. \tag{9.5}$$

The term α determines the Gaussian envelope of the pulse, and the term $j\beta t$ is a linear frequency shift during the pulse (chirp). From (9.5) follows, for the pulse width at the half-intensity points,

$$t_p = \left(\frac{2\ln 2}{\alpha}\right)^{1/2}, \tag{9.6}$$

and a bandwidth again taken at the half-power points of the pulse spectrum [2]

$$\Delta \nu_p = \frac{1}{\pi} \left[2 \ln 2 \left(\frac{\alpha^2 + \beta^2}{\alpha} \right) \right]^{1/2}. \tag{9.7}$$

Note how the frequency chirp contributes to the total bandwidth. The pulsewidth–bandwidth product is a parameter often used to characterize pulses. For Gaussian pulses, the pulsewidth-bandwidth product is given by

$$t_p \Delta \nu_p = \left(\frac{2 \ln 2}{\pi} \right) \left[1 + \left(\frac{\beta}{\alpha} \right)^2 \right]^{1/2}. \tag{9.8}$$

For the important special case $\beta = 0$ (i.e., no frequency chirp), one obtains $\psi = t_p \Delta \nu_p \approx 0.441$.

Before leaving this subject, we have to mention one other important pulse shape, namely the hyperbolic secant function. The steady-state solution of the differential equation describing the pulse envelope $I(t)$ of a cw mode-locked pulse is a function of the form [3]:

$$I(t) = I_O \text{sech}^2(t/t_p). \tag{9.9}$$

The FWHM pulsewidth of a sech^2 pulse is t_p' where $t_p' = 1.76 t_p$, and a transform-limited sech^2 pulse has a pulsewidth-bandwidth product of

$$t_p' \Delta \nu = 0.315. \tag{9.10}$$

Mode-locked lasers can have high peak powers because the power contained in the entire output of the uncoupled laser is now contained within the more intense ultrafast pulses. From (9.4) follows, for the ratio of pulse-on-to-pulse-off time, that is, the duty cycle, a value of $t_p/t_R = 1/N$, so that the peak power of the pulse is N times the average power resulting from incoherent phasing of the axial modes

$$P_p = N P_{av}. \tag{9.11}$$

Typical output-pulse formats from pulsed and cw mode-locked lasers are illustrated in Fig. 9.4. The output of a flashlamp-pumped, mode-locked, solid-state laser consists of a burst of pulses with amplitudes that fit underneath the envelope of a Q-switched pulse. A cw mode-locked laser produces a train of pulses with equal amplitude. In both cases are consecutive pulses separated by the cavity round-trip time.

A modern diode end-pumped Nd : YAG laser, cw mode-locked with an acousto-optic modulator, produces pulses of about 20 ps in duration and has an average output power of about 200 mW. If we assume a resonator length of $L = 1.2$ m we obtain, from (9.1), a PRF of 125 MHz or a pulse separation of 8 ns. The pulsewidth of 20 ps in combination with (9.4) indicates that about 400 longitudinal modes are phase-locked together ($N = 400$). From (9.11) follows a peak power of 80 W.

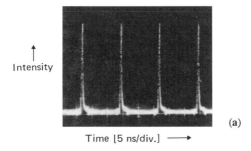

(a)

Time [5 ns/div.] ⟶

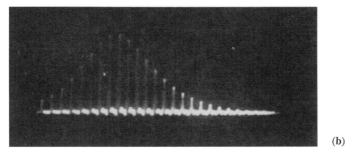

(b)

FIGURE 9.4. Output pulses from (a) a cw mode-locked and (b) a pulsed mode-locked laser.

The homogeneously broadened line in Nd : YAG is $\Delta v = 120$ GHz. Therefore, the ultimate limit in short-pulse generation in Nd : YAG is, according to (9.10), about two picoseconds. Passively cw mode-locked Nd : YAG have approached this limit. The modulation index one can achieve in active mode-locking is not large enough to support phase-locking over a very large bandwidth. Therefore, actively mode-locked lasers emit longer pulses compared to cw passively mode-locked systems.

The generation of mode-locked pulses from a laser requires that the longitudinal modes be coupled together. This is achieved experimentally by placing inside the laser cavity either an externally driven loss or phase modulator, or a passive device which exhibits saturable absorption. Details of the generation of mode-locked pulses with active and passive devices will be discussed in the following sections. Tutorial discussions of the work on mode-locked lasers with extensive references can be found in [4], [5].

9.2 Passive Mode-Locking

Originally the nonlinear absorption of saturable absorbers was employed for simultaneously Q-switching and mode-locking solid-state lasers. The saturable absorbers consisted of organic dyes that absorb at the laser wavelength. At sufficient intense laser radiation, the ground state of the dye becomes depleted, which decreases the losses in the resonator for increasing pulse intensity.

In pulsed mode-locked solid-state lasers, pulse shortening down to the limit set by the gain-bandwidth is prevented because of the early saturation of the absorber, which is a result of the simultaneously occurring Q-switching process. Shorter pulses and a much more reproducible performance are obtained if the transient behavior due to Q-switching is eliminated. In steady-state or cw mode-locking, components or effects are utilized, which exhibit a saturable absorber-like behavior, that is, a loss that decreases as the laser intensity increases. In this section, we will briefly review mode-locking by means of saturable dye absorbers, and then concentrate on cw mode-locking via the Kerr effect.

9.2.1   Liquid Dye Saturable Absorber

The distinction between an organic dye suitable for simultaneous mode-locking and Q-switching, as opposed to only Q-switching the laser, is the recovery time of the absorber. If the relaxation time of the excited-state population of the dye is on the order of the cavity round trip, that is, a few nanoseconds, passive Q-switching will occur, as described in Chapter 8. With a dye having a recovery time comparable to the duration of mode-locked pulses, that is, a few picoseconds, simultaneous mode-locking and Q-switching can be achieved. A typical example of a mode-locked laser utilizing a saturable absorber is illustrated in Fig. 9.5.

One major requirement in the resonator design of a mode-locked system is the complete elimination of reflections that can occur from components located between the two cavity mirrors. This is accomplished by employing laser rods with Brewster's angle at the ends, placing the dye cell at Brewster's angle in the resonator, and by using cavity mirrors which are wedged. Reflection from an optical surface that is parallel to the cavity mirrors will create a secondary resonator. The mode-locked pulse will be split into several pulses that will circulate inside the resonators with different round-trip times. The result is a very erratic output usually consisting of several superpositioned pulse trains or containing subsidiary pulses in the train. With all optical surfaces inside the resonator, either at Brewster angle or antireflection-coated and tilted away from the resonator axis, coupled resonator structures can be avoided and the occurrence of satellite pulses is mini-

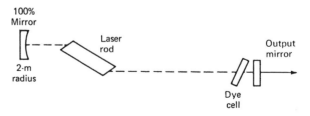

FIGURE 9.5. Mode-locking with a liquid dye saturable absorber.

mized. Similar attention must be paid to avoid back-reflection into the cavity from external components.

Despite the relatively simple construction of a passively mode-locked laser oscillator, the output will be very unpredictable unless dye concentration, optical pumping intensity, and resonator alignment are carefully adjusted. Furthermore, mixing and handling the dye solution and maintaining proper dye concentration proved cumbersome. As a result of the inherent shortcomings of pulsed passive mode-locking, this technology has been superseded by cw mode-locking.

9.2.2   Kerr Lens Mode-Locking

In recent years, several passive mode-locking techniques have been developed for solid-state lasers, whereby fast saturable absorberlike action is achieved in solids. Most of these novel optical modulators utilize the nonresonant Kerr effect. The Kerr effect produces intensity-dependent changes of the refractive index. It is generally an undesirable effect because it can lead to self-focusing and filament formation in intense beams, as explained in Section 4.5.

In contrast to the absorption in bleachable dyes, the nonresonant Kerr effect is extremely fast, wavelength-independent, and allows the generation of a continuous train of mode-locked pulses from a cw-pumped laser. For a Gaussian beam, the Kerr effect focuses the radiation toward the center, and essentially an intensity-dependent graded-index lens is formed. The action of a fast saturable absorber can be achieved if an aperture is introduced in the resonator at a position where the mode size decreases for increased intensity.

The transformation of the power-dependent change in the spatial profile of the beam into an amplitude modulation is illustrated in Fig. 9.6. Transmission through the aperture is higher for an intense beam, as compared to a low-power beam. This technique, which provides an extremely simple means for ultrashort pulse generation in tunable lasers, has been termed Kerr lens mode-locking (KLM).

The laser crystal acts as a lens with a focal length that changes with the intracavity intensity. With the assumption of a parabolic index variation and a focal length much longer than the Kerr medium we obtain, to a first approximation,

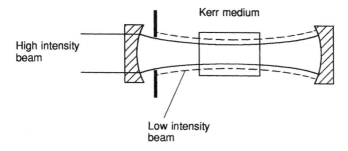

FIGURE 9.6. Intensity-dependent amplitude modulation of the resonator mode.

(see also Section 4.5)

$$f = \frac{w^2}{4\gamma I_0 l} \qquad (9.12)$$

where w is the beam waist, γ is the nonlinear refractive coefficient, I_0 is the peak intensity, and l is the length of the Kerr medium.

Assuming a waist radius w of $50\,\mu$m in a 4 mm long Ti : Al$_2$O$_3$ crystal, and a peak power of 150 kW we obtain, with $\gamma = 3.45 \times 10^{-16}$ cm^2/W, a focal length of $f = 24$ cm.

As was discussed in Section 4.5 beam focusing due to the Kerr effect can lead to catastrophic beam collapse with the beam breaking up into filaments. The critical power level at which beam collapse will set in is usually defined by

$$P_{cr} = \frac{a\lambda_0^2}{8\pi n_0\gamma}, \qquad (9.13)$$

where the factor a can take on values between 3.77 and 6.4 depending on the severity of nonlinear phase distortion. For the materials and wavelengths of interest, one finds $P_{cr} = 2$–3 MW. For example, $P_{cr} = 2.6$ MW for Al$_2$O$_3$ and BK7 glass at 0.8 μm. In KLM lasers, the peak power has to be large enough to produce a strong nonlinearity, but needs to be well below the critical power for beam collapse.

Returning to Fig. 9.6, for a hard aperture of radius r_0, the rate of change in transmission T with respect to the beam waist w for a Gaussian beam is given by

$$dT/dw = (-4r_0^2/w^3)\exp(-2r_0^2/w^2). \qquad (9.14)$$

If one assumes that the aperture size is close to the mode size $r_0 \approx w$, one obtains $\delta T \approx -(1/2w)\,\delta w$. The limiting aperture introduces a loss discrimination for low power (cw) and high peak power (mode-locked) operation which is proportional to mode size changes.

Since the mode size, w is power dependent due to the Kerr effect, the resonator losses are expressed as

$$\delta = \delta_0 - uP, \qquad (9.15)$$

where δ_0 are the fixed losses, P is the intracavity power, and u is the nonlinear loss coefficient. For the design of KLM systems, it is more convenient to use, instead of u, the small-signal relative spot size variation s, where $s \sim -u$. The parameter s, also called the Kerr lens sensitivity is an important factor for analyzing and designing KLM resonators. It is defined as

$$s = \left(\frac{1}{w}\frac{dw}{d(P/P_{cr})}\right), \qquad (9.16)$$

where w is the spot size at a given place inside the resonator. The second term is the slope of the spot size versus normalized power, taken at $P = 0$.

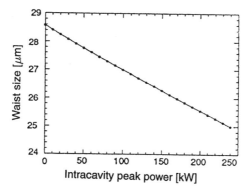

FIGURE 9.7. Beam-waist radius as a function of peak power [6].

In Fig. 9.7 the beam waist at the center of a Ti : Al$_2$O$_3$ crystal is plotted. The calculations were performed for a 100 fs pulsewidth, 3 W intracavity average power, and a 80 MHz pulse repetition rate. The beam radius decreases with a slope of approximately -1.5×10^{-5} μm/W. Introducing this value into (9.16), and normalizing to the spot size w and critical power P_{cr}, gives $s = -1.36$.

A typical resonator, used for KLM of a Ti : Al$_2$O$_3$ crystal pumped by an argon laser, consists of two plane mirrors and two focusing mirrors. The crystal is at the beam waist as shown in Fig. 9.8. An aperture is usually located close to one of the flat mirrors. As will be explained later, the spot size variation s, at the aperture as a function of the circulating power, critically depends on the position x of the Kerr medium around the focus of the beam, and on the separation z of the two focusing mirrors.

We will first consider the properties of the empty resonator illustrated in Fig. 9.8. The equivalent resonator consists of two flat mirrors with two internal lenses essentially forming a telescope with a magnification of one, as shown in Fig. 9.9. For a slightly misadjusted telescope, the lens spacing is $z = 2f - \Delta z$ where Δz measures the misadjustment. The focal length of the lens assembly is then $f_{ef} = f^2/\Delta z$, where f is the focal length of the individual lenses. The distances from the resonator mirrors and the principal planes of the lenses are L_1 and L_2. The g-parameters for a resonator comprised of two flat mirrors and an

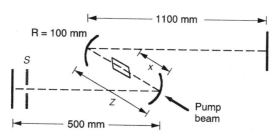

FIGURE 9.8. Typical resonator used for Kerr lens mode-locking (KLM).

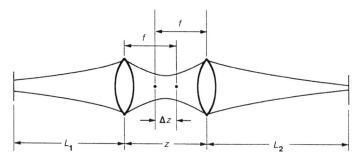

FIGURE 9.9. Equivalent resonator of the arrangement shown in Fig. 9.8 ($f = 50\,\text{mm}$, $L_1 = 500\,\text{mm}$, $L_2 = 1100\,\text{mm}$).

internal telescope are (see also Section 5.1.7)

$$g_1 = 1 - (L_2/f_{\text{ef}}) \qquad \text{and} \qquad g_2 = 1 - (L_1/f_{\text{ef}}) \qquad (9.17)$$

or

$$g_1 = 1 - (L_2\,\Delta z/f^2) \qquad \text{and} \qquad g_2 = 1 - (L_1\,\Delta z/f^2). \qquad (9.18)$$

The operating region of the passive resonator is depicted in the stability diagram (Fig. 9.10). For $\Delta z = 0$, the lenses are in focus and the resonator configuration becomes plane-parallel. As the lenses are moved closer together, a stability limit is reached for $g_2 = 0$. As both g-parameters become negative, a second zone of stable operation is reached. As Δz becomes larger, a point is reached for $g_1 g_2 = 1$, where the lens assembly focuses the beam onto one of the flat mirrors, that is, the spot size becomes zero at the mirror. If the two legs of the resonator are of unequal length, $L_1 \neq L_2$, two stability regions exist, as indicated by curve A in Fig. 9.10. Curve B gives the stability range of a resonator with $L_1 = L_2 = 850\,\text{mm}$. The two stability regions are joined and the resonator changes from plane-parallel, confocal, to concentric for increasing values of Δz (or decreasing mirror separation z).

A particularly useful and practical design procedure for KLM lasers can be found in [7]–[9]. The authors introduced a nonlinear ABCD ray matrix to treat the Gaussian beam propagation in a Kerr medium. The dependence of the spot size w on the power P is given in closed form by a set of equations. From these equations contour plots can be generated of the spot size variation s as a function of x and z for a given spot size and power of the input pump beam. The results of these calculations provide guidelines for the design of KLM resonators.

An example of such a plot is shown in Fig. 9.11. Plotted is the Kerr lens sensitivity s for the resonator characterized by curve B in Fig. 9.10. However, the resonator now contains the laser crystal which increases the optical length and also makes the resonator axis rotational unsymmetrical because of the Brewster angle.

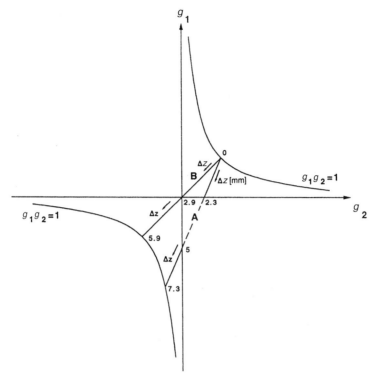

FIGURE 9.10. Stability diagram for the resonator shown in Fig. 9.9.

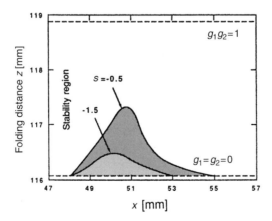

FIGURE 9.11. Contour plot of Kerr sensitivity for a resonator with $R = 100\,\text{mm}$, $L_1 = L_2 = 850\,\text{mm}$, and a 25 mm long Ti : sapphire crystal.

The plots are for the tangential plane of the resonator because δ is larger in this plane as compared to the saggital plane. The curves for δ are bound by the stability limits of the resonator in the first quadrant of the stability diagram, i.e., $g_1 \cdot g_2 = 1$ and $g_1 = g_2 = 0$. Similar curves are obtained for the third quadrant. Experiments performed on an actual system revealed that, for reliable mode-locking, a value of $|\delta| \geq 0.5$ was necessary. This defines the operating regime of the laser system indicated by the shaded areas.

It was found that the best compromise between a large δ and a reasonably stable performance is achieved for a symmetric resonator $L_1 = L_2$ and operation close to $g_1 g_2 = 0$. For this configuration, which is equivalent to a confocal resonator, the tolerance for stable operation is about ± 0.6 mm in the z-direction, with the resonator operated about 0.5–1 mm from the stability limit. From the results illustrated in Fig. 9.11 it is clear that the alignment of a KLM resonator is very critical and the tolerance in length adjustments are fractions of a millimeter.

Specific resonator designs can be analyzed by following the formulas developed in the referenced papers. The design usually starts with the assumption of a pump beam that has a certain spot size and power in the Kerr medium (defined by n_2 and P_{cr}). From this a nonlinear matrix is developed which, combined with the linear matrix of the passive elements of the resonator, describes the behavior of the system. Kerr lens mode-locking is the preferred method for the generation of femtosecond lasers. Pulses on the order of a few tens of femtoseconds have been obtained from solid-state tunable lasers. The ultimate limit of pulse duration for Ti : sapphire is somewhere in the range of ~ 3 fs, or about 1 cycle of light at 800 nm.

9.3   Active Mode-Locking

By placing inside a laser cavity either a phase modulator (FM) or an amplitude modulator (AM), driven at exactly the frequency separation of the axial modes, one can cause the laser to generate a train of mode-locked pulses with a pulse repetition rate of $\nu_m = c/2L$. Active mode-locking, performed on cw-pumped lasers such as the Nd : YAG system, is achieved by inserting into the resonator an electro-optic or acousto-optic modulator.

A cw actively mode-locked laser generates a train of equal pulses at a repetition rate typically in the range of 80–250 MHz with pulse energies in the nanojoule range. Figure 9.12 presents an example of an end-pumped Nd : YAG laser in a folded resonator containing an LiNbO$_3$ phase modulator for active mode-locking of the laser. Although the system architecture looks identical, we have to distinguish between AM and FM modulation.

9.3.1 AM Modulation

From a frequency-domain viewpoint, introducing a time-varying transmission $T(t)$ through an amplitude modulator inside the laser resonator creates sidebands

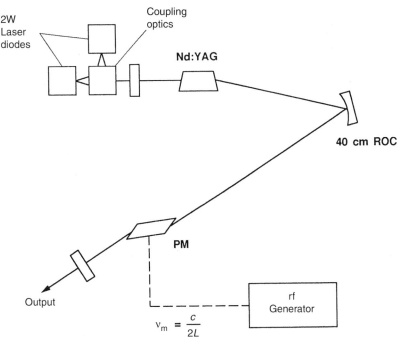

FIGURE 9.12. Schematic of an FM mode-locked Nd : YAG laser [10].

on each oscillating axial mode that overlap with adjoining axial modes. The operation can be described by assuming that the mode with the frequency ν_0, nearest
the peak of the laser gain profile, will begin to oscillate first. If a loss modulator
operating at a frequency ν_m is inserted into the resonator, the carrier frequency
ν_0 will develop sidebands at $\pm\nu_m$. If the modulating frequency is chosen to be
commensurate with the axial mode frequency separation $\nu_m = c/2L$, the coinci-
dence of the upper ($\nu_0 + \nu_m$) and the lower ($\nu_0 - \nu_m$) sidebands with the adjacent
axial mode resonances will couple the $\nu_0 - \nu_m$, ν_0, and $\nu_0 + \nu_m$ modes with a well-
defined amplitude and phase. As the $\nu_0 + \nu_m$ and $\nu_0 - \nu_m$ oscillations pass through
the modulator, they will also become modulated and their sidebands will couple
the $\nu_0 \pm 2\nu_m$ modes to the previous three modes. This process will continue until
all axial modes falling within the laser linewidth are coupled.

Viewed in the time domain, the same intracavity modulating element, with its
modulation period equal to the round-trip transit time $2L/c$, can reshape the internal circulating field distribution repeatedly on each successive round trip inside
the cavity. For example, light incident at the modulator during a certain part of
the modulation cycle will be again incident at the same point of the next cycle
after one round trip in the laser resonator. Light suffering a loss at one time will
again suffer a loss on the next round trip. Thus, all the light in the resonator will
experience loss except that light which passes through the modulator when the
modulator loss is zero (Fig. 9.13(a)). Light will tend to build up in narrow pulses

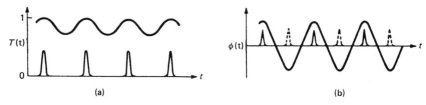

FIGURE 9.13. Mode-coupling behavior in the case of (a) AM mode-locking and (b) FM mode-locking.

in these low-loss time positions. In a general way, we can see that these pulses will have a width given by the reciprocal of the gain bandwidth since wider pulses will experience more loss in the modulator, and narrower pulses will experience less gain because their frequency spectrum will be wider than the gain bandwidth.

By following a single mode-locked pulse through one round trip around the laser cavity a comprehensive analysis of active mode-locking has been developed [11]–[13]. For steady-state mode-locking the pulse shape should be unchanged after a complete round-trip. The self-consistent solution carried out for a Gaussian pulse leads to a simple expression that shows the dependence of the mode-locked pulsewidth on the spectral linewidth, modulation frequency, depth of modulation, and saturated gain.

For an acousto-optic AM operating in the Bragg regime, as well as for electro-optic AMs, the round trip amplitude transmission is

$$T(t) \approx \cos^2(\delta_{AM} \sin \omega_m t), \tag{9.19}$$

where δ_{AM} is the modulation depth and $\omega_m = 2\pi \nu_m$ is the angular frequency of the modulation. In the ideal mode-locking case, the pulse passes through the modulator at the instant of maximum transmission. This occurs twice in every period of the modulation signal ω_m and, hence, one drives these modulators at a modulation frequency equal to half the axial mode-spacing of the laser. Expanding (9.19) at the transmission maximum gives, for the round-trip modulation function,

$$T(t) = \exp(-\delta_{AM}\omega_m^2 t^2), \tag{9.20}$$

which results in a pulse width for the AM mode-locked laser of

$$t_p(AM) = \mu \frac{(gl)^{1/4}}{(\delta_{AM}\nu_m \Delta\nu)^{1/2}}. \tag{9.21}$$

where $\mu = 0.53$ for Bragg deflection and $\mu = 0.45$ for Raman–Nath modulation, g is the saturated gain coefficient at the line center, $\Delta\nu$ is the gain bandwidth of the laser, and l is the length of the active medium. The AM mode-locked pulses have no frequency "chirp"; that is, $\beta = 0$ in (9.5).

The pulsewidth-bandwidth product for AM is

$$t_p(AM) \times \Delta\nu(AM) = 0.440. \tag{9.22}$$

From (9.21) it follows that the pulse duration of cw AM mode-locked lasers is inversely related to the product of the modulation depth and modulation frequency, $t_p(AM) \sim (\delta_{AM}\nu_m)^{-1/2}$. It follows that pulse duration can be shortened by increasing the modulation depth or the frequency of the mode locker. The pulsewidth is also inversely proportional to the gain bandwidth $\Delta\nu$. Therefore Nd : glass and Nd : YLF produce shorter pulses as compared to Nd : YAG.

As is the case with passive mode-locked systems, etalon effects due to intracavity elements will reduce the bandwidth of the system and broaden the mode-locked pulses. In a cw-pumped Nd : YAG laser, even a weak etalon effect due to the modulator or rod surfaces can decrease the effective value of $\Delta\nu$ by a large amount. Therefore, one of the most important considerations in a practical mode-locking system is the elimination of residual reflections and optical interference effects in the laser cavity.

On the other hand, it is possible to use a tilted etalon inside a mode-locked laser to deliberately lengthen the pulse width. With uncoated quartz etalons of thickness between 1 and 10 mm, good control of the pulse width can be achieved.

9.3.2 FM Modulation

Light passing through an electro-optic phase modulator will be up- or down-shifted in frequency unless it passes through at the time when the intracavity phase modulator $\delta(t)$ is stationary at either of its extrema. The recirculating energy passing through the phase modulator at any other time receives a Doppler shift proportional to $d\delta/dt$, and the repeated Doppler shifts on successive passes through the modulator eventually push this energy outside the frequency band over which gain is available from the laser medium. The interaction of the spectrally widened circulating power with the narrow laser linewidth leads to a reduction in gain for most frequency components. Thus, the effect of the phase modulator is similar to the loss modulator, and the previous discussion of loss modulation also applies here. As shown in Fig. 9.13(b), the existence of two phase extrema per period creates a phase uncertainty in the mode-locked pulse position since the pulse can occur at either of two equally probable phases relative to the modulating signal. The quadratic variation of $\delta(t)$ about the pulse arrival time also produces a frequency "chirp" within the short mode-locked pulses.

In the FM case, the internal FM introduces a sinusoidally varying phase perturbation such that the round-trip transmission through the modulator is given approximately by

$$T(t) \approx \exp(\pm j\delta_{FM}\omega_m^2 t^2), \tag{9.23}$$

where δ_{FM} is the peak phase retardation through the modulator. The $\pm$ sign corresponds to the two possible phase positions at which the pulse can pass through the modulator, as mentioned earlier. With these parameters the pulse width of phase mode-locked pulses is given by

$$t_p \text{ (FM)} = 0.54 \left(\frac{gl}{\delta_{FM}} \right)^{1/4} \frac{1}{(\nu_m \, \Delta \nu)^{1/2}}. \tag{9.24}$$

The time-bandwidth product is given by

$$t_p \text{ (FM)} \times \Delta \nu \text{ (FM)} = 0.626. \tag{9.25}$$

In an electro-optic phase modulator the phase retardation is proportional to the modulating voltage, hence $\delta_{FM} \propto P_{in}^{1/2}$, where P_{in} is the drive power into the modulator. Therefore, we obtain from (9.24) the following expression for the pulse width: $t_p(FM) \propto P_{in}^{-1/8}$, which indicates that the pulses shorten very slowly with increased modulator drive. More effective in shortening the pulses is an increase of the modulation frequency. Since $\nu_m = c/2L$, the pulse width will be proportional to the square root of the cavity length.

In order to calculate the pulse width from (9.24), we can calculate the saturated gain coefficient g by equating the loop gain with the loss in the resonator

$$2gl \approx \ln \left(\frac{1}{R} \right), \tag{9.26}$$

where R is the effective reflectivity of the output mirror and includes all losses. For a typical Nd : YAG laser with 10% round-trip loss, that is, $R = 0.9$, a resonator length of 60 cm, and a linewidth of 120 GHz, the pulse length is given by $t_p(FM) = 39(1/\delta_{FM})^{1/4}$. For $\delta_{FM} = 1$ rad, which is easily obtainable, pulses of 39 ps can be generated.

9.4 Picosecond Lasers

Today, diode-pumped cw Nd : YAG or Nd : YLF lasers, mode-locked with an acousto-optic modulator, are the standard sources for the production of picosecond pulses. A typically end-pumped and actively mode-locked neodymium laser is shown in Fig. 9.14. The output from a GaAlAs diode array is focused onto the end of the laser crystal. Spot sizes of the pump beam range from 50 μm to a few hundred micrometers. A common resonator configuration for diode-pumped actively mode-locked lasers is a three-mirror arrangement. A folded cavity geometry provides a beam waist at both the laser crystal and the modulator, as well as astigmatic compensation. In commercial mode-locked lasers, the resonator is usually folded a few times to decrease the overall length of the system. The length of the cavity is usually a compromise between the need for short pulses, which requires a short resonator and a high-modulation frequency, and the ability to slice out a single pulse with a Pockels cell.

Typical of end-pumped lasers is the highly reflective coating applied to the back face of the crystal (Fig. 9.14) to form one of the resonator mirrors. Small thermal or vibrational changes in the pump radiation readily induce relaxation

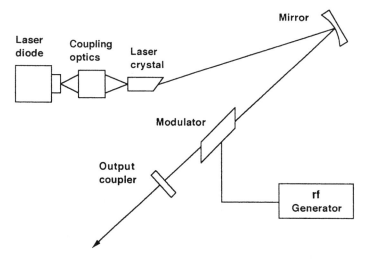

FIGURE 9.14. Typical end-pumped and actively mode-locked neodymium laser [14].

oscillations in mode-locked systems. These relaxation oscillations cause fluctuations in amplitude, accompanied by damped oscillations at frequencies in the 30 to 150 kHz range. Therefore, careful alignment and isolation is required for clean mode-locking. The modulator is usually placed close to the front mirror at the beam waist created by the folding mirror. The resonator length is typically on the order of 1 m or longer to provide a sufficiently large separation between pulses for pulse selection by an external switching device. Actively mode-locked lasers are very sensitive to cavity length detuning. In practice, cavity length changes on the order of 1 μm can cause serious degradation in pulse quality. Therefore, active length stabilization of the resonator by a feedback loop is needed to maintain long-term stability of mode-locking.

Active mode-locking of a laser can be achieved by using a tunable rf oscillator and adjusting the modulator frequency to agree with the cavity length; or, alternatively, selecting a fixed frequency and adjusting the mirror spacing.

9.4.1 AM Mode-Locking

Amplitude modulation for mode-locking cw lasers is usually performed with an acousto-optic modulator. These devices are different from the modulators employed for Q-switching lasers because they are operated at resonance. The modulator material, such as a quartz block, has parallel faces and the sound wave is reflected back and forth in the material. The length of the quartz cell is cut such that the length is equal to an integer number of half-wavelength of the sound wave. In such a standing-wave pattern, the diffraction loss of the optical beam will be modulated at twice the frequency of the sound wave since the diffraction loss reaches a maximum wherever the standing-wave pattern has a maximum. The

TABLE 9.1. Modulation depth and rf power requirements for typical mode lockers.

Frequency [MHz]	Modulation depth (δ_{AM})	Drive power [W]	Corresponding resonator length
38	0.8	5	197.4 cm
120	0.63	1	62.5 cm

transducer of the acousto-optic modulator is therefore driven at an rf frequency of $v_{rf} = c/4L$. Operation of the modulator at resonance requires considerably less rf power compared to a traveling-wave device. For Q-switching of lasers, this approach is not feasible because the high Q of the acoustic resonance prevents fast switching.

Acousto-optic mode-locking requires a precise match of the drive rf with one of the acoustic resonances of the modulator and also with the frequency spacing of the cavity modes. Since an acousto-optic mode locker operates at a fixed frequency, exact synchronism between the modulation frequency and the cavity length is achieved by mounting one of the resonator mirrors on a piezoelectric translator for cavity-length adjustment. A second feedback loop is usually required to maintain synchronism between the rf drive frequency and the resonance frequency of the high-Q modulator. This is achieved by sampling the rf reflected back from the modulator by means of a directional coupler. This reflected rf power provides an error signal for slight frequency adjustments caused by a shift of the modulator's resonance due to thermal effects.

Usually, fused quartz is used as the Bragg deflector in mode lockers because of its excellent optical quality, which ensures a low-insertion loss. A transducer, such as $LiNbO_3$, is bonded to the quartz block and launches an acoustic wave into the substrate. Such mode lockers are driven by an rf generator anywhere from 40 to 120 MHz and at power ranges from 1 to 5 W. For example, two commercially available units at the low and high ends of the frequency range have the performance characteristics, as listed in Table 9.1.

Optimum mode-locking is achieved by adjusting the cavity length such that its resonant frequency matches the frequency of the mode locker. This is usually accomplished by fixing the frequency of the mode locker and tuning the resonator length to match the mode locker's frequency. The cavity length is tuned by translating one mirror mounted on a piezoelectric translator.

Figure 9.15 shows feedback control of a laser mode-locked with an acousto-optic Bragg cell. The mode locker is driven by an rf generator whose output frequency is divided by 2. Manually, the cavity length is adjusted until a region of stable mode-locking is achieved. This is accomplished by adjusting the reference voltage of a differential operational amplifier, which then provides a dc voltage to the piezoelectric transducer. At that point, the feedback loop is closed. The train of mode-locked pulses is detected by a photodiode. Appropriate filtering of the photodiode output provides a sine wave that corresponds to the resonator round-trip

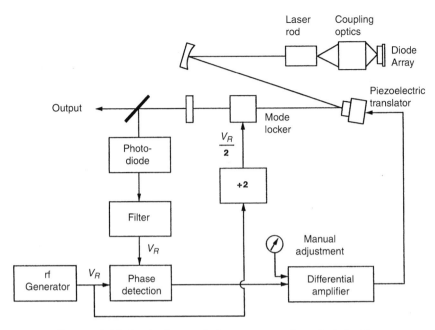

FIGURE 9.15. Feedback control of an acousto-optic mode-locked laser.

repetition frequency. The phase of this waveform is compared with the waveform of the rf generator in a phase detector. A deviation or drift between these two frequencies produces an error signal directed to the differential amplifier that, in turn, provides a dc voltage to the piezoelectric translation stage.

9.4.2   FM Mode-Locking

To achieve FM mode-locking, an intracavity electro-optic phase modulator is driven at the cavity resonance frequency. Electrically, the modulator consists of a pair of electrodes between which the crystal is placed. An inductive loop and a tuning capacitor are shunted across these electrodes and are used to achieve the desired resonant frequency.

Lithium niobate is usually the crystal employed in electro-optic modulators. With the electric field applied transversely to the optical beam, the r_{33} electro-optic coefficient can be utilized, and one obtains

$$\Delta n_z = \frac{-1}{2} r_{33} (n^e)^3 E_z, \tag{9.27}$$

where n^e is the extraordinary index of refraction. The light propagates in the x-direction, and the beam is polarized in the z-direction in the same direction as the applied electric field.

The total phase change in the crystal is $\delta_{FM} = 2\pi \, \Delta n \ell / \lambda_0$, where ℓ is the length of the crystal in the x-direction. If a voltage $V = V_0 \cos \omega_m t$ is applied across the crystal in the z-direction, the peak single-pass phase retardation of the modulator is

$$\delta_{FM} = \frac{\pi r_{33} (n^e)^3 V_0 \ell}{\lambda_0 d}, \tag{9.28}$$

where d is the dimension of the crystal in the z-direction.

For LiNbO$_3$, the materials parameters are $n^e = 2.16$ at $1.06 \, \mu m$ and $r_{33} = 30.8 \times 10^{-10}$ m/V. Typically, a phase retardation of about $\delta_{FM} = 1$ rad is achieved with 300 V across a $5 \times 5 \times 20$ mm crystal. At a frequency of a few hundred megahertz, this requires an rf power of a few watts.

We will illustrate the performance of an FM mode-locked Nd : YLF laser that has an optical design, as shown in Fig. 9.14. The Brewster cut LiNbO$_3$ FM had dimensions of $15 \times 6 \times 2$ mm and was driven by 1 W of rf power at a frequency of 200 MHz. The Nd : YLF rod was pumped by a 3 W cw laser diode. The output from the $500 \times 1 \, \mu m$ diode-emitting aperture was focused to a $300 \, \mu m$ spot at the laser crystal. The mode-locked oscillator generated 14 ps pulses at an output power of 830 mW at an output wavelength of $1.047 \, \mu m$.

In the case of an FM mode-locked laser, a simple mode-locking system can be built by allowing the laser to determine its own drive frequency, as shown in Fig. 9.16. This can be accomplished by using a high-speed photodetector to sense the first beat frequency $c/2L$ of the oscillator. This signal is amplified, phase-shifted a variable amount, and then applied to the intracavity modulator. The adjustable phase shifter compensates for the delay in the feedback loop. When the phase of the electric signal on the modulator and the signal derived from the

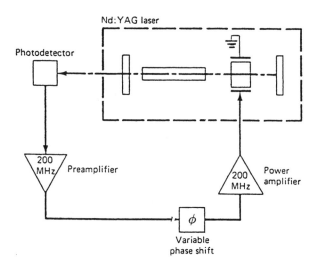

FIGURE 9.16. Direct drive mode-locking system.

laser equal an integer times π, the loop goes into regeneration and the laser is mode-locked. The resultant mode-locking system is a closed-loop oscillator using the laser cavity as the basic reference. Such a system will automatically track changes in cavity length.

9.5 Femtosecond Lasers

The goal of an all solid-state laser source for ultrafast pulse generation has motivated research efforts world-wide. Kerr lens mode-locking (KLM) has revolutionized femtosecond pulse generation in solid-state materials because the KLM technique produces the shortest pulses, and it is relatively simple to implement.

9.5.1 Laser Materials

The laser most actively explored for ultrashort pulse generation is Ti : Al_2O_3. This laser has the largest gain bandwidth and is therefore capable of producing the shortest pulses; it also provides the widest wavelength tunability. Typically, the Ti : Al_2O_3 crystal is pumped by a cw argon laser. The future trend will be to replace the argon laser with a diode-pumped, frequency-doubled Nd : YAG laser in order to create an all-solid-state source.

With regard to providing a compact, reliable femtosecond source, Cr : LiSAF (Cr : $LiSrAlF_6$) is an intriguing material because it can be pumped directly by laser diodes. Compared to Ti : Al_2O_3 systems, Cr : LiSAF does not have as wide a tuning range, but the much longer upper-state lifetime (67 μs) and a broad absorption band centered at 640 nm permit pumping with red AlGaInP diodes. Diode-pumped Cr : LiSAF lasers are currently the most compact femtosecond pulse sources.

A third material which has a broad gain bandwidth is Cr : Forsterite. This laser can be pumped with the output from a diode-pumped Nd : YAG laser to produce femtosecond pulses at around 1.3 μm. Table 9.2 lists key parameters of these three important materials for femtosecond pulse generation in solid-state lasers.

TABLE 9.2. Spectroscopic properties of broadband solid-state laser materials.

	Ti : sapphire	Cr: LiSAF	Cr: Forsterite
Peak emission cross section			
$[\times 10^{-20} \, cm^2]$	30	4.8	14.4
Emission peak [nm]	790	850	1240
Gain bandwidth [nm]	230	180	170
Upper state lifetime [μs]	3.2	67	2.7
Nonlinear coefficient γ			
$[\times 10^{-16} \, cm^2/W]$	3.2	1.5	2.0

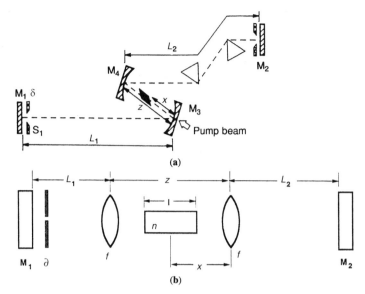

FIGURE 9.17. (a) Resonator for Kerr lens mode locking and (b) equivalent resonator.

9.5.2   Resonator Design

A resonator commonly employed for KLM is an astigmatically compensated arrangement consisting of two focusing mirrors and two flat mirrors. In order to obtain a high nonlinearity, the Kerr medium is inserted into the tightly focused section of the resonator, as shown in Fig. 9.17(a). Taking into account the astigmatism of the Brewster cut crystal and the tilted mirrors, the resonator shown in Fig. 9.17(a) has to be evaluated in the tangential and sagittal plane. The curved mirrors correspond to lenses with focal lengths f, which are different for the two planes, the same is true for the equivalent lengths l of the laser crystal. Figure 9.17(b) depicts the equivalent resonator. The function of the two prisms is to provide dispersion compensation in the resonator, as will be explained in the next subsection.

As was explained in Section 9.2.2, the operation of a KLM laser is a trade-off between output power, stability, and tolerance to the exact position of the components. An analytical treatment of nonlinear resonators has shown that for a given pump power and pump spot size, the most critical parameters are:

(a) the distance z of the two focusing mirrors;
(b) the location x of the Kerr medium with respect to the mirrors, and
(c) the spot size variation s at the aperture.

As we have seen in Section 9.2.2, the Kerr lens sensitivity s is highest near the limit of the stability range of the resonator, therefore loss-modulation efficiency has to be traded off against stable laser performance.

Because of the Brewster angle design, these resonators behave differently in the tangential and sagittal planes. This is illustrated in Fig. 9.18 which shows the change in beam size in two orthogonal directions at the flat mirror for a resonator configuration as depicted in Fig. 9.17. Since the spot size changes essentially in only one direction, a slit rather than a round aperture is usually employed for the adjustment of the proper loss modulation.

Dispersion Compensation

For ultrashort pulse generation, the round-trip time t_r in the resonator for all frequency components of the mode-locked pulse must be frequency independent, i.e., $t_r(v) = d\varphi/dv = $ constant, where φ is the phase change after one round trip. Otherwise, frequency components that experience a cumulative phase shift no longer add constructively and are attenuated. This limits the bandwidth of the pulse and leads to pulsewidth broadening. The frequency-dependent phase shift of the pulse during one round trip can be expressed in a Taylor series about the center frequency v_0,

$$\frac{d\varphi}{dv} = \varphi'(v_0) + \varphi''(v_0)\,\Delta v + \frac{1}{2}\varphi'''(v_0)\,\Delta v^2, \tag{9.29}$$

where φ', φ'', φ''' are the derivatives of the phase with respect to frequency. When φ'' is nonzero, the pulse will have a linear frequency chirp, while a nonzero third-order dispersion will induce a quadratic chirp on the pulse.

Generally speaking, in the design of the femtosecond lasers, dispersion effects have to be minimized. The two major sources of dispersion in a mode-locked laser are self-phase modulation that is part of the Kerr effect and normal dispersion in the laser crystal or any other optical component in the resonator.

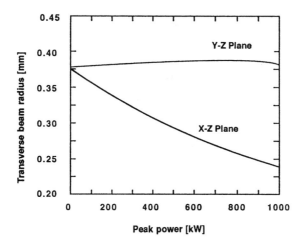

FIGURE 9.18. Spot sizes as a function of intracavity peak power for two orthogonal directions [15].

Self-Phase Modulation. Besides modifying the spatial profile of the beam leading to a self-induced quadratic index gradient, the Kerr effect also causes a phase shift among the frequency components as the pulse propagates through the crystal. The time varying phase shift $\varphi(t)$ or phase modulation produced by the pulse itself can be expressed by

$$\Delta\varphi(t) = (2\pi/\lambda)n_2 I(t)l, \tag{9.30}$$

where $I(t)$ is the intensity and l is the length of the Kerr medium. Since the Kerr effect always leads to an increase of n_0 for increasing intensity, the rising edge of the pulse experiences a medium which is getting optically denser, i.e., $dn_0/dt > 0$. The increasing index of refraction at the rising edge of the pulse delays the individual oscillations of the electric field, which is equivalent of red-shifting of the leading edge. The opposite occurs at the trailing edge of the pulse, the trailing edge is blue-shifted. Self-phase modulation will thus cause a frequency chirp of the pulse and prevent the pulsewidth from becoming transform-limited.

Normal Dispersions. The change of group velocity with frequency is usually expressed by

$$\frac{dv_g}{dv} = -v_g^2 \beta'', \tag{9.31}$$

where β'' is the group dispersion of the medium. Materials in the visible region of the spectrum have positive or normal dispersion, i.e. $\beta'' > 0$. Therefore in a laser crystal v_g decreases with increasing frequency, that is, longer wavelengths travel faster than short ones, causing a red-shift of the pulse.

Negative Dispersion. From the foregoing discussion it follows that a high-intensity mode-locked pulse is red-shifted due to self-phase modulation and normal dispersion. Positive self-phase modulation and positive group-velocity dispersion in the Kerr medium can be compensated for by a dispersive delay line based on a prism pair that intentionally introduces negative dispersion into the resonator. Although the glasses of the prisms have normal dispersion, the geometry of the ray path can be arranged such that the blue components of the pulse traverses the two prisms in a shorter time than do the red components. Although a number of prism arrangements can be devised, usually two prisms are used at minimum deviation and Brewster angle incidence at each surface. In Fig. 9.19 the entrance face of prism II is parallel to the exit face of prism I, and the exit face of prism II is parallel to the entrance of prism I. The prisms are cut so that the angle of minimum deviation is also the Brewster angle. The plane MM' normal to the rays is a plane of symmetry.

Referring to Fig. 9.19 it can be shown that the optical path that contributes to dispersion is $l = 2L\cos\beta$. Group-velocity dispersion is the second derivative of the pathlength with respect to wavelength. From [16] we obtain

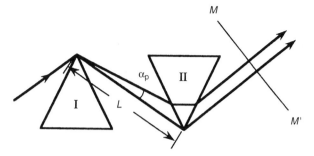

FIGURE 9.19. Dispersive delay line employing a prism pair.

$$\frac{d^2l}{d\lambda^2} = 4L \left\{ \left[\frac{d^2n_0}{d\lambda^2} + \left(2n_0 - \frac{1}{n_0^3} \right) \left(\frac{dn_0}{d\lambda} \right)^2 \right] \sin\alpha_p - 2 \left(\frac{dn_0}{d\lambda} \right)^2 \cos\alpha_p \right\}.$$

(9.32)

The second part of (9.32) is responsible for negative dispersion; therefore the first part has to be made as small as possible. The term $L\sin\alpha_p$ expresses the distance of the beam from the apex of the first prism. This term is minimized by placing the beam as close to the apex as possible. In actual systems, the incident beam is adjusted to pass at least one beam diameter inside the apex of the first prism. We introduce $L\sin\beta = 4w$, where w is the beam radius, and with $\cos\beta \approx 1$ and $2n_0 \gg 1/n_0^3$ we obtain

$$\frac{d^2l}{d\lambda^2} = 16w \left[\frac{d^2n_0}{d\lambda^2} + 2n \left(\frac{dn_0}{d\lambda} \right)^2 \right] - 8L \left(\frac{dn_0}{d\lambda} \right)^2.$$

(9.33)

For sufficiently large prism separation L, the right-hand side of (9.33) can be made negative, as illustrated by the following example. We assume a Ti : sapphire laser, with two SF10 glass prisms for dispersion compensation. With $n_0 = 1.711$, $dn_0/d\lambda = -0.0496\,\mu\mathrm{m}^{-1}$, $d^2n_0/d\lambda^2 = 0.1755\,\mu\mathrm{m}^{-2}$ at 800 nm, and a beam radius of $w = 1$ mm,

$$\frac{d^2l}{d\lambda^2} = 0.294 - 0.0197L \;(\mathrm{cm}) \quad [1/\mathrm{cm}].$$

(9.34)

Therefore, for a prism separation larger than about 15 cm, negative group-velocity dispersion is obtained.

Self-Starting of KLM. The Kerr nonlinearity is usually not strong enough for the cw mode-locking process to self-start. In order to initiate KLM, usually a strong fluctuation must be induced by either perturbing the cavity or by adding another nonlinearity to the system. The simplest method to start KLM in a laboratory set-up is to slightly tap one of the resonator mirrors. Disturbing the cavity mirrors will sweep the frequencies of competing longitudinal modes, and strong amplitude modulation due to mode-beating will occur. The most intense mode-beating pulse will be strong enough to initiate mode-locking.

Several methods for the self-starting of KLM that are practical enough to be implemented on commercial lasers have recently been developed. A simple approach for starting KLM is to mount one of the resonator mirrors on a PZT and introduce a vibration on the order of a few micrometers at a frequency of tens of hertz.

In synchronous pumping, the Kerr material is pumped from a source that produces already mode-locked pulses. For example, a Ti : Al$_2$O$_3$ crystal can be pumped with a frequency-doubled Nd : YAG laser which is actively mode-locked.

Summary

This chapter deals with the formation of ultrashort pulses by longitudinal mode-locking and the techniques employed for the generation of pulses with durations less than 100 picoseconds and as short as a few femtoseconds.

In typical lasers the phases of different longitudinal modes vary randomly with time. This causes the intensity of the laser output to fluctuate producing an irregular, spiking emission. When the longitudinal modes are forced to oscillate with their phases locked, the modes interfere and produce short pulses in the pico- or femto-second regime. The longitudinal modes of a laser oscillator can be locked in phase by means of a suitable nonlinear process or by the action of an active element that modulates the loss in the resonator.

In passive mode-locking the intensity of the radiation in combination with a passive nonlinear element generates a periodic modulation that leads to a fixed phase relationship of the longitudinal modes. In active mode-locking, a rf signal is applied to a phase or frequency modulator at exactly the frequency interval of the longitudinal modes.

The output of a mode-locked laser consists of a train of pulses with a separation equal to the round-trip time of the resonator. The laser radiation can therefore also be understood in terms of a pulse that propagates back and forth between the mirrors of the resonator. The pulse width is equal to the round-trip time divided by the number of modes locked together. Therefore the shortest pulses are obtained from solid-state lasers, such as Ti : sapphire, that have a large gain bandwidth and can support many longitudinal modes.

The generation of femtosecond pulses requires a nonlinear process in the resonator that has an extremely fast response time. Most femtosecond lasers utilize the Kerr effect in a solid that produces an intensity-dependent change of the refractive index. The Kerr medium acts as a lens with an intensity-dependent focal length. Dispersion of various elements in the resonator causes a phase shift of individual modes, thus reducing the number of modes that can be mode-locked. In the design of femtosecond lasers, dispersion effects have to be minimized.

References

[1] J.A. Fleck: *Phys. Rev. B* **1**, 84 (1970).

[2] R. Bracewell: *The Fourier Transform and its Applications* (McGraw-Hill, New York), 1965.

[3] H.A. Haus, J.G. Fujimoto and E.P. Ippen: *IEEE J. Quantum Electron.* **QE-28**, 2086 (1992).

[4] S.L. Shapiro (ed.): *Ultrashort Light Pulses*. Topics Appl. Phys., Vol. 18 (Springer, Berlin), 1977.

[5] W. Kaiser (ed.) *Ultrashort Laser Pulses*, 2nd ed., Topics Appl. Phys., Vol. 60 (Springer, Berlin), 1993.

[6] F. Salin, J. Squier, G. Mourou, M. Piché, and N. McCarthy: *OSA Proc. Ad. Solid-State Lasers* **10**, 125 (1991).

[7] V. Magni, G. Cerullo, and S. DeSilvestri: *Opt. Commun.* **101**, 365 (1993).

[8] V. Magni, G. Cerullo, and S. DeSilvestri: *Opt. Commun.* **96**, 348 (1993).

[9] G. Cerullo, S. DeSilvestri, V. Magni, and L. Pallaro: *Opt. Lett.* **19**, 807 (1994).

[10] C.J. Flood, D.R. Walker, and H.M. van Driel: *Opt. Lett.* **20**, 58 (1995).

[11] D.J. Kuizenga and A.E. Siegman: *IEEE J. Quantum Electron.* **QE-6**, 694, 709 (1970).

[12] A.E. Siegman and D.J. Kuizenga: *Appl. Phys. Lett.* **14**, 181 (1969).

[13] A.E. Siegman and D.J. Kuizenga: *Opto-Electr.* **6**, 43 (1974).

[14] J.L. Dallas: *Appl. Opt.* **33**, 6373 (1994).

[15] D.K. Negus, L. Spinelli, N. Goldblatt, and G. Feuget: *OSA Proc. Adv. Solid-State Lasers* **10**, 120 (1991).

[16] R.L. Fork, O.E. Martinez, and J.P. Gordon: *Opt. Lett.* **9**, 150 (1984).

Exercises

1. Consider a continuously operating laser that is restricted to a single transverse mode of oscillation but which is oscillating in $N + 1$ longitudinal modes that are exactly equally spaced from each other in frequency by the amount Ω Hz. Assume that they all have equal electric field amplitude E_0. Since these modes must all satisfy the same boundary conditions on a cavity mirror, they can be assumed to have equal phase intervals from one mode to the next. Let us set that to zero. Each mode can be written as an oscillating electromagnetic field that satisfies Maxwell's equations. The electric field of each mode is given as

$$E_p = E_0 \exp(i(k_p z + \omega_p t)),$$

where p is an integer from zero to N. The total output electric field of this laser is the sum of these fields. Find an expression for the electric field as a function of time at some arbitrary plane, say where $z = 0$. Then show that the output intensity is given by

$$I = E_0^2 \left[\frac{\sin^2 \left(\frac{1}{2}(N\Omega t) \right)}{\sin^2 \left(\frac{1}{2}(\Omega t) \right)} \right].$$

If there are peaks of intensity in the output what is the intensity of the laser at a peak? How does this compare to the intensity you would have if the modes were spaced by random frequency differences? What is Ω in terms of cavity parameters and what is the interval in time between the peaks? (*Hint:* This is the derivation of (9.9). See the discussion of mode-locking in *Laser Fundamentals* by William T. Silfvast, Cambridge University Press, Cambridge, UK, 1996, ISBN 0-521-55424-1, if you can't do this problem yourself. To do it yourself you will need to look up the summation of a geometric series with a finite limit. It is not hard once you do that.)

2. In Problem 1 you found an expression for the intensity as a function of time for a mode-locked laser. Use that expression and find the full width of the peaks at half-maximum (FWHM). If the cavity of the laser has an optical length of 15 cm, and the number of modes that are locked together (e.g., have constant intermode frequency spacing) in $N + 1 = 100$, what is the duration at FWHM of the intensity maxima?

10

Nonlinear Devices

10.1 Nonlinear Optics

10.2 Harmonic Generation

10.3 Parametric Oscillators

10.4 Raman Laser

10.5 Optical Phase Conjugation

References

Exercises

Nonlinear optical devices, such as harmonic generators and parametric oscillators, provide a means of extending the frequency range of available laser sources. In 1961, Franken and coworkers detected ultraviolet light at twice the frequency of a ruby laser beam when this beam was propagated through a quartz crystal [1]. This experiment marked the beginning of an intense investigation into the realm of the nonlinear optical properties of matter.

Frequency conversion is a useful technique for extending the utility of high-power lasers. It utilizes the nonlinear optical response of an optical medium in intense radiation fields to generate new frequencies. It includes both elastic (optical-energy-conserving) processes, such as harmonic generation, and inelastic processes (which deposit some energy in the medium), such as stimulated Raman or Brillouin scattering.

There are several commonly used elastic processes. Frequency doubling generates a single harmonic from a given fundamental high-power source. The closely related processes of sum- and difference-frequency generation also produce a single new wavelength but require two high-power sources. These processes have been used to generate high-power radiation in all spectral regions, from the ultraviolet to the far-infrared. Optical parametric oscillators and amplifiers generate two waves of lower frequency from a pump source. They are capable of generating a range of wavelengths from a single frequency source, in some cases spanning the entire visible and near-infrared regions.

As far as inelastic processes are concerned, the Raman process can be utilized in solid-state lasers for the generation of additional spectral output lines. The strongest interaction is for the output shifted toward a longer wavelength (first Stokes shift), but at sufficiently high-pump intensities additional lines at longer, as well as shorter, wavelengths with respect to the pump wavelength will appear (Stokes and anti-Stokes lines).

Although it produces a small wavelength shift, stimulated Brillouin scattering is mainly of interest for the realization of phase-conjugating mirrors. The application of phase conjugation, or wavefront reversal, via stimulated Brillouin scattering, offers the possibility of minimizing thermally induced optical distortions [2] which occur in solid-state laser amplifiers.

10.1   Nonlinear Optics

Nonlinear optical effects are analyzed by considering the response of the dielectric material at the atomic level to the electric fields of an intense light beam. The propagation of a wave through a material produces changes in the spatial and temporal distribution of electrical charges as the electrons and atoms react to the electromagnetic fields of the wave. The main effect of the forces exerted by the fields on the charged particles is a displacement of the valence electrons from their normal orbits. This perturbation creates electric dipoles whose macroscopic manifestation is the polarization. For small field strengths this polarization is proportional to the electric field. In the nonlinear case, the reradiation comes from dipoles whose amplitudes do not faithfully reproduce the sinusoidal electric field that generates them. As a result, the distorted reradiated wave contains different frequencies from that of the original wave.

On a microscopic scale, the nonlinear optical effect is rather small even at relatively high-intensity levels, as the following example illustrates. We will compare the applied electric field strength of a high-intensity beam to the atomic electric field which binds the electrons to the nucleus. The magnitude of the atomic field strength is approximately equal to $E_{at} = e/4\pi\epsilon_0 r^2$ where e is the charge of an electron, ϵ_0 is the permitivity of free space, and r is the radius of the electron orbit. With $r = 10^{-8}$ cm one obtains $E_{at} = 10^9$ V/cm. A laser beam with a power density of $I = 100\,\mathrm{MW/cm^2}$ generates an electric field strength in the material of about $E = 10^5$ V/cm according to $I = n_0 E^2 (\mu_0/\epsilon_0)^{-1/2}$, where n_0 is the refractive index and $(\mu_0/\epsilon_0)^{1/2}$ is the impedance of free space. To observe such a small effect, which is on the order of 10^{-4}, it is important that the waves add coherently on a macroscopic scale. This requires that the phase velocities of the generated wave and the incident wave are matched.

In a given material, the magnitude of the induced polarization per unit volume **P** will depend on the magnitude of the applied electric field **E**. We can therefore expand **P** in a series of powers of **E** and write

$$P = \epsilon_0 \chi^{(1)} E + \epsilon_0 \chi^{(2)} EE + \epsilon_0 \chi^{(3)} EEE + \cdots, \qquad (10.1)$$

where ϵ_0 is the permitivity of free space and $\chi^{(1)}$ is the linear susceptibility representing the linear response of the material. The two lowest-order nonlinear responses are accounted for by the second- and third-order nonlinear susceptibilities $\chi^{(2)}$ and $\chi^{(3)}$. The expression of the polarization in anisotropy materials is rather complicated, for the induced polarization depends not only on the magnitude of the electric vector but on the magnitude and direction of all vectors that characterize the electromagnetic field. Normally $\chi^{(1)} \gg \chi^{(2)} \gg \chi^{(3)}$ and therefore the nonlinear effects will be negligible unless the electric field strength is very high.

The linear term in (10.1) represents three separate equations, one for each value of the ith Cartesian coordinate of **P**, that is,

$$\mathbf{P}_i = \chi_{ij}\mathbf{E}_j \qquad (i, j = 1, 2, 3). \tag{10.2}$$

The linear polarization tensor, which is rank 2, has an array of nine coefficients. The linear susceptibility, related to the refractive index through $\chi = n_0^2 - 1$ and to the dielectric constant $\epsilon = \epsilon_0(1 + \chi)$, is responsible for the linear optical properties of the medium such as refraction, dispersion, absorption, and birefringence. In the linear regime optical properties are independent of light intensity and the wavelength of radiation does not change.

The nonlinear optical susceptibilities of second and third order which describe three- and four-wave mixing processes, respectively, give rise to a large variety of optical phenomena since each electric field can have a different frequency, propagation vector, polarization, and relative phase. Therefore **P** can contain many different product terms of the various interacting fields. In the following overview only interactions that are utilized in the nonlinear devices described in the remainder of this chapter are described.

10.1.1 Second-Order Nonlinearities

The second-order susceptibility $\chi^{(2)}$ is responsible for second harmonic generation, sum and difference frequency generation, and optical parametric amplification. Sections 10.2 and 10.3 describe devices based on these nonlinear phenomena. The above-mentioned second-order nonlinear effects are produced by two waves, which interact to produce a third wave. Conservation of momentum and conservation of photon energy is always required in these processes. The optical fields of the three waves are coupled to one another through the second-order susceptibility. The coupling provides the mechanism for the exchange of energy among the interacting fields. The second-order nonlinear susceptibility in a Cartesian coordinate system can be written as

$$\mathbf{P}_j = \epsilon_0\chi_{ijk}^{(2)}\mathbf{E}_j\mathbf{E}_k, \qquad i, j, k = 1, 2, 3. \tag{10.3}$$

The tensor $\chi_{ijk}^{(2)}$ in general has 27 independent coefficients. By taking into account symmetry conditions, the second-order susceptibility can be expressed by a two-dimensional 3×6 tensor, commonly known as the d-tensor. The nonlinear coefficients ranging from d_{11} to d_{36} will be frequently used throughout this chapter.

Most nonlinear crystals are described by only a few d-coefficients. For example, the nonlinear crystal KDP has only three nonvanishing terms (d_{14}, d_{25}, d_{36}). Furthermore, for those crystals of main interest there is usually one predominant coefficient associated with a beam propagation direction which yields maximum harmonic power. In centrosymmetric crystals the $\chi^{(2)}$ tensor is actually zero; therefore second-order nonlinear processes are only possible in crystals that lack inversion symmetry. Using the d-coefficients, the nonlinear polarization is given by

$$\mathbf{P} = \epsilon_0 d \mathbf{E}\mathbf{E}, \tag{10.4}$$

where d has the dimension of meter per volt.

Sum- and Difference-Frequency Generation

Two waves of comparable intensity and frequency ω_1 and ω_2 are initially present, and a wave at the third frequency ω_3 is created through the nonlinear interaction

$$\mathbf{P}(\omega_3) = \epsilon_0 \chi^{(2)} \mathbf{E}(\omega_1)\mathbf{E}(\omega_2), \tag{10.5}$$

where $\omega_3 = \omega_1 + \omega_2$. Similarly, a wave at the difference frequency $\omega_3 - \omega_2 = \omega_1$ may be generated ($\omega_3 > \omega_2 > \omega_1$). Sum frequency generation is described in section 10.2.5. A very common technique to produce the third harmonic of a neodymium laser is, first, the generation of the second harmonic and then sum frequency generation by mixing the second harmonic (ω_2) with the unconverted portion of the fundamental beam (ω_1).

Second Harmonic Generation

This is a special case of frequency mixing with the initial waves having a common frequency, i.e., $\omega = \omega_1 = \omega_2$ and $\omega_3 = 2\omega$ in (10.5). The generation of a second harmonic can be readily shown if we describe the applied field by $E = E_0 \sin(\omega t)$. The induced polarization, according to (10.1) takes the form

$$\mathbf{P} = \epsilon_0[\chi^{(1)}\mathbf{E}\sin(\omega t) + \chi^{(2)}\mathbf{E}^2 \sin^2(\omega t) + \cdots, \tag{10.6a}$$

which can be written as

$$\mathbf{P} = \epsilon_0[\chi^{(1)}\mathbf{E}\sin(\omega t) + 1/2\chi^{(2)}\mathbf{E}^2(1 - \cos 2\omega t) + \cdots. \tag{10.6b}$$

In the last expression, the presence of the second term shows that a wave having twice the frequency of the fundamental can be formed in a nonlinear medium. Second harmonic generation is described in Section 10.2.

Parametric Amplification

Parametric amplification refers to the coupled growth of two waves in the presence of a strong wave, which will be referred to as the pump. Energy and momentum are conserved if $\omega_3 = \omega_2 + \omega_1$, where ω_3 is the frequency of the input pump beam and ω_2 and ω_1 are the frequencies of the resulting signal and idler beams. Parametric oscillators are described in Section 10.3.

10.1.2   Third-Order Nonlinearities

Third-order optical nonlinearities involve the nonlinear susceptibility tensor $\chi^{(3)}$ in (10.1). This term governs third harmonic generation, optical Kerr effect, Raman effect, and Brillouin scattering. In general, $\chi^{(3)}$ will couple together four frequency components, that is, three fields interact to produce a fourth field,

$$\mathbf{P}(\omega_4) = \epsilon_0 \chi^{(3)} \mathbf{E}(\omega_3) \mathbf{E}(\omega_2) \mathbf{E}(\omega_1), \tag{10.7}$$

where $\omega_4 = \omega_3 + \omega_2 + \omega_1$. $\chi^{(3)}$ is a four-rank tensor with 81 elements, because an array of 3×27 coefficients couple a single component of the polarization $\mathbf{P}$ to three independent components of three electric field vectors.

The third-order optical nonlinearity is thus a four-photon process. There can be up to three different input laser frequencies, but several of the frequencies can also be the same. In a lossless medium the susceptibility coefficients of $\chi^{(3)}$ are real. In this case, the primary nonlinear optical effects are the generation of new frequency components and the intensity-dependent change in the refractive index (Kerr effect). In third-order interactions involving absorption, the imaginary part of $\chi^{(3)}$ describes Raman and Brillouin scattering.

Third Harmonic Generation

The third harmonic generated by an intense beam at frequency ω can be described by (10.7) with $\omega = \omega_1 = \omega_2 = \omega_3$ and $\omega_4 = 3\omega$. The expression of a sinusoidal electric field raised to the third power will include a term $\sin^3(\omega t) = 3/4 \sin(\omega t) - 1/4 \sin(3\omega t)$ which illustrates the creation of a third harmonic. The possibility of third-harmonic generation exists in principle, although it suffers from practical drawbacks. Typical values of $\chi^{(3)}$ are orders of magnitude smaller than $\chi^{(2)}$ coefficients found in nonlinear crystals. Therefore the preferred method of third-harmonic generation is the utilization of two second-order effects. (Second-harmonic generation, followed by sum-frequency mixing, is explained in Section 10.2.5.)

Optical Kerr Effect

The third-order nonlinearity can lead to an intensity-dependent refractive index known as the optical Kerr effect. With $\omega_2 = -\omega_1$, (10.7) transforms to

$$\mathbf{P}(\omega) = [\chi^{(3)} \mathbf{E}_{av}^2] \mathbf{E}(\omega), \tag{10.8}$$

where $\mathbf{E}_{av}$ is the average electric field strength. The tensor $\chi^{(3)}$ describes the effect of an intense light beam upon the propagation path of the beam. The self-induced effect manifests itself as an intensity-dependent contribution to the index of refraction

$$n_2 = n_0 + (12\pi/n_0)\chi^{(3)}\mathbf{E}_{av}^2. \tag{10.9}$$

The effect of an intensity-dependent refractive index is exploited in Kerr lens mode-locking (KLM), which is described in Section 9.2.2. In high-intensity beams the nonlinear refractive index can be the source of optical damage because it can lead to self-focusing and small-scale ripple growth. Both phenomena are discussed in Sections 4.3 and 4.5.

Stimulated Raman Scattering

This nonlinear process involves the interaction of a laser beam with the molecular vibrations in gases or liquids. In the presence of a strong external electric field, the charge distribution of a diatomic molecule is polarized. The molecule acquires a dipole that can interact with the driving field. The external field exerts a force on the molecule proportional to E^2. This implies that there must be two input photons driving the interaction. Therefore two electric fields in (10.7) have the same frequency and represent the pump beam ω_p which drives the interaction to provide amplification at $\omega_p - \omega_0$, where ω_0 is the frequency of the molecular vibration wave. In accordance with (10.7) three waves produce a fourth wave via the third-order polarization, and energy at the pump frequency can be transferred to the signal frequency $\omega_p - \omega_0$. In addition, nonlinear mixing processes create additional frequency components at $\omega_p - j\omega_0$ and $\omega_p + j\omega_0$ where $j = 1, 2, 3, \ldots$. Raman lasers employed to shift the wavelength of solid-state lasers are described in Section 10.4.

Stimulated Brillouin Scattering

In this process an optical beam interacts with an acoustic wave in the medium. The acoustic wave is associated with the propagation of pressure in the medium, leading to periodic density fluctuations. Electrostriction provides the coupling mechanism between the acoustic wave and electromagnetic wave, that is, a local compression of the medium in response to the strength of the electromagnetic field. An incident laser beam can scatter with the periodic index variations associated with a propagating acoustic wave. The interaction is stimulated because the acoustic and scattered waves can grow as the pump beam is depleted. The acousto-optic interaction with the laser pump beam depends on the intensity of the pump, just as in stimulated Raman scattering; therefore the polarizability is a function of E^2. The four-wave mixing process is characterized by two electromagnetic fields with the same frequency ω_p interacting with an acoustic wave of frequency ω_a to produce a fourth wave at $\omega_p - \omega_a$. The frequency of the acoustic wave is determined by the conservation of energy and momentum. The stimulated wave is usually radiated backward toward the incoming pump beam. The acoustical frequency for a backward wave is $\omega_a = 2\omega_p(n_0 v_a/c)$, where v_a is the speed of sound in the medium. Since v_a is very small compared to the speed of light in the medium c/n_0, the frequency shift is very small.

Acoustic waves occur at frequencies orders of magnitude smaller than molecular vibrations. Brillouin scattering is therefore frequency shifted from the incident radiation by a much smaller amount as compared to Raman scattering. From an engineering point of view, interest in Brillouin scattering is not related to the frequency shift which occurs in the interaction, but in the phase-conjugate reflection of the pump beam. Brillouin scattering is the mechanism for phase conjugation discussed in Section 10.5.

10.2   Harmonic Generation

In this section, we will review the basic theory and discuss system parameters and material properties which affect harmonic generation. The subject of nonlinear optics has been treated in several books [2]–[6] and tutorial review articles [7], [8]. For a general introduction to crystal optics the reader is referred to standard texts [9], [10].

10.2.1   Basic Equations of Second-Harmonic Generation

The process of harmonic generation by an incident wave of frequency ω_1 must be viewed as a two-step process. First, a polarization wave at the second harmonic $2\omega_1$ is produced and has a phase velocity and wavelength in the medium which are determined by $n_{1\omega}$, the index of refraction for the fundamental wave, that is, $\lambda_p = c/2\nu_1 n_{1\omega}$. The second step is the transfer of energy from the polarization wave to an electromagnetic wave at frequency $2\nu_1$. The phase velocity and the wavelength of this electromagnetic wave are determined by $n_{2\omega}$, the index of refraction for the doubled frequency, that is, $\lambda_2 = c/2\nu_1 n_{2\omega}$. For efficient energy transfer it is necessary that the two waves remain in phase, which implies that $n_{1\omega} = n_{2\omega}$. Since almost all materials have normal dispersion in the optical region, the radiation will generally lag behind the polarization wave. The phase mismatch between the polarization wave and the electromagnetic wave for collinear beams are usually expressed as the difference in wave number, $\Delta k = 2k_\omega - k_{2\omega}$, or

$$\Delta k = \frac{4\pi}{\lambda_1}(n_{1\omega} - n_{2\omega}). \tag{10.10}$$

If Maxwell's equations are solved for a coupled fundamental and second-harmonic wave propagating in a nonlinear medium, then the ratio of the power generated at the second-harmonic frequency, to that incident at the fundamental, is given by [2]

$$\frac{P_{2\omega}}{P_\omega} = \tanh^2\left[lK^{1/2} \left(\frac{P_\omega}{A}\right)^{1/2} \frac{\sin \Delta kl/2}{\Delta kl/2} \right], \tag{10.11}$$

where

$$K = 2Z^3 \omega_1^2 d_{eff}^2, \tag{10.12}$$

l is the length of the nonlinear crystal, A is the area of the fundamental beam, Z is the plane-wave impedance $Z = \sqrt{\mu_0/\varepsilon_0 \varepsilon} = 120\pi/n_0 [V/A]$, ω_1 is the frequency of the fundamental beam, and d_{eff} is the effective nonlinear coefficient of the nonlinear polarizability tensor $\chi^{(2)}$ in (10.1). The dimension of d_{eff} in (10.12) is given in the MKS system and includes ε_0, the permittivity of free space, thus $d_{eff} [As/V^2]$. Some authors exclude ε_0 from the d coefficient, in this case $d [As/V^2] = 8.855 \times 10^{-12} d [m/V]$. The conversion from the cgs system to MKS units becomes $d [As/V^2] = 3.68 \times 10^{-15} d [esu]$.

For low conversion efficiencies, (10.11) may be approximated by

$$\frac{P_{2\omega}}{P_\omega} = l^2 K \frac{P_\omega}{A} \frac{\sin^2 (\Delta k l/2)}{(\Delta k l/2)^2}. \tag{10.13}$$

For a given wavelength and a given nonlinear material, K is a constant. The conversion efficiency, therefore, depends on the length of the crystal, the power density, and the phase mismatch. For a crystal of fixed length, the second-harmonic power generation is strongly dependent on the phase mismatch expressed by the $sinc^2$ function, as is illustrated in Fig. 10.1. In this case, a variation of Δk was obtained by changing the crystal temperature. The harmonic power is at maximum

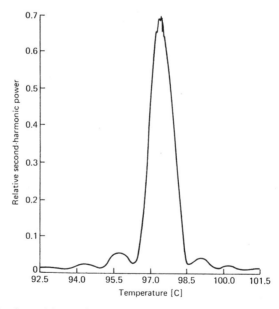

FIGURE 10.1. Second-harmonic generation as a function of temperature in a $Ba_2 NaNb_5 O_{15}$ crystal employed to frequency-double an Nd : YAG laser.

when $\Delta k = 0$, that is, at the exact phase-matching temperature. For a fixed Δk, the second-harmonic power as a function of distance l along the crystal grows and decays with a period of $\Delta k l/2 = \pi$. Half of this distance has been termed the coherence length l_c. It is the distance from the entrance face of the crystal to the point at which the second-harmonic power will be at its maximum value. This parameter is not to be confused with the coherence length l_c of the laser beam, which was defined in Chapter 5.

For normal incidence the coherence length is given by

$$l_c = \pi/\Delta k. \tag{10.14}$$

Expressing the phase mismatch Δk in terms of coherence length in (10.13) one obtains

$$\frac{P_{2\omega}}{P_\omega} = l_c^2 K \frac{4}{\pi^2} \frac{P_\omega}{A} \sin^2 \left(\frac{\pi l}{2l_c}\right). \tag{10.15}$$

The oscillatory behavior of (10.15) is shown in Fig. 10.2 for several values of l_c. For the ideal case $l_c = \infty$, the second-harmonic conversion efficiency is proportional to the square of the crystal length, at least in the small-signal approximation

$$\frac{P_{2\omega}}{P_\omega} = l^2 K \frac{P_\omega}{A}. \tag{10.16}$$

Clearly, if the crystal is not perfectly phase-matched ($l_c = \infty$), the highest second-harmonic power we can expect to generate will be the signal obtained after the beam propagates one coherence length, no matter how long the crystal

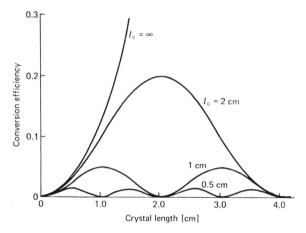

FIGURE 10.2. Second-harmonic power conversion efficiency as a function of distance l from the entrance surface of a CDA crystal. Parameter is the coherence length l_c ($K = 1.3 \times 10^{-9}\,\mathrm{W}^{-1}$; $\lambda = 1.06\,\mu\mathrm{m}$; $I = 100\,\mathrm{MW/cm}^2$).

is. The decrease in harmonic power, for example, between l_c and $2l_c$, is explained by a reversal of the power flow. Instead of power being coupled from the polarization wave into the electromagnetic wave, it is coupled from the electromagnetic wave into the polarization wave: that is, the power is coupled back into the input beam. Thus we find that the power oscillates back and forth between the harmonic and the fundamental wave.

In almost all practical cases the coherence length is limited by the beam divergence and the bandwidth of the laser beam and by angular and thermal deviations of the crystal from the phase-matching angle and temperature.

10.2.2   Index Matching

With typical dispersion values in the visible region, the coherence length in most crystals is limited to about $10\,\mu$m. For this reason the intensity of second-harmonic power is small. Only if n_1 can be made substantially equal to n_2 will relatively high efficiencies of frequency-doubled power be obtained.

An effective method of providing equal-phase velocities for the fundamental and second-harmonic waves in the nonlinear medium utilizes the fact that dispersion can be offset by using the natural birefringence of uniaxial or biaxial crystals. These crystals have two refractive indices for a given direction of propagation, corresponding to two allowed orthogonally polarized beams; by an appropriate choice of polarization and direction of propagation it is often possible to obtain $\Delta k = 0$. This is termed phase matching or index matching.

We shall restrict our discussion to that of uniaxial crystals. These crystals, to which the very important nonlinear crystals KDP, and its isomorphs, and LiNbO$_3$ belong, have an indicatrix that is an ellipsoid of revolution with the optic axis being the axis of rotation, as shown in Fig. 10.3. The two directions of polarization and the indices for these directions are found as follows: We draw a line

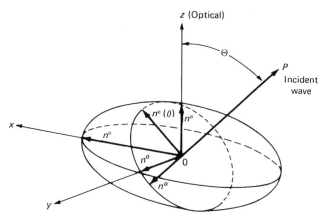

FIGURE 10.3. Indicatrix ellipsoid of a uniaxial crystal. Also shown is a cross section perpendicular to the light propagation direction P.

through the center of the ellipsoid in the direction of beam propagation (line OP in Fig. 10.3). Then we draw a plane perpendicular to the direction of propagation. The intersection of this plane with the ellipsoid is an ellipse. The two axes of this ellipse are parallel to the two directions of polarization, and the length of each semiaxis is equal to the refractive index in that direction.

We now examine how the indices of refraction vary when the direction of propagation is changed. We notice that for the direction of polarization perpendicular to the optic axis, known as the ordinary direction, the refractive index is independent of the direction of propagation. For the other direction of polarization, known as the extraordinary direction, the index changes between the value of the ordinary index n_0 when OP is parallel to z, and the extraordinary index n_e when OP is perpendicular to z. When the wave propagation is in a direction Θ to the optic axis, the refractive index for the extraordinary wave is given by [9]

$$n^e(\Theta) = \frac{n^o n^e}{[(n^o)^2 \sin^2 \Theta + (n^e)^2 \cos^2 \Theta]^{1/2}}, \qquad (10.17)$$

where the superscripts "o" and "e" refer to the ordinary and extraordinary rays.

Changing the point of view, we look now at the shape of the wavefronts for these two rays instead of their direction and polarization. Assuming that the input is a monochromatic point source at O, the expanding wavefront for the o-ray is spherical, whereas the spreading wavefront for the e-ray is an ellipsoid. This property of crystals is described by the index surface, which has the property that the distance of the surface from the origin along the direction of the wave vector is the refractive index. For a uniaxial crystal this surface has two sheets—a sphere for ordinary waves polarized perpendicular to the optic axis with index n^o, and an ellipsoid for extraordinary waves with index $n^e(\Theta)$. By definition the optic axis is that direction at which the o- and e-rays propagate with the same velocity. If the value of $n^e - n^o$ is larger than zero, the birefringence is said to be positive, and for $n^e - n^o$ smaller than zero the birefringence is negative; the corresponding crystals are called positive or negative uniaxial. Figure 10.4 shows a cross section of the index surface of a negative uniaxial crystal (for the moment we consider only the solid lines n_1^o and n_1^e). The complete surfaces are generated by rotating the given sections about the z-axis. The wavefront velocity v and the refractive index n_o are related by $v = c/n_o$, where c is the velocity of light.

Both refractive indices n^o and n^e are a function of wavelength. Figures 10.4 and 10.5 illustrate how the dependence of the refractive index on beam direction, wavelength, and polarization can be utilized to achieve angle-tuned phase matching. The dashed lines in these figures show the cross section of the index surfaces n_2 at the harmonic frequency. As can be seen, the negative uniaxial crystal has sufficient birefringence to offset dispersion, and the matching condition can be satisfied for a beam deviating from the z-axis by the angle Θ_m.

The directions for phase-matched second-harmonic generation are obtained by considering the intersection of the index surfaces at the fundamental and harmonic frequencies. As was mentioned earlier, frequency doubling may be considered

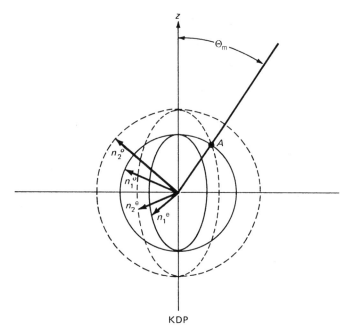

FIGURE 10.4. Cross section of the index surface of a negative uniaxial crystal for two different wavelengths.

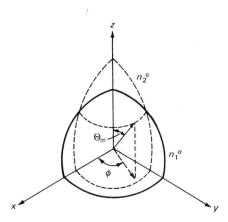

FIGURE 10.5. Direction for phase-matched second-harmonic generation (type-I) in a uniaxial crystal, where Θ_m is the phase-matching angle measured from z and ϕ is the azimuth angle measured from x.

as a special case, where two incident waves with electric fields $\mathbf{E}_m$ and $\mathbf{E}_n$ are identical waveforms. There are two types of processes in harmonic generation, depending on the two possible orientations for the linear polarization vectors of the incident beams. In the type-I process both polarization vectors are parallel: in the type-II process the polarization vectors are orthogonal.

In a negative uniaxial crystal there are two loci where the index surfaces intersect and $\Delta k = 0$ [11]

$$n_2^e(\Theta_m) = n_1^o \quad \text{type-I,} \tag{10.18a}$$

$$n_2^e(\Theta_m) = \tfrac{1}{2}[n_1^e(\Theta_m) + n_1^o] \quad \text{type-II,} \tag{10.18b}$$

first, in a symmetrical cone at Θ_m (type-I) about the optic axis where two o-rays at ω are matched to an e-ray at 2ω; second, in a cone at Θ_m (type-II) where an o-ray and an e-ray at ω are matched to an e-ray at 2ω.

The harmonic power is not independent of the azimuthal angle of the phase-matched direction (Fig. 10.5). In general, d_{eff} is a combination of one or several coefficients of $\chi^{(2)}$, and the angles Θ and ϕ that define the direction of the wave propagation vector. For example, for KDP and its isomorphs and type-I index matching, one obtains

$$d_{\text{eff}} = d_{14} \sin 2\phi \sin \Theta_m. \tag{10.19}$$

The phase-matching angle Θ_m is obtained by combining (10.17) and (10.18a)

$$\sin^2 \Theta_m = \frac{(n_1^o)^{-2} - (n_2^o)^{-2}}{(n_2^e)^{-2} - (n_2^o)^{-2}}. \tag{10.20}$$

For frequency doubling of 1.05 μm radiation, the beam propagation vector is at $\Theta_m = 41°$ with respect to the optical axis of the KDP crystal and the azimuth angle is at $45°$ in type-I phase-matching. The fundamental beams, $E_j = E_k$ in (10.3), are polarized perpendicular to the optical axis (ordinary direction) and the frequency doubled beam is polarized parallel to the optical axis of the crystal (extraordinary direction). The orientation of the beam propagation vector, polarization, and crystal axes is illustrated in Fig. 10.6(a) for the case of KDP.

For type-II phase-matching in KDP one obtains

$$d_{\text{eff}} = d_{14} \cos 2\phi \sin 2\Theta_m. \tag{10.21}$$

In this case the angle between the beam direction and the crystal axis is $\Theta_m = 59°$ and $\phi = 90°$ with respect to the x-axis. A single linearly polarized fundamental beam incident on the crystal may be equally divided into o- and e-rays by orienting the polarization vector at $45°$ with respect to the x-axis. The polarization vector of the harmonic beam is rotated $45°$ from the fundamental beam with the polarization parallel to the optical axis (see Fig. 10.6(b)).

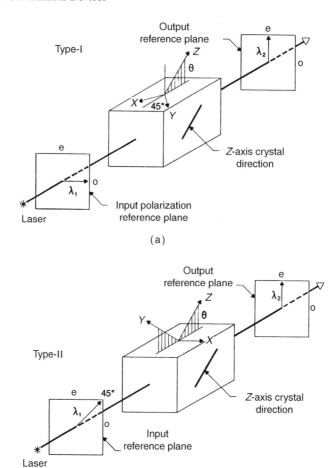

FIGURE 10.6. Crystal and electric vector orientation for harmonic generation in KDP and its isomorphs for (a) Type-I and (b) Type-II phase-matching [12].

Comparing type-I and type-II phase-matching, we find that type-I is more favorable when Θ_m is near 90°, whereas type-II leads to a higher d_{eff} when Θ_m lies near 45°.

Critical Phase-Matching

The reason for not achieving the optimum second harmonic output from a particular set-up is in many cases the divergence of the pump beam. For second-harmonic generation in a negative crystal we find from (10.10) that $\Delta k = 0$ if $n_1^o = n_2^e(\Theta_m)$, which is exactly true only at $\Theta = \Theta_m$. In Section 10.1.2 it will be shown that there is a linear relationship between small deviations $\delta\theta$ from the phase-matching angle and Δk. The change of Δk as a function of $\delta\theta$ can be large enough to limit the conversion efficiency in real devices. Consider as an example second-harmonic generation in KDP at 1.064 μm. The linear change in Δk with Θ is sufficiently great to restrict the deviation from the phase-matched direction to approximately 1.2 mrad if the coherence length is to be greater than 1 cm. Phase-matching under these unfavorable conditions is termed "critical phase-matching."

Also if phase-matching is accomplished at an angle Θ_m other than $90°$ with respect to the optic axis of a uniaxial crystal, there will be double refraction. Therefore, the direction of power flow (Poynting vector) of the fundamental and second harmonics will not be completely collinear but occur at a small angle.

For a negative uniaxial crystal and type-I phase-matching, this angle is given by [13]

$$\tan \varrho = \frac{(n_1^o)^2}{2} \left(\frac{1}{(n_2^e)^2} - \frac{1}{(n_2^o)^2} \right) \sin 2\Theta. \tag{10.22}$$

The angle ϱ has the effect of limiting the effective crystal length over which harmonic generation can take place. The beams completely separate at a distance of order

$$l_a = D/\varrho \tag{10.23}$$

called the aperture length where D is the beam diameter. Of course, at only a fraction of this distance the reduction of conversion efficiency because walk-off becomes noticeable and has to be taken into account.

For weakly focused Gaussian beams, the aperture length can be expressed as

$$l_a = w_0 \sqrt{\pi}/\varrho, \tag{10.24}$$

where w_0 is the fundamental beam radius.

Noncritical Phase-Matching

If the refractive indices can be adjusted so that $\Theta_m = 90°$, by variation of a parameter such as the temperature or chemical composition of the crystal, the dependence of Δk on beam divergence or angular misalignment $\delta\theta$ is due to a much smaller quadratic term, instead of a linear relationship as is the case in critical phase-matching.

For example, second-harmonic generation in LiNbO$_3$ occurs at $90°$ for 1.064 μm radiation provided the crystal is at a temperature of $\approx 107°$ C (for a crystal doped with 5% MgO). The allowable departure from the phase-matched direction is greater than 26 mrad if the coherence length is to be 1 cm. For these reasons, $90°$ phase-matching is often called noncritical phase-matching. Provided that n_1^o and n_2^e are nearly equal and $d(n_1^o - n_2^e)/dT \neq 0$, noncritical phase-matching can be achieved by temperature tuning the crystal.

In addition, at $\Theta_m = 90°$ there are no walk-off effects due to double refraction.

TABLE 10.1. Properties of important nonlinear materials (data are for 1064 to 532 nm conversion).

Material	Phase-matching type	Effective nonlinear coefficient $[10^{-12}\,\mathrm{m/V}]$	Refractive index n_0	Phase matching angle Θ_m	Walk-off angle ϱ	Angular sensitivity $\beta_\Theta\,(\mathrm{mr\text{-}cm})^{-1}$
KDP[a]	II	0.78	1.49	58.9	1.45°	2.53
KTP	II	3.18	1.74	24.3°	0.26°	0.22
BBO	I	1.94	1.65	22.8°	3.19°	11.1
LiNbO$_3$[b]	I	4.7	2.23	90°	0	0.12

[a] Conversion of 1.054 μm to 527 nm.
[b] With 5% MgO at 107° C.

10.2.3   Parameters Affecting the Doubling Efficiency

Second-harmonic conversion depends on parameters that are related to the laser source, such as power density, beam divergence, and spectral linewidth, and parameters associated with the harmonic generator, such as the value of the nonlinear coefficient and angular, thermal, and spectral sensitivity to deviations from the exact phase-matching condition.

The most critical laser parameters are power density and beam divergence. In general, one finds that Nd : YAG lasers have sufficiently narrow linewidths for efficient harmonic generation. In these lasers attention is focused mainly on obtaining a diffraction limited beam, i.e., TEM$_{00}$-mode operation. Nd : glass lasers and tunable lasers require longitudinal mode selection for efficient harmonic conversion. However, the large Nd : glass lasers used in fusion research employ Nd : YLF oscillators which automatically solve the spectral linewidth problem.

The most important crystal parameters are the nonlinear index and the angular sensitivity which are listed in Table 10.1 for several important materials.

Power Density

From (10.13) it follows that the conversion efficiency is proportional to the power density of the fundamental beam, whereas the harmonic power itself is quadratically proportional to the fundamental power. At conversion efficiencies above 20%, the second-harmonic generation starts to deviate markedly from the linear relationship of (10.13) because of depletion of the fundamental beam power. At these high efficiencies, (10.11) should be used.

Figure 10.7 shows the experimental data and predicted performance for a harmonic generator comprised of a 10 mm long KTP crystal. The fundamental beam is derived from a Q-switched diode-pumped Nd : YAG laser at 1.064 μm. The laser had a maximum pulse energy of 40 mJ in a 12 ns pulse. At the peak intensity of 175 MW/cm^2 for the pump beam, the conversion efficiency approached 80%. The solid curve is obtained from (10.11) if one introduces the following

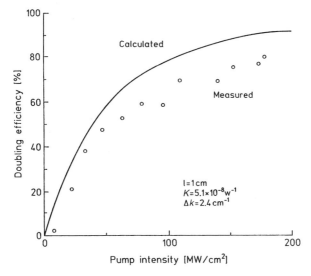

FIGURE 10.7. Second-harmonic conversion efficiency versus power density for a KTP crystal pumped by a Q-switched Nd : YAG laser at 1.06 μm.

system parameters: crystal length $l = 10\,\text{mm}$, effective nonlinear coefficient $d_{\text{eff}} = 3.2\,\text{pm/V}$, and fundamental wavelength $\lambda_1 = 1.064\,\mu\text{m}$. The fundamental beam was a weakly focused Gaussian beam with a spot radius of $w_0 = 0.75\,\text{mm}$. The beam area is therefore $A = \pi w_0^2 = 1.8 \times 10^{-2}\,\text{cm}^2$.

As will be demonstrated in the next section, phase-mismatch can be expressed as $\Delta k = \beta_\Theta(\Delta\Theta/2)$, where β_Θ is the angular sensitivity and $\Delta\Theta$ is the beam divergence of the fundamental beam. From Table 10.1 follows $\beta_\Theta = 0.2\,(\text{mrad cm})^{-1}$ for KTP at 1.06 μm and type-II phase-matching. The beam divergence of the weakly focused beam was about 22 mrad from which follows $\Delta k = 2.4\,\text{cm}^{-1}$. The value of K listed in Fig. 10.7 is obtained by introducing the parameters for KTP from Table 10.1 into (10.12).

Walk-off reduces the region of overlap between the beams, thus reducing the effective gain length of the crystal. From (10.19) we obtain for the aperture length $l_a = 290\,\text{mm}$ if we use the beam walk-off angle from Table 10.1. Since the aperture length is large compared to the crystal length, double refraction has only a negligible effect on the conversion efficiency.

The theoretical doubling efficiency calculated using (10.11) gives quite acceptable agreement with the experimental data. As expected, the theory predicts higher efficiency than can be achieved in practice due to the imperfect nature of the pump beam and crystal quality.

Beam Divergence

When collinear phase-matched second-harmonic generation is used, the light waves will have a small but finite divergence. It is necessary to consider the mismatch Δk of the wave vector for small deviations $\delta\Theta$ from the phase-matched direction. An expansion for $n_1^o - n_2^e(\Theta)$ taken for a direction close to the perfect phase-matching direction Θ_m yields

$$n_1^o - n_2^e(\Theta) = \frac{\partial[n_1^o - n_2^e(\Theta)]}{\partial\Theta}(\Theta - \Theta_m). \tag{10.25}$$

The expression given in (10.17) for the dependence of n_2^e on the angular direction in the crystal can be very well approximated by [14]

$$n_2^e(\Theta) = n_2^o - (n_2^o - n_2^e)\sin^2\Theta. \tag{10.26}$$

Introducing (10.26) into (10.25) gives

$$n_1^o - n_2^e(\Theta) = \delta\Theta(n_2^o - n_2^e)\sin 2\Theta_m, \tag{10.27}$$

where we note that $\partial n_1^o/\partial\Theta = 0$ and $\partial n_2^o/\partial\Theta = 0$.

Introducing (10.27) into (10.10) yields

$$\Delta k = \beta_\Theta\,\delta\Theta' = \beta_\Theta(\Delta\Theta/2) \tag{10.28}$$

where

$$\beta_\Theta = \frac{4\pi}{\lambda_1 n_1^o}(n_2^o - n_2^e)\sin 2\Theta_m \tag{10.29}$$

is a material constant and expresses the angular sensitivity of the crystal, and $\Delta\Theta$ is the full angle of the pump beam. The expression for β_Θ has been divided by n_1^o, because $\delta\Theta'$ is the angular misalignment measured external to the crystal.

Under noncritical phase-matching conditions ($\Theta_m = 90°$), we can make the approximation $\sin 2(90° + \delta\Theta) \approx 2\delta\Theta$ and, instead of (10.29), we obtain

$$\beta_\Theta = \frac{4\pi}{\lambda_1(n_1^o)^2}(n_2^o - n_2^e)\,\Delta\Theta. \tag{10.30}$$

Properties of Nonlinear Crystals

Table 10.1 lists parameters for several important nonlinear materials for 1064 to 532 nm conversion. Frequency doubling of neodymium lasers is one of the major applications of harmonic generation.

Since KDP is rather easily grown from a water solution, crystals with cross sections of 30 cm square have been produced. Because of the availability of large aperture harmonic crystal assemblies, KDP is the material of choice for the large Nd : glass lasers employed in fusion research. Despite the lower nonlinear coefficient of KDP compared to other nonlinear crystals, conversion efficiencies as high as 80% have been obtained. A high-conversion efficiency is actually the result of material properties as well as pump source characteristics, the high-peak power, narrow spectral bandwidth, small beam divergence, and clean spatial and temporal beam profiles obtained from Nd : glass lasers employed in fusion research, make possible such high harmonic conversions even in crystals with modest non-linearity.

The material β-BaB$_2$O$_4$ (BBO) is a new nonlinear optical crystal that possesses excellent properties for nonlinear frequency conversion in a spectral range that extends from the ultraviolet to the mid-infrared. This material has a moderately large nonlinear coefficient, a large temperature tolerance, low absorption, and a very high damage threshold. The principal shortcoming of BBO is the low-angular tolerance of 0.5 mr-cm, which requires a diffraction-limited beam for efficient frequency doubling. BBO is of particular interest for frequency doubling into the blue region.

The crystals of lithium niobate are transparent in the region 0.42 to 4.2 μm. Temperature sensitivity of birefringence is such that, by varying the temperature, phase-matching can be achieved at 90° to the optical axis. Unfortunately, LiNbO$_3$ is particularly susceptible to photorefractive damage from propagation from visible or ultraviolet radiation. The relatively low-damage threshold has severely limited practical applications for this material.

The emergence of Periodically Poled LiNbO$_3$ (PPLN), which will be discussed in Section 10.2.2, has attracted renewed interest in this material. In PPLN, the largest nonlinear index of LiNbO$_3$ can be utilized. This allows efficient nonlinear conversion at modest power levels which are well below the damage threshold.

At the present time, the leading candidate material for nonlinear experiments with solid-state lasers is KTP since the crystal has a large nonlinear coefficient, high-damage threshold, and large angular and temperature acceptance ranges. Its major drawbacks are the limited size (about 1–2 cm^3) and the high cost associated with the difficult growth process. Although a few specific characteristics of other materials are better, KTP has a combination of properties that make it unique for nonlinear conversion processes. Figure 10.8 shows the crystal orientation for the phase-match condition of the type-II interaction at 1.06 μm. The phase-match direction in KTP for second-harmonic generation of 1.06 μm radiation results in a walk-off angle of 4.5 mrad between the fundamental and second-harmonic beam.

KDP is superior to any other nonlinear crystal with regard to availability in large sizes combined with an excellent optical quality. Crystals with diameters as large as 27 cm have been fabricated. Crystals grown from the melt such as

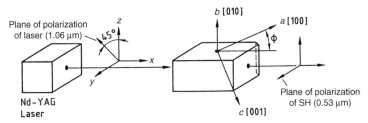

FIGURE 10.8. Orientation of KTP for type-II interaction at 1.06 μm. $\phi = 26°$ for hydrothermally and $\phi = 21°$ for flux-grown material.

LiNbO$_3$ or KTP are generally hard, chemically stable and can easily be polished and coated. Solution grown crystals such as KDP and its isomorphs are soft and hygroscopic. These crystals need to be protected from the atmosphere.

10.2.4 Intracavity Frequency Doubling

In the previous section we discussed frequency doubling by placing a nonlinear crystal in the output beam of the laser system. Frequency doubling a cw-pumped laser system in this manner results in an unacceptably low harmonic power because large conversion efficiencies require power densities that are not available from a cw-pumped laser. One obvious solution to this problem is to place the nonlinear crystal inside the laser resonator, where the intracavity power is approximately a factor $(1 + R)/(1 - R)$ larger than the output power. The power is coupled out of the resonator at the second-harmonic wavelength by replacing the output mirror by one that is 100% reflective at the fundamental and totally transmitting at the second harmonic. Functionally, the second-harmonic crystal acts as an output coupler in a manner analogous to the transmitting mirror of a normal laser. Normally the transmitting mirror couples out power at the laser frequency, whereas the nonlinear crystal inside the laser couples out power at twice the laser frequency.

Intracavity harmonic generation produces a beam of harmonic power in two directions. If it is desired to obtain all the harmonic power in a single output beam, it is necessary to employ a dichroic mirror to reflect one of the beams back in the same direction as the other. Figure 10.9 shows a technique that combines into a single output beam the second-harmonic power which is generated in both directions by the frequency-doubling crystal. The frequency-doubled beam has a polarization which is rotated 90° with respect to the polarization of the fundamental beam. The dichroic mirror M$_1$ is designed to reflect the 1.06 μm beam

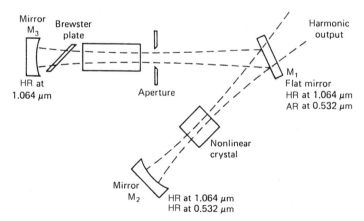

FIGURE 10.9. Three-mirror, folded cavity configuration allows harmonic power generated in two directions to be combined into a single output beam.

completely and transmit virtually all the orthogonally polarized $0.53\,\mu$m beam. Mirror M_2 is a 100% reflector for both 1.06 and $0.53\,\mu$m mirrors. In this way the forward and reverse green beams are combined into one.

The steady-state condition for intracavity doubling can be determined if we equate the round-trip saturated gain of the laser to the sum of the linear and non-linear losses [15]

$$\frac{2g_0 l^*}{1 + I/I_s} = \delta + K'I, \tag{10.31}$$

where g_0 is the unsaturated gain coefficient, l^* is the length of the laser medium, I is the power density in the laser rod, and I_s is the saturation power density of the active material. All linear losses occurring at the fundamental frequency are lumped together into the parameter δ, the quantity $K'I$ is the nonlinear loss. The nonlinear coupling factor K', defined by

$$I(2\omega) = K'I^2(\omega), \tag{10.32}$$

is related to K in (10.12); it is $K' = (I_{\mathrm{crys}}/I_{\mathrm{rod}})l^2 K$, where $(I_{\mathrm{crys}}/I_{\mathrm{rod}})$ accounts for different power densities in the laser rod and nonlinear crystal in the case of focused beams, and l is the crystal length.

From the theoretical treatment of intracavity doubling, it follows that a maximum value of second-harmonic power is found when

$$K'_{\mathrm{max}} = \frac{\delta}{I_s}. \tag{10.33}$$

The magnitude of the nonlinearity required for optimum second-harmonic production is proportional to the loss, inversely proportional to the saturation density, and independent of the gain. Thus, for a given loss, optimum coupling is achieved for all values of gain and, hence, all power levels. As we recall from Chapter 3, in a laser operating at the fundamental wavelength the optimum output coupling of the front mirror is gain-dependent. The usefulness of (10.33) stems from the fact that by introducing the material parameters for a particular crystal into (10.12) and combining with (10.33) the optimum length of the nonlinear crystal is obtained. For example, from (10.12) we obtain $K = 5.1 \times 10^{-8}\,\mathrm{W}^{-1}$ for KTP and $\lambda = 1.064\,\mu$m. With a saturation flux of $2.9\,\mathrm{kW/cm}^2$ for Nd : YAG and assuming a 2% linear cavity loss, we obtain from (10.33) a nonlinear coupling factor of $K'_{\mathrm{max}} = 0.69 \times 10^{-5}\,\mathrm{cm}^2/\mathrm{W}$. Focusing the intracavity beam to achieve a flux enhancement of $(I_{\mathrm{crys}}/I_{\mathrm{rod}}) = 24$, a crystal 2.4 cm long would be required to optimally couple the second harmonic from the laser.

As an example, the performance of a typical intracavity doubled Nd : YAG laser is illustrated in Fig. 10.10. The output at the harmonic wavelength is compared to the TEM$_{00}$ output available at $1.064\,\mu$m from the same laser. The Nd : YAG laser employed in these experiments was cw-pumped by diode arrays and intracavity doubled with KTP. After measuring the output at $1.064\,\mu$m, a KTP crystal was inserted in the folded resonator (similar to the arrangement shown in Fig. 10.9)

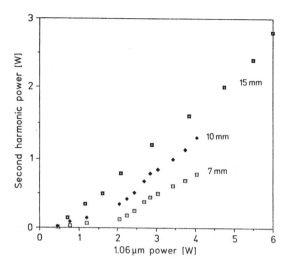

FIGURE 10.10. Comparison of fundamental and second harmonic output from a cw-pumped intracavity frequency-doubled Nd : YAG laser for different KTP crystal lengths.

and the output coupler was replaced with a mirror of high reflectivity at 1.064 and 0.53 μm. For the short crystals, the nonlinear coupling factor K' is too small and far from optimum. A higher output at 532 nm would require an increase of intensity by tighter focusing of the beam. However, this makes alignment very critical and sacrifices mechanical and thermal stability of the laser. A longer nonlinear crystal is therefore a better solution. However, a 15 mm long crystal is about the longest crystal available for KTP.

From the theory presented in [15] it follows that if the loss due to the inserted second-harmonic crystal is small compared to the total internal loss, the value of K' is the same for fundamental and second-harmonic output coupling. Also, the maximum second-harmonic power equals the fundamental power obtainable from the same laser.

In most systems, the second-harmonic power is considerably below the output power that can be achieved at the fundamental wavelength. Insertion losses of the nonlinear crystal, or any other additional intracavity element, are often the reason for the poor performance, or in the case of Nd : YAG, thermally induced depolarization losses can have a major effect.

10.2.5 Third-Harmonic Generation

Efficient conversion of the output of a laser to the third harmonic is possible by invoking a two-step process, namely second harmonic generation followed by frequency mixing. In the first crystal ("doubler") some fraction of the fundamental radiation is converted to the second harmonic, followed by a second crystal ("tripler"), in which unconverted fundamental radiation is mixed with the second harmonic to produce the third harmonic.

The optimum situation is achieved if, for three photons at the fundamental frequency ω entering the first crystal, two photons are converted to create a photon at 2ω, and the third photon at ω is combined in the second crystal with the 2ω photon to produce one 3ω photon. The $1 = 1$ mix of an ω and a 2ω photon at the exit of the doubling crystal represents a 67% conversion efficiency. Given this ratio, in a lossless and perfectly matched crystal, a complete conversion of the fundamental beam to the third harmonic is theoretically possible.

Third-harmonic generation is very important for frequency up-conversion of high-energy Nd : glass lasers employed in fusion research. In these lasers the frequency converters are constructed from two type-II potassium dihydrogen phosphate (KDP) crystals arranged in tandem. As illustrated in Fig. 10.6(b), in conventional type-II doubling in KDP the polarization vector of the incoming beam is oriented at 45° with respect to the x-axis. In this case an equal number of photons are incident along the ordinary and extraordinary directions that may combine one for one to create a second harmonic photon in the extraordinary direction.

In order to optimize the production of the third harmonic, the doubling crystal is oriented at an angle of $\tan^{-1}(1/\sqrt{2}) = 35°$ with respect to the x-axis of the KDP crystal. This ensures that two photons in the ordinary direction are incident for every photon in the extraordinary direction. Each e-photon will combine with one o-photon to produce an e-photon at 2ω, leaving one o-photon at ω unconverted. The doubler output of one o-photon at ω and one e-photon at 2ω provides the input to the frequency mixing second crystal. This situation is depicted in Fig. 10.11(a).

Third-harmonic generation has been analyzed starting with the general equation governing frequency mixing [16]

$$dE_1/dz = -jK_1E_3E_2^* \exp(-j\,\Delta k \cdot z) - \tfrac{1}{2}\gamma_1 E_1,$$

$$dE_2/dz = -jK_2E_3E_1^* \exp(-j\,\Delta k \cdot z) - \tfrac{1}{2}\gamma_2 E_2, \qquad (10.34a)$$

$$dE_3/dz = -jK_3E_1E_2 \exp(j\,\Delta k \cdot z) - \tfrac{1}{2}\gamma_3 E_3.$$

Here the E_j's are the complex electric vectors of waves propagating in the z-direction with frequencies ω_j, where $\omega_3 = \omega_1 + \omega_2$. The electric field of wave j is the real part of $E_j \exp(i\omega_j t - ik_j z)$, and the phase mismatch $\Delta k = k_3 - (k_1 + k_2)$ is proportional to the deviation $\Delta\theta$ of the beam path from the phase-matching direction. The γ_j's are absorption coefficients. For tripling, $\omega_2 = 2\omega_1$, $\omega_3 = 3\omega_1$, $K_2 \simeq 2K_1$, and $K_3 \simeq 3K_1$ where

$$K_1 = \frac{\omega_1 d_{36}}{2c\varepsilon_0}(n_{2\omega}n_{3\omega})^{-1/2}\sin r\theta_m, \qquad (10.34b)$$

$r = 1$ or 2 for type-I or type-II interactions and θ_m is the corresponding phase-matching angle. The solution of such an analysis for the phase-matched case $(\Delta k = 0)$ of a type-II KDP crystal is shown in Fig. 10.11(b). The curves have

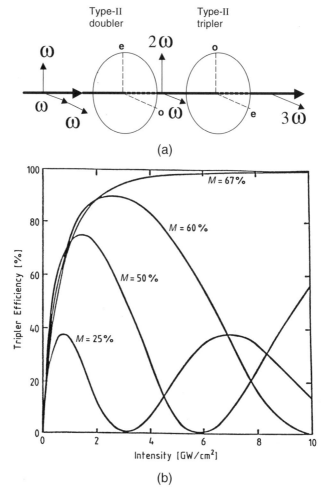

FIGURE 10.11. Frequency tripling. (a) Orientation of polarization vectors for optimum third-harmonic generation. (b) Tripling efficiency of a 9 mm thick phase-matched KDP type-II crystal as a function of total input intensity, for various percentages M of the second harmonic in the input. A small absorption of $0.04\,\text{cm}^{-1}$ is included for the fundamental [16].

been calculated assuming a value of $K_1 = 1.4 \times 10^{-6}\text{V}^{-1}$, an absorption coefficient of $\gamma = 0.09\,\text{cm}^{-1}$ at $1.054\,\mu\text{m}$, and a crystal thickness of 9 mm.

The parameter M is the ratio of second-harmonic power to total power in the tripler

$$M = P_{2\omega}/(P_\omega + P_{2\omega}). \tag{10.35}$$

If one ignores the small absorption in the doubling crystal, then M is essentially the conversion efficiency of the doubler. If the input photons at ω and 2ω are matched $1:1$, then $P_{2\omega} = 2P_\omega$ and $M = 0.67$. In principle, complete conversion of the input beams to the third harmonic can be achieved.

A deviation from the optimum photon ratio results in a reduction of the peak conversion efficiency, which is approximately proportional to the ratio of the actual M value and $M = 0.67$. Also, as can be seen from Fig. 10.11(b) the crystal thickness and beam intensity have to be carefully matched to avoid back conversion of energy. For values other than $M = 0.67$, after a ray has propagated an optimum distance z_{opt} into the crystal, the number of photons in one of the components depletes to zero, and as z increases beyond z_{opt} the mixing process reverses and the third-harmonic radiation reconverts.

10.3   Parametric Oscillators

As was mentioned at the beginning of this chapter, two beams with different frequencies incident on a nonlinear crystal will generate a traveling polarization wave at the difference frequency. Provided the polarization wave travels at the same velocity as a freely propagating electromagnetic wave, cumulative growth will result. For reasons that will soon become clear, the two incident beams are termed "pump" and "signal" waves having a frequency of ν_p and ν_s, and the resulting third wave is termed an "idler" wave with frequency ν_i. Under proper conditions, the idler wave can mix with the pump beam to produce a traveling polarization wave at the signal frequency, phased such that growth of the signal wave results. The process continues with the signal and idler waves both growing, and the pump wave decaying as a function of distance in the crystal.

Since each pump photon with energy $h\nu_p$ is generating a photon at the signal ($h\nu_s$) and idler frequency ($h\nu_i$), energy conservation requires that

$$\frac{1}{\lambda_p} = \frac{1}{\lambda_s} + \frac{1}{\lambda_i}. \tag{10.36}$$

In order to achieve significant parametric amplification, it is required that at each of the three frequencies the generated polarization waves travel at the same velocity as a freely propagating electromagnetic wave. This will be the case if the refractive indices of the material are such that the k vectors satisfy the momentum-matching condition $k_p = k_s + k_i$. For collinearly propagating waves this may be written as

$$\frac{n_p}{\lambda_p} - \frac{n_s}{\lambda_s} - \frac{n_i}{\lambda_i} = 0, \tag{10.37}$$

where n_s, n_i, and n_p are the refractive indices at the signal, idler, and pump frequency, respectively.

Since the three indices of refraction depend on the wavelength, the direction of propagation in the crystal, and on the polarization of the waves, it is generally possible, by using birefringence and dispersion, to find conditions under which (10.37) is satisfied.

Tunability is a fundamental characteristic of all parametric devices. With the pump providing input at the fixed wavelength λ_p, small changes of the refractive index around the phase-matching condition will change the signal and idler wavelengths such that a new phase-matching condition is achieved. Tuning is possible by making use of the angular dependence of the birefringence of anisotropic crystals.

Figure 10.12 illustrates different configurations that make use of the parametric-interaction process of three waves. The simplest device is a nonresonant configu-

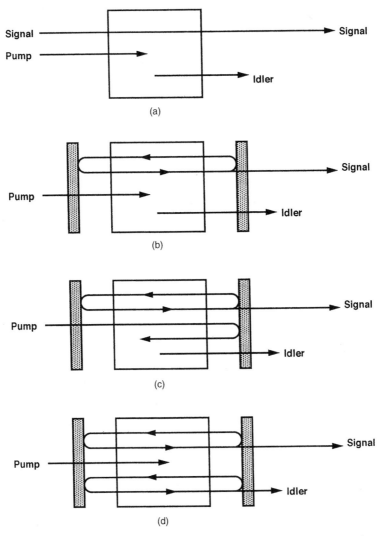

FIGURE 10.12. Configurations for parametric interactions. (a) Optical parametric amplifier (OPA). (b) Singly resonant optical parametric oscillator (SRO). (c) Singly resonant oscillator with pump beam reflected. (d) Doubly resonant OPO (DRO).

ration, namely an optical parametric amplifier (OPA) exhibited in Fig. 10.12(a). In this case, a pump beam and a signal beam are present at the input. If the output of a Q-switched laser is focused into the crystal and if the intensity of the pump is sufficiently high, and phase-matching conditions are met, gain is obtained for the signal wave and at the same time an idler wave is generated. Typically, an OPA is used if the signal obtained from an optical parametric oscillator (OPO) is too weak and further amplification is desired.

The most common optical parametric device is the singly resonant oscillator depicted in Figs. 10.12(b) and (c). In this device, the crystal is located inside a resonator that provides feedback at either the signal or idler frequency. In the example illustrated, the pump beam enters through a mirror that is highly transmitting at the pump wavelength and highly reflective for the signal wavelength. In Fig. 10.12(b) the opposite mirror, which is the output coupler, has typically 80–95% reflectivity for the signal wavelength, and high transmission for the idler and pump beam. Only the signal wavelength is resonant in the cavity, and a small fraction is coupled out through the front mirror. In the configuration in Fig. 10.12(c) the pump beam is reflected at the output mirror and makes one return pass through the crystal. Since the input mirror through which the pump enters is highly transmissive for this wavelength, no resonance condition is set up for the pump wavelength. However, threshold is lowered, because on the return path the pump beam provides gain for the resonant signal wave.

Figure 10.12(d) depicts a doubly resonant oscillator (DRO), which provides feedback at both the signal and idler wavelengths. The double-resonance condition, in which both the signal and the idler waves are simultaneously resonant within the optical cavity, lowers the threshold significantly. However, this advantage of a DRO is off-set by a reduction in stability and tunability. Maintaining the doubly resonant condition in a common resonator requires that the pump be single frequency and stable, and that the OPO cavity length be actively controlled to maintain the resonance condition. The considerably lower threshold of a DRO makes it possible to obtain parametric gain at the low-power densities achievable under cw conditions.

10.3.1   Performance Modeling

Of greatest interest for the design of parametric devices are simple models that describe gain, threshold, phase-matching, and conversion efficiency as a function of device and input parameters.

For the amplifier depicted in Fig. 10.12(a) a parametric gain coefficient for the amplification of the signal wave can be defined [17]

$$g = \sqrt{\kappa I_p}, \tag{10.38}$$

where I_p is the pump flux and κ is a coupling constant,

$$\kappa = \frac{8\pi^2 d_{\mathrm{eff}}^2}{\lambda_s \lambda_i n_s n_i n_p \varepsilon_0 c}. \tag{10.39}$$

The effective nonlinear coefficient d_{eff} connects the pump, signal, and idler fields. The parameters n_p, n_s, n_i and λ_s and λ_i are the refractive indices of the three waves, and the wavelengths of the signal and idler waves, respectively.

The single-pass power gain of the parametric amplifier in the high-gain limit can be approximated by

$$G = \tfrac{1}{4}\exp(2gl), \tag{10.40}$$

where l is the length of the crystal.

A phase-mismatch between the waves can be expressed by

$$\Delta k = k_p - k_s - k_i, \tag{10.41}$$

where $k_j = 2\pi n_j/\lambda_j$, with $j = $ p, s, i, are the propagation constants of the three waves.

In the presence of phase-mismatch, the gain coefficient is reduced according to

$$g_{eff} = \left[g^2 - \left(\tfrac{1}{2}\Delta k\right)^2\right]^{1/2}. \tag{10.42}$$

The reduction of gain resulting from momentum or phase-mismatch is clearly evident. Maximum gain is achieved for $\Delta kl = 0$. Typical values for the coupling constant are on the order of $\kappa = 10^{-8}$/W, as will be discussed later. Therefore, a power density of at least $100\,\mathrm{MW/cm^2}$ is required to obtain a gain coefficient of $g = 1\,\mathrm{cm^{-1}}$ and a modest power gain of $G = 1.8$ in a 1 cm long crystal. At a $1\,\mu\mathrm{m}$ wavelength, the propagation constant is $k \approx 10^5$/cm in a material with an $n = 1.7$ refractive index. In order to minimize the effect of phase-mismatch, we require from (10.42) that $\Delta k/2 < g$. This means that the propagation constants have to be phase-matched to better than 10^{-5}/cm.

From these very basic considerations, one can already draw several conclusions that govern the design of parametric devices. The optimum configuration of a parametric converter depends critically on the pump intensity in the nonlinear crystal. There is a strong incentive to operate at the highest attainable levels for a given pump source. However, the practical and acceptable pump intensity depends strongly on the optical damage threshold of the crystal and its coatings. The importance of a high nonlinear coefficient d_{eff} is also clearly evident from these equations, as is the detrimental effect of phase-mismatch.

Figure 10.13 shows an optical schematic of an OPO designed to generate output at the eye-safe wavelength of $1.6\,\mu\mathrm{m}$. The OPO contains a 15 mm long KTP crystal as the nonlinear material which is pumped by a Q-switched Nd : YAG laser at $1.064\,\mu\mathrm{m}$. The maximum output energy of the pump laser was 10 mJ, produced in a 50 ns pulse. The crystal was positioned to achieve type-II noncritical phase-matching for a pump wavelength of $1.06\,\mu\mathrm{m}$ (Nd : YAG) and a signal wavelength near $1.6\,\mu\mathrm{m}$; the idler wavelength was therefore near $3.2\,\mu\mathrm{m}$. Noncritical phase-matching maximizes the effective nonlinear coefficient and essentially eliminates walk-off. The crystal was positioned inside an optical cavity formed by a pair

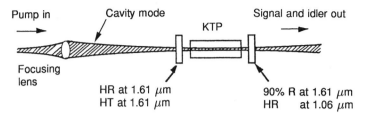

FIGURE 10.13. Plane-parallel resonator for optical parametric oscillator, showing mode-matching of focussed pump to OPO cavity mode.

of plane-parallel mirrors. The input mirror, through which the pump enters, was antireflection coated at the pump wavelength ($1.06\,\mu$m) and highly reflecting at the signal wavelength ($1.6\,\mu$m); while the output mirror was highly reflecting at the pump wavelength and 10% transmitting at the signal wavelength. The laser output was focused by a 100 cm focal length lens, carefully positioned so as to mode-match the waist of the pump to the cavity mode of the OPO resonator.

In this noncritically phase-matched configuration the wavelengths are $\lambda_s = 1.61\,\mu$m, $\lambda_i = 3.2\,\mu$m, $\lambda_p = 1.06\,\mu$m, and the respective indices of refraction are $n_s = 1.7348$, $n_i = 1.7793$, and $n_p = 1.7474$. The effective nonlinear coefficient of KTP for parametric conversion (as well as type-II harmonic generation) can be approximated as follows [18]

$$d_{\text{eff}} \approx (d_{24} - d_{15}) \sin 2\phi \sin 2\theta - (d_{15} \sin^2 \phi + d_{24} \cos^2 \phi) \sin \theta, \quad (10.43)$$

which reduces to $|d_{\text{eff}}| = 0.15 d_{15} + 0.84 d_{24}$ for $\theta = 90°$ and $\phi = 23°$. With the values of $d_{24} = 3.64$ pm/V and $d_{15} = 1.91$ pm/V, one obtains $d_{\text{eff}} \approx 3.3$ pm/V. Using these parameters we calculate from (10.39) $\kappa = 1.2 \times 10^{-8}$ W^{-1}.

Saturation and Pump Beam Depletion

When the Q-switched pump pulse is incident on the nonlinear crystal, the signal and ideal waves are amplified from the initial noise level. The number of round trips in the optical cavity necessary to amplify the signal and idler waves, multiplied by the cavity round-trip time, determines the delay in achieving threshold. This build-up time, required to achieve parametric oscillation, causes a temporal compression of the OPO output with respect to the pump pulse. Above threshold, after a short transition period during which a steady-state condition is established, the pump power is limited at the threshold value. Any pump input power above threshold is divided into power at the signal and idler beams. Since $\nu_3 = \nu_1 + \nu_2$, it follows that for each input pump photon above threshold, one photon at the signal and idler wavelengths is generated.

The oscilloscope trace in Fig. 10.14, taken from the OPO described on the previous pages, shows the dynamics of signal generation very nicely. The dashed curve is the input pump pulse. The two solid curves display the signal pulse and the depleted pump pulse at the output of the OPO. During the early part of the

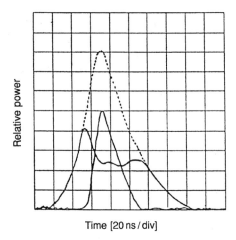

Time [20 ns / div]

FIGURE 10.14. Oscilloscope trace showing the depleted pump and the generated OPO output at 1.61 μm (solid curves) and the input pump (dashed curve).

pump pulse, there is a transient period during which the oscillation builds up from noise. After the transient period, the pump beam is clamped at its threshold value until the pump power falls below threshold.

Conversion Efficiency

In the plane-wave approximation, the conversion efficiency of an SRO for the ideal case of perfect phase-matching and zero losses is given by [19]

$$\eta_{\text{par}} = \sin^2 gl, \tag{10.44}$$

where g and l have been defined in (10.38), (10.40). Using this expression, total conversion of the pump can be achieved in theory. As the point of total conversion is exceeded, power starts to couple back into the pump field at the expense of the signal and idler fields and the conversion efficiency decreases.

The conversion efficiency in (10.44) is defined as the ratio of the sum of signal and idler power to pump power. The energy or power between the signal and idler beams is divided according to the ratio of the photon energies, that is,

$$\frac{h\nu_s}{h\nu_i} = \frac{\lambda_i}{\lambda_s}. \tag{10.45}$$

From this follows the ratio of the energy of the signal compared to the total energy converted by the OPO

$$\frac{h\nu_s}{h\nu_s + h\nu_i} = \frac{1}{1 + \lambda_s/\lambda_i}. \tag{10.46}$$

In the so-called degenerate mode, for which $\lambda_s = \lambda_i = 2\lambda_p$, each pump photon generates two photons at twice the pump wavelength. In this case, two orthogonally polarized beams at the same wavelength are emitted from the OPO.

Phase-Matching

In order to achieve significant gain in the parametric device, the pump, signal, and idler waves have to be phase-matched according to (10.37). In a medium without dispersion, all waves propagate with the same velocity, hence $\Delta k = 0$. In reality, Δk is not zero for different wavelengths, and just as in harmonic generation, dispersion is compensated for by birefringence. The index of refraction at a given wavelength is a function of the direction of propagation in the crystal as well as orientation of polarization.

When the signal and idler waves are both ordinary rays, one has type-I phase-matching. When either one is an ordinary ray, it is referred to as type-II phase-matching.

In a uniaxial crystal, if the signal and idler are ordinary rays (type I) with indices of refraction n_s and n_i, then the index of refraction for the pump wave necessary to achieve phase-matching is

$$n_p = (\lambda_p/\lambda_s)n_s = (\lambda_p/\lambda_i)n_i. \qquad (10.47)$$

In uniaxial crystals, the index of refraction n_o for an extraordinary wave propagating at an angle θ to the optic axis is given by

$$\frac{1}{n(\theta)^2} = \frac{\cos^2\theta}{(n^o)^2} + \frac{\sin^2\theta}{(n^e)^2}, \qquad (10.48)$$

where n^o and n^e are the ordinary and extraordinary indices of refraction, respectively. The index n_0 is thus limited by n^o and n^e. If n_p falls between n^o and n^e, then an angle θ_m exists for which $n_0 = n_p$. Propagation with $\Delta k = 0$ results, and all three waves travel at the phase-matching angle Θ_m with respect to the optical axis.

Phase-matching with a propagation direction normal to the optic axis ($90°$ phase-matching) is preferred if possible. In that direction, the double refraction angle is zero and the nonlinear interaction is not limited by the effective gain length but by the crystal's physical length. Non-$90°$ phase-matching does, however, allow angle tuning.

Tuning curves for parametric oscillators can be determined by solving the phase-matching equations (10.36), (10.37) for signal and idler frequencies at a given pump frequency as a function of the tuning variable. The most common tuning method is varying the direction of propagation with respect to the crystal axes. To carry out the calculations, the indices of refraction must be known over the entire tuning range. For most nonlinear crystals of interest, the indices can be obtained from the Sellmeier equation.

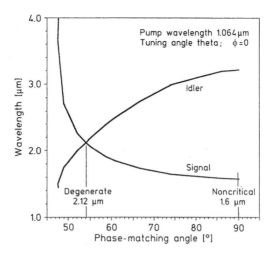

FIGURE 10.15. Tuning range of an OPO pumped by a 1.06 μm neodymium laser and employing a type-II KTP crystal.

Figure 10.15 exhibits the calculated phase-matching angles for a KTP optical parametric converter pumped at $1.064\,\mu\text{m}$. The two curves correspond to the signal and idler wavelengths for type-II phase-matching with $\phi = 0$. The polarization of the pump wave is along the y-axis (o-wave) of the crystal as is that of the signal wave. The idler wave is polarized in the x–z-plane (e-wave). The degenerate point at $2.12\,\mu\text{m}$ corresponds to a phase-matching angle of $\theta = 54°$.

Figure 10.16 illustrates a configuration of an intracavity doubled OPO designed to produce output in the ultraviolet/blue wavelength region. The device generates radiation between 760 and 1040 nm, which is internally doubled to produce tun-

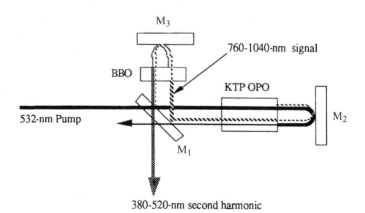

FIGURE 10.16. Layout of intracavity OPO, showing the different pump wavelengths involved: the 532 nm pump, the 760–1040 nm signal, and the 380–520 nm intracavity doubled signal [20].

able output from 380 to 520 nm. The OPO employs a 15 mm long KTP crystal, antireflection coated at 911 nm, cut for normal incidence at $\phi = 0°$ and $\theta = 69°$. The angles correspond to type-II phase-matching for a signal wavelength of 911 nm. The pump source is a diode-pumped Nd : YAG laser that is frequency doubled by a 1 cm long KTP crystal. The waist of the pump beam is positioned at the center of the 5 cm long OPO cavity.

The 532-nm pulse enters the OPO cavity through a flat dichroic turning mirror (M_1; highly reflecting at 800–950 nm and highly transmitting at 440–540 nm). The 532-nm pulse passes through the OPO crystal and is reflected off the flat rear cavity mirror (M_2; highly reflecting at 532 and 880–950 nm), to leave the cavity through mirror M_1 after making a second pass through the OPO crystal.

All mirrors have minimal reflectivity at the idler wavelength, making the OPO singly resonant. The intracavity 911-nm flux is reflected by the turning mirror (M_1) to pass through a 3 mm long BBO type-I doubling crystal. The final cavity mirror (M_3) is highly reflective at both 880–950 and 410–480 nm, which allows the doubled output to be extracted through the turning mirror (M_1) after making another pass through the doubling crystal. The use of two-pass nonlinear generation in both the doubler and the OPO significantly improves the conversion efficiency of these processes. The advantage of this configuration is that the KTP crystal is not exposed to the blue-ultraviolet flux, which could cause damage problems.

The output energies obtained at the various wavelengths are shown in Fig. 10.17 plotted as a function of electrical input energy to the diode arrays of the Nd : YAG pump laser.

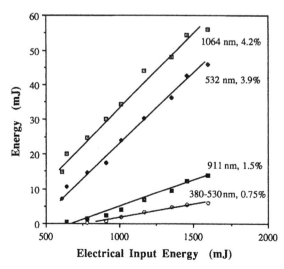

FIGURE 10.17. Output at various wavelengths of the OPO depicted in Fig. 10.16 as a function of electrical input to the Nd : YAG pump laser.

10.3.2   Quasi-Phase-Matching

In this subsection we will briefly review the concept of quasi-phase-matching (QPM), and summarize the properties of Periodically Poled LiNbO$_3$ (PPLN). Although QPM is applicable to any nonlinear process, the subject is covered here because of the importance of PPLN and its application in OPOs for the generation of mid-infrared radiation.

PPLN is the first commercially available crystal in which efficient nonlinear conversion processes, such as harmonic generation or parametric interactions, are not based on birefringence phase-matching but on a periodic structure engineered into the crystal. We will briefly explain this technique, termed quasi-phase-matching, by referring to Section 10.1.1. An in-depth treatment of this subject can be found in [21], [22].

As illustrated in Fig. 10.2 and expressed in (10.13) in a phase-matched condition $\Delta k = 0$ the harmonic power increases with the square of the interaction length l. In the situation of a fixed phase mismatch Δk, energy flows back and forth sinusoidally between the fundamental and harmonic beams with a period of $\Delta k \ell / 2 = \pi$ as the waves propagate through the crystal. Half of this period is the coherence length l_c given by (10.14) which is the maximum distance over which the second harmonic can grow. In a non-phase-matched condition, this distance is on the order of 2–20 μm for conversion processes of interest. The objective of birefringence phase-matching is to make this distance as long as possible by taking advantage of the wavelength and the polarization dependence of the refractive indices in a crystal. By arranging proper balance between dispersion and birefringence the technique of phase-matching increases l_c by a factor of 10^3.

Even before the concept of birefringence phase-matching was invented, it was proposed that a periodic structure in the crystal, which corrects the phase of the propagating beams each time it reaches π, would enable continued energy flow from the fundamental to the harmonic beam. An implementation of this concept is a crystal where the sign of the nonlinear coefficient is reversed after each distance l_c. In this case the relative phase between the waves is inverted after the conversion has reached its maximum. Therefore, on average, the proper phase relationship between the beams is maintained, and the second-harmonic power increases with the square of crystal length similar to the birefringence phase-matched case. The nonlinear coefficient d_Q for quasi-phase-matching is, however, reduced as compared to the phase-matched interaction according to [21],

$$d_Q = \frac{2}{\pi} d_{\text{eff}}. \qquad (10.49)$$

A phase reversal is the equivalent of slicing a crystal in thin wafers and stacking the wafers by rotating alternate wafers by 180°. The periodicity Λ of this structure is twice that of l_c, that is,

$$\Lambda = 2l_c. \qquad (10.50)$$

Because of the micrometer size thickness of alternating layers, the practical realization of quasi-phase-matching had to wait until it was possible to engineer a periodic phase reversal into a monolithic crystal. Only recently has the technique of periodically reversed polarization domains in ferroelectric crystals, combined with advances in lithography, made it possible to realize this concept. A reversal of the ferroelectric domains corresponds to a sign reversal of the nonlinear coefficient. In this process, standard lithography produces a patterned electrode with a period between 5 and 30 μm on the surface of a ferroelectric crystal such as LiNbO$_3$. A high-voltage pulse is applied to the crystal which is sandwiched between the patterned electrode and a uniform electrode. The high-electric-field strength of the voltage pulse permanently reverses the sign of the nonlinear coefficient in a pattern determined by the electrode structure.

This electric-field poling technique is employed to reproducibly manufacture periodically poled lithium niobate (PPLN) suitable for applications in infrared parametric oscillators and second-harmonic generation.

We will now derive an expression for the quasi-phase-matching condition in OPOs for periodically poled crystals. Parametric conversion requires energy conservation and momentum-matching conditions, as given by (10.36) and (10.37). A momentum mismatch expressed as a phase mismatch between the pump, signal, and idler waves has been defined in (10.41). In the frequency domain, a periodic structure can be represented by a grating wave vector

$$k_g = 2\pi/\Lambda, \tag{10.51}$$

where Λ is the period of the grating. In the presence of a grating structure in the crystal, the equation for phase mismatch (10.41) includes, as an additional term, the grating wave vector k_g [22]

$$\Delta k = k_p - k_s - k_i - k_g, \tag{10.52}$$

the first three terms being the conventional phase-matching condition. It is assumed that all wave vectors are collinear with the grating vector.

The objective of high parametric conversion is to eliminate the phase-mismatch caused by dispersion by selecting the appropriate crystal orientation, temperature, and polarization such that $k = 0$ is achieved. The grating vector in (10.52) provides an additional adjustable parameter that is independent of inherent material properties.

Differences in the three wave vectors k_p, k_s, k_i can be compensated for by an appropriate choice of the grating vector k_g such that $\Delta k = 0$ can be achieved. Introducing (10.51) into (10.52) and with $\Delta k = 0$ the grating period under quasi-phase-matching conditions is given by

$$\Lambda_g = \frac{2\pi}{k_p - k_s - k_i}. \tag{10.53}$$

Equation (10.53) can also be expressed by

$$\frac{1}{\Lambda_g} = \frac{n_p}{\lambda_p} - \frac{n_s}{\lambda_s} - \frac{n_i}{\lambda_i}.$$ (10.54)

This equation is the equivalent of (10.37) which was derived for the birefringence phase-matching case. If one substitutes Δk in (10.10) with k_g from (10.51) one obtains the condition for harmonic generation

$$\Lambda_g = \frac{\lambda_1}{2(n_{2\omega} - n_{1\omega})}.$$ (10.55)

The fact that the domain thickness is a free parameter, which can be customized for a particular nonlinear process, offers significant advantages over birefringent phase-matching. For example, quasi-phase-matching permits wavelength selection over the entire transmission window of the crystal, it allows utilization of the largest nonlinear coefficient, and it eliminates problems associated with walk-off since all interactions are noncritical.

For example, in $LiNbO_3$ interactions with all waves polarized parallel to the crystal optic axis utilizes the largest nonlinear coefficient $d_{33} = 27 \, pm/V$. PPLN permits noncritical phase-matching with this coefficient. Birefringent phase-matching requiring orthogonally polarized beams can only be accomplished with the smaller coefficient $d_{31} = 4.3 \, pm/V$. Therefore, PPLN has a parametric gain enhancement over single-domain material of $(2d_{33}/\pi d_{31})^2 \approx 20$.

10.4   Raman Laser

The Raman laser, which is based on stimulated Raman scattering (SRS), can provide access to wavelengths not directly available from solid-state lasers. In its basic form, the Raman laser consists of a high pressure gas cell and resonator optics. If this completely passive device is pumped by a high-power laser, a fraction of the laser wavelength is shifted to a longer wavelength. The particular wavelength shift depends on the gas in which SRS takes place. Also, if the power level of the laser is increased, additional spectral lines will appear at longer, as well as shorter, wavelengths with respect to the pump wavelength.

The basic Raman effect is an inelastic light-scattering process. The energy levels of interest for Raman scattering are shown in Fig. 10.18. An incident photon $h\nu_p$ is scattered into a photon $h\nu_s$ whereas the difference in energy $h(\nu_p - \nu_s) = h\nu_R$ is absorbed by the material. In Fig. 10.18, u is the upper state of the molecule, and i and f are the initial and final states. In principle, the excitation of the material may be a pure electronic excitation or a vibrational or rotational excitation of a molecule. Solid-state-laser-pumped Raman lasers typically employ gases such as hydrogen or methane; therefore levels i and f are the vibrational levels of the ground state of the molecule. The upper level u can be a real state or a "virtual" upper state. The frequency ν_s is called a Stokes frequency and is lower than the

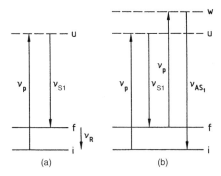

FIGURE 10.18. Raman process. Generation of (a) first Stokes light and (b) first anti-Stokes light.

incident light frequency v_p. The difference between v_p and v_s,

$$v_p - v_s = v_R, \tag{10.56}$$

is the Raman shift which is characteristic of the material in which the Raman process is observed.

If the system is in an excited state to begin with, it may make a transition downward while the light is scattered. In that case the scattered light contains anti-Stokes frequencies that are larger than the incident frequency.

In the stimulated Raman effect, the pump laser at frequency v_p excites molecules to level u, and if a population inversion exists between levels u and f, it can produce lasing action. In this case the radiation v_s becomes amplified, while the pump radiation v_p loses energy. The process has typical laser characteristics, such as pump energy threshold, exponential gain, and narrow linewidth. The emission in Fig. 10.18(a) is called the first Stokes line, usually written S_1. If a high-power laser is focused into a Raman medium, additional lines will appear at the output. Actually, a single laser frequency interacting with molecules will produce a "comb" of frequencies, each separated from its neighbor by the frequency spacing v_R. These additional lines will be to the left and right of the wavelength scale with regard to the laser pump wavelength.

The additional lines are produced by parametric four-wave mixing of the various waves propagating in the Raman medium. As an example, Fig. 10.18(b) illustrates the generation of one such line, having a wavelength shorter than the pump wavelength. This so-called anti-Stokes line is the result of the interaction of v_p and v_{s1} both propagating in the same direction. The parametric four-wave mixing process does not require a population inversion between w and i, therefore there is no well-defined threshold. The simplest way of looking at this interaction is that the two frequencies beat together to produce polarization (induced dipole moments in the molecules) at the difference frequency. This polarization then modulates the laser–molecule interaction and produces light beams at the side frequencies.

Stokes lines have lower frequency (longer wavelength) and anti-Stokes lines have higher frequency (shorter wavelength). In each case the line is labeled first, second, and so on, by counting the number of frequency shifts from the pump laser.

SRS can be described as a nonlinear interaction involving the third-order nonlinear susceptibility χ^3. At a medium's Raman resonance, the third-order susceptibility reduces to the peak Raman susceptibility χ_R'', where the double prime indicates the imaginary part of the total susceptibility.

The growth of the electric field E_2 at the Stokes wavelength and depletion of the pump field E_1 is governed by the equations

$$\frac{\partial E_1}{\partial z} = -\frac{\omega_p}{2cn_p}\chi_R''|E_2|^2 E_1, \tag{10.57}$$

$$\frac{\partial E_2}{\partial z} = \frac{\omega_s}{2cn_s}\chi_R''|E_1|^2 E_2, \tag{10.58}$$

where $\omega_p - \omega_s = \omega_R$ are the frequencies, n_s and n_p are the indices of refraction, and c is the velocity of light. For a constant pump field, the Stokes field grows exponentially with a power gain given by

$$P_s(l) = P_s(0)\exp(g_s l), \tag{10.59}$$

where g_s is the gain coefficient and l is the interaction length in the Raman medium

$$g_s = \frac{\omega_s \chi_R''|E_p|^2}{n_s c} = \frac{4\pi \chi_R'' I_p}{\lambda_S n_s n_p \varepsilon_0 c}. \tag{10.60}$$

If one expresses the third-order Raman susceptibility, χ_R'', in terms of spontaneous Raman scattering cross section, $d\sigma/d\Omega$, one obtains

$$g_s = \frac{\lambda_p \lambda_S^2 N(d\sigma/d\Omega) I_p}{n_s^2 hc\pi \,\Delta\nu_R}, \tag{10.61}$$

where λ_S is the Stokes wavelength, I_p is the pump intensity, N is the number density of molecules, h is Planck's constant, and $\Delta\nu_R$ is the full-width, half-maximum Raman linewidth.

From these equations it follows that the gain for a single-pass Raman medium is proportional to the incident intensity, the active media cross section (which includes pressure and linewidth dependencies), and the length of the Raman cell.

The threshold of a Raman laser is usually defined as the gain required to achieve an output power at the Stokes wavelength that is of the same order as the pump radiation. For example, to achieve a 1 MW Stokes shifted power output, one requires a gain length product of $g_s l = 36$ in the Raman medium in order for the radiation to build up from the initial spontaneous noise level which is $P_s = h\nu_s \,\Delta\nu_s \approx 10^{-19}$ W in the visible. Table 10.2 summarizes the data for the most important Raman media.

TABLE 10.2. Stokes shift and Raman
scattering cross section for several gases.

Medium	ν_R [cm^{-1}]	$\dfrac{d\sigma}{d\Omega}\left[\dfrac{cm^2}{Ster}\right]$
H$_2$	4155	8.1×10^{-31}
CH$_4$	2914	3.0×10^{-30}
N$_2$	2330	3.7×10^{-31}
HF	3962	4.8×10^{-31}

The maximum theoretical conversion efficiency of a Raman laser is

$$\eta_{Ram} = \frac{\nu_p - \nu_R}{\nu_p}. \tag{10.62}$$

If one introduces into (10.62) the numbers given for ν_R in Table 10.2 it is obvious that the conversion efficiency can be very high. For example, a frequency-doubled Nd : YAG laser, Raman shifted with CH$_4$, provides an output at 630 nm. With $\nu_R = 2914\,\mathrm{cm}^{-1}$ and $\nu_p = 18{,}797\,\mathrm{cm}^{-1}$ one obtains $\eta_{Ram} = 84\%$.

A rigorous mathematical treatment of SRS can be found in [24], [25].

The simplest Raman laser is based on the single-pass emission in a gas-filled cell, as shown in Fig. 10.19(a). The output beam quality is similar to that of the input pump. Whereas the optics may be simply designed to prevent optical damage to the windows at focal intensities high enough to produce significant energy conversion, many nonlinear processes may occur to limit the conversion efficiency at high energies. Copious second Stokes and anti-Stokes production may occur, as well as stimulated Brillouian scattering and optical breakdown. The single-pass cell does not provide discrimination against these other nonlinear processes at high-energy inputs.

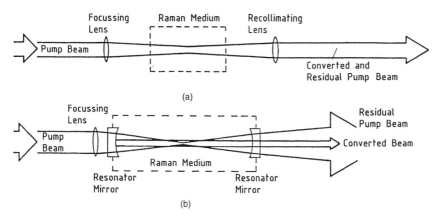

FIGURE 10.19. Raman laser configuration: (a) single pass cell and (b) Raman resonator.

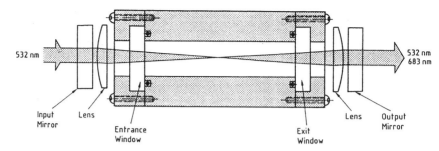

FIGURE 10.20. Schematic diagram of high-energy Raman cell [23].

By using mirrors at each end, feedback can be selectively enhanced at only the first Stokes wavelength. A Raman laser utilizing a resonator is shown schematically in Fig. 10.19(b). The mirror coatings selected allow all pump light to pass into the cavity, while inducing resonator action at the Raman shifted frequency. The resonator transforms a multimode pump laser beam into a nearly diffraction-limited output beam with a slightly narrower pulse width. The high-quality beam is a result of keeping the pump intensity below single-pass threshold so that only multiple reflections of the lowest-order mode will achieve sufficient intensity to reach the stimulated scattering threshold.

A state-of-the-art hydrogen gas Raman laser that converts the frequency doubled output of an Nd : YAG laser from 532 to 683 nm with an efficiency of up to 40% is illustrated in Fig. 10.20. The Raman laser consisted of the gas-pressure cell and the concentric resonator configuration comprised of two flat mirrors and a pair of plano-convex lenses. The resonator length was 20 cm. The entrance mirror

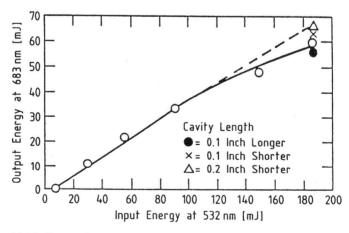

FIGURE 10.21. Concentric resonator output energy for different pump input energies and cavity lengths. The solid line corresponds to the concentric position. (Hydrogen pressure: 1000 psi, output reflector 50% at 683 nm, 20 cm resonator.)

has high transmission at 532 nm and high reflection at 683 nm. The exit mirror has high transmission for the pump wavelength at 532 nm and 50% reflectance at the first Stokes wavelength. The reflectance for the second Stokes (954 nm) and first anti-Stokes (436 nm) wavelengths was only a few percent at both mirrors.

Figure 10.21 exhibits the output versus input for the resonator at the exact length for the concentric geometry (solid line) and for slight variations in length [23].

10.5 Optical Phase Conjugation

Optical phase conjugation, also referred to as wave-front reversal, has been demonstrated in solids, liquids, and gaseous media using a number of nonlinear optical interactions, such as three- and four-wave mixing, stimulated Brillouin scattering (SBS), stimulated Raman scattering (SRS), and photon echoes.

From the standpoint of solid-state laser engineering, phase conjugation via SBS is particularly important because it provides a means to correct optical aberrations produced by a laser amplifier stage. Comprehensive descriptions of the principle of phase conjugation can be found in [26], [27].

Basic Considerations. From a mathematical point of view, phase-conjugation can be explained by considering an optical wave of frequency ω moving in the $(+z)$-direction,

$$E_1(x, y, z, t) = A(x, y) \exp\{j[kz + \phi(x, y)] - j\omega t\}, \tag{10.63}$$

where E_1 is the electric field of the wave with wavelength $\lambda = 2\pi/k$. The transverse beam profile is given by the function $A(x, y)$ and the phase factor $\phi(x, y)$, indicating how the wave deviates from a uniform, ideal plane wave. In particular, the phase factor carries all the information about how the wave is aberrated.

If the beam given by (10.63) is incident upon an ordinary mirror, it is reflected upon itself and the sign of z changes to $(-z)$, all other terms of the equation remain unchanged. However, if the beam is incident upon a phase-conjugate mirror it will be reflected as a conjugated wave E_2 given by

$$E_2(x, y, z, t) = A(x, y) \exp\{j[-kz - \varphi(x, y)] - j\omega t\}. \tag{10.64}$$

In addition to the sign change of z, the phase term has changed sign, too. The conjugate beam corresponds to a wave moving in the $(-z)$-direction, with the phase $\varphi(x, y)$ reversed relative to the incident wave. We can think of the process as a reflection combined with phase- or wave-front reversal. The phase-reversal expressed by (10.64), for example, means that a diverging beam emitted from a point source, after reflection at a phase-conjugate mirror, will be converging and be focused back to the point of origin.

A practical application of phase-reversal in a laser system is depicted in Fig. 10.22. Shown is an oscillator which produces an output with a uniform flat wavefront that is distorted in the amplifier medium. An ordinary mirror merely

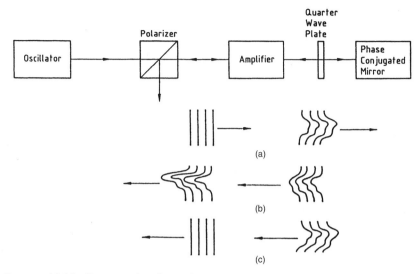

FIGURE 10.22. Compensation for optical phase distortions caused by an amplifying medium using nonlinear optical phase-conjugation. (a) Original wave, (b) ordinary reflection, and (c) phase conjugated reflection.

inverts the distortion as it reflects the beam, thereby keeping the distortion fixed with respect to the propagation direction. With a second pass through the amplifier, the distortion is essentially doubled. A phase-conjugate mirror, on the other hand, reverses the wavefronts relative to the wave-propagation direction; hence, the same region of the amplifier that originally created the distortion compensates for it during the second pass through the amplifier.

However, it should be noted that although the wavefront is conjugated, the polarization state of the backward-going field is not conjugated. As far as the polarization vector is concerned, the SBS mirror behaves like an ordinary mirror.

As already mentioned, SBS of laser radiation in liquids or gases is the preferred approach for designing phase-conjugate mirrors. SBS involves the scattering of light by sound waves, i.e., acoustic phonons or pressure/density waves. The incoming laser radiation generates an ultrasound wave by electrostriction corresponding to a spatial density modulation in the material. This modulation changes the refractive index and, therefore, a refractive index grating is created in the material.

The resultant Stokes scattered light is down-shifted in frequency by a relatively small amount ($\approx 10^9$ to 10^{10} Hz). The gain of the Stokes-shifted wave is generally the highest in a direction opposite to that of the incident beam. The efficiency of the SBS process (defined as the ratio of the Stokes-shifted, backward-going energy or power, to the incident optical energy or power) can be as high as 90%. SBS has been studied extensively in liquids, solids, and gases. The steady-state gain factor of SBS is generally large, making SBS the dominant simulated scattering process in many substances.

Phase-conjugate mirrors can be employed to compensate thermal lensing in double-pass amplifiers if the beam intensity is high enough to provide sufficient reflectivity from the SBS process. An SBS mirror in its most common form consists of a lens that focuses the incident beam into the bulk nonlinear material. Figure 10.22 illustrates the most common configuration of compensating thermally induced lensing in amplifiers with an SBS mirror. The output from the oscillator passes through the amplifiers, is reflected by the SBS mirror, and propagates a second time through the amplifier chain. Output coupling is realized by a polarizer and a 90° polarization rotation carried out by a $\lambda/4$ wave plate which is passed twice. Conjugate reflectivities in this type of configuration are typically 50% to 80%, so that the efficiency of the amplifier stage is reduced by a fraction that depends on the ratio of energy extraction between the first and second pass in the amplifiers.

Material Properties. Pressurized gases such as CH_4 and SF_6, or liquids such as CCl_4, acetone, and CS_2 are usually employed to provide efficient SBS for a $1.06\,\mu$m pump. SBS materials can be characterized by a gain coefficient and an acoustic-decay time. The gain coefficient g determines the threshold and slope of the reflectivity curve. Materials with a lower gain show a higher threshold and a slower increase of the reflectivity with increased pumping.

The acoustic-decay time τ establishes criteria for the coherence or linewidth of the pump beam. For efficient interaction with the sound-wave grating, a coherent interaction on the order of the acoustic decay time is required. An inverse bandwidth of the pump pulse, which is larger than the acoustic-phonon damping time, describes the steady-state condition, that is, $(\Delta v)^{-1} > \tau_B$. Maximum gain is achieved if this condition is satisfied. Most calculations, as well as the simple relationships for gain and threshold given below, are for this case.

Table 10.3 lists steady-state SBS gain coefficient g and acoustic decay time τ_B of representative gases and fluids.

Quite noticeable is the longer phonon lifetime in gases compared to liquids. Single longitudinal mode Q-switched lasers have a pulse length of 10 to 20 ns;

TABLE 10.3. Steady-state SBS gain coefficients and acoustic decay time of representative gases and fluids.

Medium	g [cm/GW]	τ_B [ns]
CH_4 (30 atm)	8	6
CH_4 (100 atm)	65	17
SF_6 (20 bar)	14	17
SF_6 (22 atm)	35	24
CCL_4	6	0.6
Acetone	20	2.1
CS_2	130	5.2

therefore the coherence requirement for steady-state conditions, stated above, will not be met in gases. SBS in gases will be highly transient, while in liquids it will reach a steady state. This has the important implication that the effective gain for gases is actually considerably lower than the values in Table 10.3. The effective gain in a transient SBS process can be as much as 5 to 10 times less than the value given in the table. The lower SBS threshold and higher gain is the reason, while most current systems employ liquids as the nonlinear medium. Also, liquid cells are not as long as gas-filled cells. Liquid cells are typically 10–15 cm long, whereas gas cells are between 30–100 cm in length.

Gain and Threshold. Reflectivity is an important parameter in designing an SBS mirror since it greatly influences the overall laser-system efficiency. The calculation of the reflectivity of SBS mirrors requires the solution of coupled differential equations for the incident and reflected light waves and the sound wave grating. The calculations of the scattering processes require a great numerical effort [28].

The reflectivities of SBS mirrors depend on the excitation condition and the material properties of the nonlinear medium. Reflectivity versus pump power or energy can be described by a threshold condition and a nonlinear power relationship. Threshold depends mainly on materials properties and cell geometry, whereas the increase of reflectivity is a function of the pump-beam power.

The gain of an SBS process is a function of the frequency v_0 of the pump radiation, the phonon lifetime τ_B, the electrostrictive coefficient γ, the velocity of light and sound c and v, and the refractive index n_0 and density ρ_0 of the material [28],

$$g_B = \frac{v^2 \gamma^2 \tau_B}{c^2 v_0 n_0 \rho_0}. \tag{10.65}$$

The threshold pump power for SBS is a parameter of practical importance. If one measures reflectivity versus pump power input, one observes that at a particular input power, reflectivity changes over many orders of magnitude for just a few percent increase in power. In this regime of rapid signal growth, a threshold is usually defined if the gain has reached exp(30).

The gain of the backward reflected Stokes wave over a length l of the nonlinear medium is $\exp(g_B I_P l)$ for a plane wave of pump intensity I_P. If the threshold P_{th} is defined as the condition when the gain reaches exp(30), then one obtains [29]

$$P_{th} = 30A/g_B l, \tag{10.66}$$

where A is the area of the pump beam.

For a Gaussian pump beam, one obtains [30]

$$P_{p,th} = \frac{\lambda_p}{4g_B} \left\{ 1 + [1 + 30/\arctan(l/b)]^{1/2} \right\}^2, \tag{10.67}$$

where b is the confocal parameter defined in (5.8) and l is the length of the SBS cell. If the cell is long compared to the confocal parameter $l \gg b$, the lowest threshold is obtained, since $\arctan(l/b) = \pi/2$ and (10.67) reduces to

$$P_{th} = 7.5\frac{\lambda_p}{g_B}. \tag{10.68}$$

The pump power has to be well above the threshold of the nonlinear reflectivity curve. Otherwise, the fidelity of the reflected signal becomes small because the parts of the beam away from the central peak, with energies below this threshold value, are not reflected. The useful power range of the incident radiation is limited by the SBS threshold at the lower limit and by optical breakdown at the upper end.

A typical curve for SBS reflectivity versus input energy is shown in Fig. 10.23. These experimental data were obtained by focusing a Q-switched Nd : YAG laser into a CCl_4 cell with an 80 mm long focal length lens. Both threshold and the slope of reflectivity depend on the coherence properties of the beam. There is a minimum requirement on the coherence or linewidth of the incident beam such that a coherent spatial and temporal interaction with the sound-wave grating is possible.

First, the coherence length of the beam should exceed the longitudinal sound-wave extension. If the pump beam has a Gaussian spatial profile, the sound wave has a longitudinal extent given approximately by the confocal parameter. Second, the coherence time (i.e., the inverse of the linewidth, $1/\Delta\nu$), of the pump beam should exceed the response time τ_B of the acoustic phonons. Typically, τ_B is on the order of nanoseconds for liquids, and tens of nanoseconds for pressurized gases. Therefore, high-efficiency SBS requires linewidth narrowing or preferably single-longitudinal-mode operation.

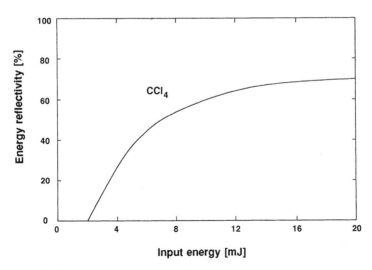

FIGURE 10.23. SBS energy reflectivity of CCL_4 as a function of pump energy for a Q-switched Nd : YAG laser [31].

Summary

One usually assumes a linear relationship between the induced electric polarization and the electric field strength. The term *nonlinear optics* refers to those phenomena in which the polarization per unit volume is no longer proportional to the electric field strength. Nonlinear effects can be described by a power-series expansion of the induced polarization and electric field. The nonlinear response of the medium can give rise to an exchange of energy between electromagnetic waves of different frequencies (second harmonic generation, frequency mixing, optical parametric intractions), or lead to an exchange of energy between electromagnetic waves and phonons in transparent media (stimulated Raman or Brillouin scattering). In the first group of interactions that take place between electromagnetic waves, crystals with large nonlinear coefficients are utilized as the nonlinear medium, whereas high-pressure gas or a liquid is used as the nonlinear medium for Raman or Brillouin scattering.

Second-harmonic generation takes place when two photons combine and a photon at twice the energy is created; in this case a wave at frequency ν_1 is converted to a wave at frequency $2\nu_1$. In frequency mixing two waves at frequencies ν_1 and ν_2 produce a wave at frequency ν_3, such that $\nu_1 + \nu_2 = \nu_3$. In an optical parametric oscillator the reverse process takes place, namely, a strong pump beam at frequency ν_3 produces radiation at two lower frequencies $\nu_3 = \nu_1 + \nu_2$. In all these processes the interaction of the various beams and the flow of energy among them is critically dependent on a velocity match of the interacting waves in the nonlinear crystal. Dispersion causes waves at different frequencies to propagate at different velocities. Taking advantage of the birefringence of nonlinear crystals is the most common approach to compensate for the effects of dispersion. In a birefringent crystal the index of refraction and therefore the propagation velocity of a wave at a given frequency depends on the propagation direction within the crystal and its sense of polarization.

Besides the aforementioned nonlinear processes, stimulated Raman scattering in a high-pressure gas cell is exploited in solid-state laser technology to expand the frequency range. Stimulated Raman scattering is an inelastic scattering process and involves interaction of light with the vibrational levels of the ground state of the gas molecule. An incident pump photon is scattered, while the difference energy is absorbed in the gas raising the vibrational level of the molecule to a higher level. The frequency shift is relatively large and is determined by the different vibrational frequencies of the gas. At very-high pump levels additional spectral lines will appear at longer as well as shorter wavelengths with respect to the pump wavelength. Stimulated Brillouin scattering involves scattering of laser radiation by sound waves. The incoming laser radiation generates an ultrasound wave by electrostriction corresponding to a density modulation in the material. This modulation changes the refractive index, and a refractive index grating is created in the material. Stimulated Brillouin scattering is the nonlinear effect exploited in phase conjugation, which provides a means to correct for optical aberrations produced in a laser amplifier stage.

References

[1] P.A. Franken, A.E. Hill, C.W. Peters, and G. Weinreich: *Phys. Rev. Lett.* **7**, 118 (1961).

[2] N. Bloembergen: *Nonlinear Optics* (Benjamin, New York), 1965.

[3] F. Zernike and J.E. Midwinter: *Applied Nonlinear Optics* (Wiley, New York), 1973.

[4] R.L. Sutherland: *Handbook of Nonlinear Optics* (Dekker, New York), 1996.

[5] R.W. Boyd: *Nonlinear Optics* (Academic, San Diego, CA), 1992.

[6] D.L. Mills: *Nonlinear Optics*, 2nd ed. (Springer, Berlin), 1998.

[7] D.A. Kleinman: in *Laser Handbook* (edited by. F.T. Arecchi, E.O. Schulz-DuBois) (North-Holland, Amsterdam), 1972, Vol. 2, pp. 1229–1258.

[8] D. Hon: in *Laser Handbook* (edited by M. Stitch) (North-Holland, Amsterdam), 1979, Vol. 3, pp. 421–456.

[9] M. Born and E. Wolf: *Principles of Optics* (Macmillan, New York), 1964.

[10] N.F. Nye: *Physical Properties of Crystals* (Clarendon, Oxford), 1960.

[11] M.V. Hobden: *J. Appl. Phys.* **38**, 4365 (1967).

[12] Lasermetrics, Electro-Optics Div., Englewood, NJ, Data Sheet 8701, *Optical Harmonic Generating Crystals* (February 1987).

[13] G.D. Boyd, A. Ashkin, J.M. Dziedzic, and D.A. Kleinman: *Phys. Rev.* **137**, 1305 (1965).

[14] W.F. Hagen, P.C. Magnante: *J. Appl. Phys.* **40**, 219 (1969).

[15] R.G. Smith: *IEEE J. Quantum Electron.* **QE-6**, 215 (1970).

[16] R.S. Craxton: *IEEE J. Quantum Electron* **QE-17**, 1771 (1981); *Opt. Commun.* **34**, 474 (1980).

[17] R.L. Byer: in *Quantum Electronics: A Treatise* (edited by H. Rabin and C.L. Tang) (Academic Press, New York), 1973, Vol. I, Pt. B, pp. 587–702.

[18] R.C. Eckardt, H. Masuda, Y.L. Fan, and R.X. Byer: *IEEE J. Quantum Electron.* **QE-26**, 922 (1990).

[19] J.E. Bjorkholm: *IEEE J. Quantum Electron.* **QE-7**, 109 (1971).

[20] L.R. Marshall, A. Kaz, and O. Aytur: *Opt. Lett.* **18**, 817 (1993).

[21] M.M. Fejer, G.A. Magel, D.H. Jundt, and R.L. Byer: *IEEE J. Quantum Electron.* **QE-28**, 2631 (1992).

[22] L.E. Myers, R.C. Eckardt, M.M. Fejer, and R.L. Byer: *J. Opt. Soc. Am. B* **12**, 2102 (1995).

[23] D.G. Bruns and D.A. Rockwell: *High-Energy Raman Resonator* (Hughes Aircraft Comp., Culver City, CA). Final Report, Report FR-81-72-1035, (1981).

[24] G.L. Eesley: *Coherent Raman Spectroscopy* (Pergamon, New York) 1981; A. Owyoung: CW stimulated Raman spectroscopy, in *Chemical Applications of Nonlinear Raman Spectroscopy* (edited by B. Harvey) (Academic, New York), 1981, pp. 281–320.

[25] W. Kaiser and M. Maier: Stimulated Rayleigh, Brillouin and Raman spectroscopy, in *Laser Handbook*, Vol. II (edited by F.T. Arecchi and E.O. Schulz-DuBois), (North Holland, Amsterdam), 1972.

[26] B.Ya. Zel'dovich, N.F. Pilipetsky, and V.V. Shkunov: *Principles of Phase Conjugation*, Springer Ser. Opt. Sci., Vol. 42 (Springer, Berlin), 1985.

[27] D.M. Pepper: Nonlinear optical phase conjugation, in *Laser Handbook*, Vol. 4, pp. 333–485, (edited by M.L. Stitch and M. Bass) (North-Holland, Amsterdam), 1985.

[28] R. Menzel and H.J. Eichler: *Phys. Rev. A* **46**, 7139 (1992).

[29] I.D. Carr, D.C. Hanna: *Appl. Phys. B* **36**, 83 (1985).

[30] D. Cotter, D.C. Hanna, and R. Wyatt: *Appl. Phys.* **8**, 333 (1975).

[31] R. Menzel and D. Schulze: *Proc. Int'l Summer School on Applications of Nonlinear Optics*, Prague (1993).

Exercises

1. Use (10.17) and derive an expression for the type-I phase-matching angle for second harmonic generation in a positive uniaxial crystal.

2. Consider the third harmonic generation system shown below:

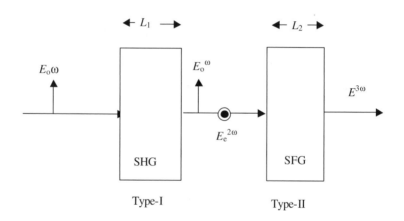

where SHG stands for second harmonic generator and SFG stands for sum frequency generator. If the input beam is of the form

$$I_\omega(r, t) = I_{0\omega} \exp\left(-\left(\frac{2r^2}{w^2} + \frac{t^2}{T^2}\right)\right).$$

(a) What is the beam radius and pulse width of the third harmonic output?

(b) Why might one prefer to use a second harmonic generator and a sum frequency generator to produce a third harmonic generator rather than a single third harmonic generator?

Appendix A

Conversion Factors and Constants

In this Appendix we have listed some of the most frequently used conversion factors and constants.

Physical Constants

$h = 6.626 \times 10^{-34}\,\mathrm{J\,s}$	Planck constant
$e = 1.602 \times 10^{-19}\,\mathrm{A\,s}$	Charge of an electron
$k = 1.381 \times 10^{-23}\,\mathrm{J\,K^{-1}}$	Boltzmann constant
$c = 2.998 \times 10^{8}\,\mathrm{m\,s^{-1}}$	Speed of light in vacuum
$\varepsilon_0 = 8.854 \times 10^{-12}\,\mathrm{A\,s\,V^{-1}\,m^{-1}}$	Permittivity of free space
$\mu_0 = 1.257 \times 10^{-6}\,\mathrm{V\,s\,A^{-1}\,m^{-1}}$	Permeability of free space
$Z_0 = \sqrt{\mu_0/\varepsilon_0} = 376.7\,\Omega$	Impedance of free space
$g = 9.81\,\mathrm{m\,s^{-2}}$	Acceleration due to gravity

Force Conversions

$$1\,\text{dyne} = 1\,\mathrm{g\,cm\,s^{-2}}$$
$$1\,\mathrm{N} = 1\,\mathrm{kg\,m\,s^{-2}} = 10^{5}\,\text{dyne}$$
$$1\,\text{kp} = 9.81\,\mathrm{kg\,m\,s^{-2}}$$

Energy Conversions

$$1\,\mathrm{J} = 1\,\mathrm{W\,s}$$
$$1\,\text{cal} = 4.19\,\mathrm{J}$$
$$1\,\text{eV} = 1.60 \times 10^{-19}\,\mathrm{J}$$
$$1\,\text{erg} = 10^{-7}\,\mathrm{J}$$

Pressure Conversions

$$1\,\text{Pa} = 1\,\mathrm{N\,m^{-2}}$$
$$1\,\text{MPa} = 145.1\,\text{psi} = 9.87\,\text{atm}$$
$$1\,\text{atm} = 1.013\,\text{bar} = 1.033\,\mathrm{kp\,cm^{-2}}$$
$$1\,\text{bar} = 10^{6}\,\mathrm{dyne\,cm^{-2}} = 10^{5}\,\mathrm{N\,m^{-2}}$$
$$1\,\text{torr} = 133.3\,\mathrm{N\,m^{-2}}$$

Conversion of English Units into the MKS System

1 in = 2.540 cm

1 gal = 3.785 ltr

1 lb = 0.453 kg

1 atm = 14.7 psi

1 Btu = 1055.8 W s

$T [^\circ C] = \frac{5}{9}(T [^\circ F] - 32)$

$\Delta T [^\circ C] = \frac{5}{9} \Delta T [^\circ F]$

$1 \, cal \, cm^{-1} \, C^{-1} \, s^{-1} = 242 \, Btu/hr \, ft \, F$

$1 \, cal \, g^{-1} = 1.8 \, Btu/lb$

Conversion of Angles

$1^\circ = 17.45 \, mrad$, $1' = 0.29 \, mrad$, $1'' = 4.85 \, \mu rad$

$1 \, rad = 57^\circ \, 17' \, 45''$, $1 \, mrad = 3' \, 26''$

Conversion of Wavenumber n [cm^{-1}] to Energy

$E = n \, hc$

$hc = 1.986 \times 10^{-23} \, W \, s \, cm$

$n = 1/\lambda$

Conversion of Linewidth Given in Wavelength ($\Delta\lambda$) or Wavenumber (Δn) to Bandwidth ($\Delta\nu$)

$\Delta\nu/\nu = \Delta\lambda/\lambda = \Delta n/n$

$\Delta\nu = c \Delta n$

$\Delta\lambda = \Delta n \lambda^2 = \Delta\nu \lambda^2/c$

Amplifier Gain

$g \, (dB) = 10 \log(E_2/E_1)$

Optical Units

Wavenumber	$n[1/cm] \approx 10^4/(\lambda \, [\mu m])$
Frequency	$\nu[Hz] \approx 3 \times 10^{14}/(\lambda \, [\mu m])$
Photon energy	$E[J] \approx 1.987 \times 10^{-19}/(\lambda \, [\mu m])$
Photon energy	$E[eV] \approx 1.24/(\lambda \, [\mu m])$

Conversion of Transmission, T, to Optical Density, D

$D = \log(1/T);$ $T = 10^{-D}.$

Conversion of the Nonlinear Refractive Index n_2 (esu) to Nonlinear Refractive Coefficient γ (cm^2/W)

$$\gamma \ (\text{cm}^2/\text{W}) = \frac{4.19 \times 10^{-3}}{n_0} n_2 \ (\text{esu}).$$

Appendix B

Definition of Symbols

A_{21}	Einstein coefficient (Chapter 1)
A^*	area
a	aperture radius
b	confocal parameter
B_{21}, B_{12}	Einstein coefficients (Chapter 1)
B	beam breakup integral (Chapter 4)
B_r	brightness (Chapter 4)
c, c_0	speed of light
C	capacitance (Chapter 6)
C_T, C_r, C_ϕ, C_B	combination of material constants (Chapter 7)
d	thickness (Chapters 5, 9)
d, d_{14}, d_{eff}	coefficient of nonlinear polarization
D, D_0	beam diameter
e	electron charge
E	Young's modulus (Chapter 7)
E_1, E_2, E_3, E_0	energy levels (Chapter 1)
E_{st}	stored energy
ΔE	energy difference
E_{EL}	electrical input energy
E_{SC}	scale factor energy (Chapter 8)
E_{in}, E_{out}, E_{opt}	input-, output-, optical energy (Chapters 3, 4)
E_{ex}	flashlamp explosion energy
E_i, E_0, E_s	input-, output-, saturation fluence
E_1, E_2, E_3, E	electric field strength (Chapter 10)
E_j, E_z	electric field strength (Chapter 8)
$f, f', f'', f_x, f_y, f_{ef}$	focal length
F	finesse of interferometer (Chapter 5)
f	pulse repetition rate (Chapter 8)
$g_1, g_2, g(n_1), g(R_1), g(R_2)$	degeneracy of energy levels (Chapters 1, 2, 3)

---

*Symbols whose definitions are not followed by chapter numbers are used throughout the book.

$g(v, v_0), g(v_0), g(v), g(v_s, v_0)$	atomic lineshape function (Chapters 1, 3)
g_0, g	gain coefficient
G, G_0	gain
g, g_1, g_2	parameters of resonator stability (Chapters 5, 9)
hv	photon energy
h	surface heat transfer coefficient (Chapter 7)
h	Planck constant (Chapter 6)
$I, I_L, I_R, I_{circ}, I_{out}, I_{IN}, I_P, I_0$	power density of laser beam
I_{av}, I_w	power density of pump radiation (Chapter 3)
I_{ASE}	fluorescence flux (Chapter 4)
I_{mn}, I_{pl}	intensity distribution of radial modes
I_{ac}	acoustic power density
I_S	saturation power density
i, i_P, i_S	current
J_{ST}, J_{th}	energy density (Chapters 2, 4)
K_0	flashlamp parameter (Chapter 6)
K	thermal conductivity
kT	thermal energy
K, K'	nonlinear coupling factor (Chapter 10)
K_1, K_2, K_3	combination of parameters (Chapter 10)
$k_p, k_s, k_i, k_W, k_{2W}, \Delta k$	wave number
$l, l(r), l_0$	length of active medium
Δl	pathlength difference (Chapter 7)
L, L_O, L_1, L_2	distances within resonator
l_C, l_a	coherence-, aperture length
L	inductance (Chapter 6)
l, m, n, p	higher order modes (Chapter 5)
M	beam quality figure of merit (Chapter 5)
M	power ratio (Chapter 10)
m	magnification of unstable resonator (Chapter 5)
M_{ac}	acoustic figure of merit (Chapter 8)
M_r, M_ϕ	thermal lensing sensitivity (Chapter 7)
$n_0, n_S, n_i, n_p, n_x, n_y$	refractive index
$\Delta n_0, \Delta n(r), \Delta n_\phi, \Delta n_r$	refractive index difference
$n_{1\omega}, n_{2\omega}, n_{3\omega}$	index at harmonic frequencies
n^0, n^e	ordinary-, extraordinary refractive index
$n(r), n_r, n_\phi$	index in radial-, tangential direction

dn/dT	temperature dependence of index
$dn/d\lambda$	wavelength dependence of index
$n(I)$	intensity dependence of index
n_2	nonlinear refractive index (Chapters 4, 9)
n_{g}, n_1	ground state-, lower laser level population density
n_2, n	upper laser level population density
n_{tot}	total population density
$n_{\mathrm{i}}, n_{\mathrm{t}}, n_{\mathrm{f}}, n_{\mathrm{oo}}$	initial-, threshold-, final-, steady-state population density
Δn	perturbation of population density (Chapter 3)
$N_1, N_2, N_0, N_{\mathrm{tot}}, N(\nu), \Delta N$	population of energy levels
N, N_{eq}	Fresnel number (Chapter 5)
N	number of axial modes (Chapter 9)
N	number of flashlamp firings (Chapter 6)
N_{ph}	number of photons (Chapter 5)
N	number density of molecules (Chapter 10)
$P_{\mathrm{in}}, P_{\mathrm{out}}, P_{\mathrm{th}}, P, P_{\mathrm{opt}}$	input-, output-, threshold-, intracavity-, optimum power
$P_{\mathrm{av}}, P_{\mathrm{p}}, P_{\mathrm{cw}}$	average-, peak-, cw power
$P_{\omega}, P_{2\omega}$	fundamental-, harmonic power
P_{ac}	acoustic power (Chapter 8)
$P_{\mathrm{a}}, P_{\mathrm{h}}$	absorbed-, dissipated power
$P_{\mathrm{cr}}, P_{\mathrm{f}}, P_{\lambda}$	critical-, fluorescence-, spectral power
$P_{\mathrm{avail}}, P_{\mathrm{e}}$	available-, incident power
$P_{\mathrm{iklm}}, p_{\mathrm{ij}}$	photoelastic tensor, coefficients
$\mathbf{P}$	induced polarization (Chapter 10)
p	gas pressure (Chapter 6)
$p, p_{\mathrm{n}}, p_{\mathrm{L}}$	resonator modes (Chapters 1, 5)
Q	resonator quality factor (Chapter 3)
Q	heat per volume (Chapter 7)
q	parameter (Chapter 7)
$R_1, R_2, R, R_{\mathrm{opt}}, R_{\mathrm{th}}, R'$	reflectivity
$r, r_0, r_{\mathrm{L}}, r_{\mathrm{R}}$	radius
$R(z), R_1, R_2, R, R_{\mathrm{eff}}$	radius of curvature of resonator mirrors (Chapter 5)
R_{d}	electrical resistance (Chapter 6)
r_{ij}	electro-optic coefficients
R_{S}	thermal shock parameter (Chapter 7)

r_P	pump intensity distribution
$R(\lambda, T)$	flashlamp radiation
s	strain (Chapter 8)
s	Kerr lens sensitivity (Chapter 9)
S, S_1, S_2, S_0	combination of material constants (Chapter 7)
$T, T_{opt}, T_n, T_o, T_i, T_{max}$	transmission
t_p, t_R	pulsewidth, rise time
$T, T(r), T(r_0), T(o), T_F$	temperature
t	time
t_r	round trip time of photons
t_{LC}	time constant
u	nonlinear loss coefficient
V	volume (Chapters 3, 8)
$V, V_0, V_z, V_{1/2}$	voltage (Chapters 6, 8)
v, v_s, v_g	velocity
W	blackbody radiation (Chapter 1)
w	width
W_p	pump rate
W_{13}, W_{03}	pump parameter (Chapter 1)
W_{av}, W_s	energy density (Chapter 3)
$w_0, w, w_1, w_2, w_P, w_{pl}, w_m, w_n$	beam radius
x, y, z	distance
Z_0, Z	impedance
z	parameter
$\alpha_0, \alpha_0(v_s), \alpha_0(E)$	absorption coefficient
α	damping factor of current pulse (Chapter 6)
α	pulse shape factor (Chapter 9)
α	thermal expansion coefficient (Chapter 7)
α_P	angle (Chapters 6, 9)
β_0	propagation constant (Chapter 8)
β	combination of parameters (Chapter 4)
β_θ	angular sensitivity of doubling crystal (Chapter 10)
β	frequency chirp (Chapter 9)
β''	group dispersion in solids (Chapter 9)
χ	susceptibility (Chapter 10)
χ	heat deposition (Chapter 7)
$\delta, \delta_m, \delta_0, \delta_{depol}$	resonator losses
δ_{AM}, δ_{FM}	modulation depth

Δ_q, $\Delta(2p + l)$, $\Delta(m + n)$	frequency separation of axial modes (Chapter 5)
$\varepsilon(\lambda, T)$	emissivity (Chapter 6)
ε	resonator losses including output coupling (Chapters 3, 8)
ε_0, ε	permittivity (Chapter 10)
ϕ, ϕ_0, $\phi(r)$, $\phi(x, t)$	photon density
ϕ	polar coordinate, azimuth angle
γ	factor determining three- or four-level laser
γ	nonlinear refractive coefficient (Chapters 4, 9)
γ	electrostrictive coefficient (Chapter 10)
η_Q, η_S, η_d	quantum-, quantum defect-, differential quantum efficiency
η_P, η_a, η_{sys}	pump-, absorption-, system efficiency
η_c, η_B, η_E	coupling-, overlap-, extraction efficiency
η_{St}, η_{ASE}, η_{EQ}	Q-switch efficiencies
η_h	fractional heat load (Chapter 7)
η_{amp}, η_{par}, η_{Ram}	amplifier-, parametric-, Raman efficiency
η, η'	combination of efficiency factors
φ, $\Delta\varphi$, $d\varphi$, $\varphi(t)$	phase angle
κ	combination of constants (Chapter 10)
κ	Boltzmann factor (Chapter 2)
λ, λ_0, λ_1, λ_2, λ_P, λ_s, λ_i, λ_L	wavelength
$\Delta\lambda$	wavelength difference
Λ	periodicity of grating (Chapter 10)
μ	factor determining Bragg deflection (Chapter 9)
μ_0	permeability of free space
ν	Poisson's ratio (Chapter 7)
ν_0, ν_s, ν_p, ν_L, ν_i, ν_R, ν_m, ν_{rf}	frequency
$\Delta\nu$, $\Delta\nu_L$, $\Delta\nu_c$, $d\nu$	bandwidth, linewidth
θ, θ_c, θ_D, θ', θ_m, $\Delta\theta$	angle
ρ	walk-off angle (Chapter 10)
ρ	normalized beam radius (Chapter 5)
ρ_0	mass density (Chapters 8, 10)
$\rho(\nu)$	radiation density per unit frequency (Chapter 1)
σ	stimulated emission cross section
σ_s	slope efficiency of laser output versus input
σ_r, σ_ϕ, σ_z, σ_{max}	thermal stress (Chapter 7)

σ_{12}, σ_{gs}, σ_{es}	absorption cross section
σ	Stefan–Boltzmann constant (Chapter 1)
τ_f, τ_{21}	spontaneous emission lifetime of upper laser level
τ_c	decay time of photon density in resonator
τ_B	phonon lifetime (Chapter 10)
τ_R	decay time of relaxation oscillations (Chapter 3)
τ_{ij}	spontaneous emission lifetime between states E_i and E_j
ω, ω_0, ω_p, ω_1, ω_2, ω_3, ω_4, ω_s	frequency
$\Delta\omega$	bandwidth
Ω	solid angle beam divergence (Chapter 4)
Ψ	pulsewidth–bandwidth product (Chapter 9)
ζ, ζ_{max}, $\zeta(t, \phi)$	cavity losses (Chapter 8)

Appendix C

Partial Solutions to the Exercises

Chapter 1

5. (page 43) For simplicity we treat the case of equal degeneracies. We want to find the temperature when

$$N_2 A_{21} = B_{21} \rho N_{21},$$

where the photon density is given by the Planck law, Eq. 1.2. Thus, the equation above becomes

$$A_{21} = (B_{21}) \frac{8\pi \nu^2}{c^3} \frac{h\nu}{\exp\left(\dfrac{h\nu}{kT}\right) - 1}.$$

We know that $A_{21}/B_{21} = 8\pi \nu^2 h\nu/c^3$ from Eq. 1.20, so we can write

$$1 = \frac{1}{\exp\left(\dfrac{h\nu}{kT}\right) - 1}$$

or

$$\exp\left(\frac{h\nu}{kT}\right) = 2.$$

Thus, $h\nu/kT = \ln 2$ and so $T = h\nu/k \ln 2$.

Chapter 2

4. (page 76) For a Lorentzian lineshape,

$$\sigma_{21} = \frac{A_{21}\lambda_0^2}{4\pi^2 n^2 \Delta \nu}.$$

The radiative lifetime for R_1 line is then

$$\tau_{21} = \frac{1}{A_{21}} = \frac{\lambda_0^2}{4\pi^2 n^2 \Delta v \sigma_{21}} = \frac{6943^2 \times 10^{-20}}{4\pi^2 \times 1.76^2 \times 330 \times 10^9 \times 2.5 \times 10^{-20} \times 10^{-4}}$$

$$= 4.778 \text{ ms,}$$

and so the nonradiative lifetime is obtained from the fact that both radiative and nonradiative processes go on in parallel. Since the fluorescent lifetime is 3.0 ms, we use

$$\frac{1}{\tau_f} = \frac{1}{\tau_r} = \frac{1}{\tau_{nr}} \Rightarrow \tau_{nr} = \frac{\tau_f \tau_r}{\tau_r - \tau_f} = \frac{4.778 \times 3.0}{4.778 - 3.0} = 8.062 \text{ ms.}$$

The fluorescence quantum efficiency is the ratio of the excited ions that emit radiatively to the total number of ions that decay in any given period. In other words it is the ratio of $\frac{1}{\tau_r}$ to the total lifetime $\frac{1}{\tau_r} + \frac{1}{\tau_{nr}}$. In other words

$$\eta_Q = \frac{\tau_f}{\tau_r} = \frac{3}{4.778} = 0.63.$$

Chapter 3

5. (pages 119–120) Let $T = 1 - R$, and use the approximations that for highly reflecting mirrors $1 + R \approx 2$ and $\ln(1 - T) = -T$. Then from Eq. 3.56, we have

$$P_{\text{out}} = AI_S \frac{T}{2} \left(\frac{2g_0 l}{\delta + T} - 1 \right).$$

Now differentiating this with respect to T:

$$\frac{dP_{\text{out}}}{dT} = AI_S \left[g_0 l \left(\frac{1}{\delta + T} - \frac{T}{(\delta + T)^2} \right) - 1/2 \right].$$

By letting $\frac{dP_{\text{out}}}{dT} = 0$ we find the value of T that maximizes P_{out}. We find

$$g_0 l \left(\frac{1}{\delta + T} - \frac{T}{(\delta + T)^2} \right) - 1/2 = 0$$

$$\Rightarrow \frac{g_0 l (\delta + T)}{(\delta + T)^2} - \frac{g_0 l T}{(\delta + T)^2} - 1/2 = 0$$

$$\Rightarrow \frac{g_0 l \delta}{(\delta + T)^2} - 1/2 = 0$$

$$\Rightarrow \delta + T = \sqrt{2 g_0 l \delta}$$

$$\Rightarrow T = \left(\sqrt{2 g_0 l / \delta} - 1 \right) \delta.$$

Substituting this value of T back into P_{out}, we have

$$P_{out} = A I_S \frac{1}{2} \left(\sqrt{2g_0 l/\delta} - 1 \right) \delta \cdot \left(\frac{2g_0 l}{\sqrt{2g_0 l/\delta}} - 1 \right)$$

$$= A I_S \frac{1}{2} \left(\sqrt{2g_0 l/\delta} - 1 \right) \left(\sqrt{2g_0 l/\delta} - 1 \right) \delta$$

$$= A I_S \frac{1}{2} \left(\sqrt{2g_0 l/\delta} - \sqrt{\delta} \right)^2$$

$$= A I_S \frac{1}{2} \left[\sqrt{2g_0 l} \cdot \left(1 - \sqrt{\delta/2g_0 l} \right) \right]^2$$

$$= A I_S g_0 l \cdot \left(1 - \sqrt{\delta/2g_0 l} \right)^2 .$$

Chapter 4

3. (page 148) For Q-246 Nd : glass we find $I_s = h\nu/\sigma\tau$ from the data in the text. It is 1.8×10^8 W/m^2. The saturated gain of the amplifier given by Eq. 4.11 is to be 10, and the small signal gain, $G_0 = \exp(g_0 l) = 20$. Since we will use Eq. 4.11, we assume a square pulse with duration $\tau_p \ll \tau$. Thus

$$E_s = I_S \tau_p \quad \text{and} \quad E_i = I_{in} \tau_p.$$

We now insert these expressions into Eq. 4.11 and write

$$G = 10 = \frac{I_S}{I_{in}} \ln \left\{ 1 + \left[\exp \left(\frac{I_{in}}{I_S} \right) - 1 \right] 20 \right\} .$$

This gives I_{in}/I_S as 0.137, and therefore $I_{in} = 2.6 \times 10^7$ W/m^2.

The output intensity is 10 times the input intensity.

Chapter 5

2. (page 185) The resonator is

$R_1 = 1000$ $R_2 = -1000$ cm

This means that $g_1 = 0.9$ and $g_2 = 1.1$, so that $g_1 g_2 = 0.99$, indicating a stable resonator.

Now use Eqs. 5.29 and 5.30 to find the mode sizes on each mirror. $w_1 = 1.876$ mm and $w_2 = 1.697$ mm. Eq. 5.13 gives the beam waist radius as $w_0 = 1.258$ mm, and Eq. 5.1 gives L_1 as 5.5 m so that the waist is located 4.5 m to the right of mirror 2.

Since the beam is largest near mirror 1, that is where to put the laser rod for maximum TEM$_{00}$ mode size in the gain medium.

Chapter 6

1. (page 243) For an input energy of $E_0 = 10$ J we use Eq. 6.3 to find the explosion limit as

$$E_{ex} = (1.2 \times 10^4) l \, dt_p^{1/2},$$

where $L = 6$ cm and $d = 0.4$ cm, $t_p = 300$ μs. The explosion energy is then $E_{ex} = 498.8$ J. The expected lifetime (number of pulses) for 10 J pulses can be calculated using Eq. 6.4,

$$N \cong \left(\frac{E_{ex}}{E_0} \right)^{8.5}$$

as 2.76×10^{14} pulses.

This result does not agree with that in the example. The calculated expected lifetime is larger than that in the example by 7 orders of magnitude. In this case the input energy is only 2% of the explosion energy. There is another example in *Solid-State Laser Engineering*, a large linear lamp employed to pump ruby and Nd:glass laser rod; the lamp is 16.5 cm long and has a bore diameter of 13 mm. The lamp has explosion energy of 8450 J for a 1-ms-long pulse. If the lamp is operated at an input energy of 2000 J, a flashlamp life of 150 (103 shots can be expected using Eq. 6.4. In this case there is good agreement between the calculated result and that in the example. So the formula seems to work.

Consider that there are other mechanisms for lamp failure besides explosion. The lamp jacket can be eroded by the plasma in the discharge after many shots. This weakens the jacket and lowers the explosion limit as use continues. The erosion also clouds the lamp jacket, making the jacket absorptive to the pump light. The lamp becomes progressively less efficient as a pump source. Thus, while not at its originally calculated explosion limited life, the lamp must be replaced after only 10 million shots because it is no longer as good a pump source as it was, and its current explosion limit after 10 million shots is less than when it was new.

Chapter 7

1. (page 278) The material constant of the Nd:YAG laser crystal required for this calculation are thermal expansion coefficient $\alpha = 7.5 \times 10^{-6}/°C$,

thermal conductivity $K = 0.14$ W/cm°C,

Poisson's ratio, $\nu = 0.25$,

the refractive index $n_0 = 1.82$, and

$dn/dT = 7.3 \times 10^{-6}/°C$

Now substitute these parameters into Eqs. 7.20 and 7.21:

$$\Delta n_r = -\frac{1}{2}n_0^3 \frac{\alpha Q}{K} C_r r^2,$$

$$\Delta n_\phi = -\frac{1}{2}n_0^3 \frac{\alpha Q}{K} C_\phi r^2,$$

where C_r and C_ϕ are functions of the elasto-optic coefficients of Nd : YAG,

$$C_r = \frac{(17\nu - 7)P_{11} + (31\nu - 17)P_{12} + 8(\nu + 1)P_{44}}{48(\nu - 1)}$$

$$C_\phi = \frac{(10\nu - 6)P_{11} + 2(11\nu - 5)P_{12}}{32(\nu - 1)}.$$

Here P_{11}, P_{12}, and P_{44} are -0.029, 0.0091, and -0.0615, respectively. Evaluating C_r, C_ϕ, Δn_r, Δn_ϕ yields

$$C_r = 0.017, C_\phi = -0.0025,$$

$$\Delta n_r = (-2.8 \times 10^{-6}) Q r^2,$$

$$\Delta n_\phi = (0.4 \times 10^{-6}) Q r^2.$$

Eq. 7.34 is used to calculate the induced focal length considering the combined effects of the temperature- and stress-dependent variation of the refractive index and the distortion of the end-face curvature of the rod. It is

$$f = \frac{KA}{P_a} \left(\frac{1}{2}\frac{dn}{dT} + \alpha C_{r,\phi} n_0^3 + \frac{\alpha r_0 (n_0 - 1)}{l} \right)^{-1}.$$

Ignoring the end effects, assuming the dimension of Nd : YAG laser rod used is 6 mm in diameter and 100 mm in length, and assuming the Nd : YAG laser rod is uniformly pumped so the rod cross-section area is 0.2826 cm², the focal length can be expressed as

$$f = \frac{KA}{P_a} \left(\frac{1}{2}\frac{dn}{dT} + \alpha C_{r,\phi} n_0^3 \right)^{-1}.$$

If the total heat dissipated in the rod is 1000 W, the focal length of this laser rod can be calculated as $f_r = 8.95$ cm and $f_\phi = 11.18$ cm. These are very short focal lengths, and clearly the rod is astigmatic.

Chapter 8

2. (pages 306–307)

(a) If the motor speed is 24,000 rpm and the misalignment tolerance is 1 mrad, then the time for switching is the time it takes the mirror to move through an angle of 1 mrad or

$$t = 10^{-3}/(2\pi \times 24000/60) \text{ sec} = 3 \times 10^{-7} \text{ sec}.$$

(b & c) If instead of a mirror one used a 40-90-45 prism as the rotating cavity reflector, the effect of the prism is to make the light require two round trips to reflect back on itself. This is because the roof prism is a retro-reflector in the plane perpendicular to its roof ridge. As indicated in the figure below, the roof prism makes the cavity seem twice as long and therefore makes it seem to have 0.5 mrad of alignment tolerance. In this figure the paths have been displaced from each other slightly to show the two round trips more clearly. However, the roof prism rotating about an axis parallel to the ridge can align with the output coupler, perhaps coated on the far end of the rod, over a wide range of angles (see sketch below). It is therefore preferable to rotate the prism about an axis perpendicular to the roof top ridge. In this manner one gains the faster Q-switching time made possible by the rooftop reflector while gaining some insensitivity to misalignments in the plane perpendicular to both the ridge and the rotation axis.

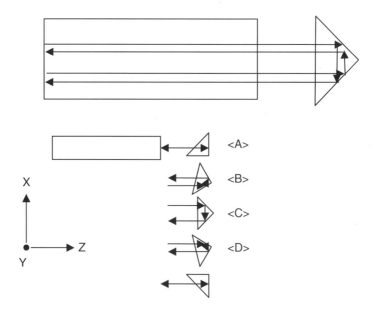

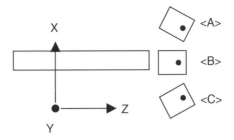

Chapter 9

2. (page 338) The intensity as a function of time is from the preceding problem

$$I(t) = I_{\max} \frac{\sin^2\left(\frac{N\Omega}{2} t_{\text{HWHM}}\right)}{\sin^2\left(\frac{\Omega}{2} t_{\text{HWHM}}\right)}.$$

To find the pulse width we first find the width at the base of each pulse, or in other words, when the pulse amplitude first becomes zero after the peak. This occurs when the numerator in the preceding equation is first zero. That is when

$$\frac{N\Omega}{2} \frac{t_{\text{base}}}{2} = \pi.$$

So $t_{\text{base}} = 4\pi/N\Omega = 2\pi/N(2\pi c/2L) = 4L/Nc = 2t_{\text{round trip}}/N$ and is exactly correct. Thus, the pulse duration at the base for the described laser is 4×10^{-11} sec.

The time at which the half maximum is reached is the time that causes

$$I_{\max} \frac{\sin^2\left(\frac{N\Omega}{2} t_{\text{HWHM}}\right)}{\sin^2\left(\frac{\Omega}{2} t_{\text{HWHM}}\right)} = \frac{I_{\max}}{2}.$$

This is a transcendental equation in t_{HWHM} and must be solved numerically for precise results. When it is solved numerically, the result for t_{FWHM} is the result in Eq. 9.4 but divided by 1.128. Thus Eq. 9.4 is off by about ~13% in this problem.

Another approximation is to consider that since the half width at half maximum time should be much less than the interpulse period we can approximate it by finding the time when

$$\sin\left(\frac{N\Omega}{2} t_{\text{HWHM}}\right) = \frac{1}{\sqrt{2}}.$$

Here the approximation is that the numerator in the expression for I changes much more quickly than the denominator. This is so, but the denominator does change from 0 to some finite value necessitating the numerical solution for a precise answer. However, the approximation may be useful. We have in this approximation

$$t_{\text{HWHM}} = \frac{2}{N\Omega} \sin^{-1}\left(\frac{1}{\sqrt{2}}\right) = \frac{2}{N\Omega} \frac{\pi}{4} = \frac{\pi}{2N\frac{2\pi c}{2L}} = \frac{1}{4N} t_{\text{roundtrip}}.$$

The full width at half maximum is twice this number or $t_{\text{round trip}}/2N$ and for the cavity described we have 0.5×10^{-11} sec. In this approximation instead of 1.13 we have 2.00 in the denominator. While it is not precisely correct, it gives a useful estimate of the pulse duration at t_{FWHM}.

Chapter 10

2. (page 386)

(a) The intensity going in is

$$I_\omega(r, t) = I_{0\omega} \exp\left(-\left(\frac{2r^2}{w^2} + \frac{t^2}{T^2}\right)\right).$$

This squared is proportional to the intensity going out of the SHG as

$$I_{2\omega}(r, t) \propto (I_{0\omega})^2 \exp\left(-2\left(\frac{2r^2}{w^2} + \frac{t^2}{T^2}\right)\right).$$

When this light and the fundamental light interact in the SFG to produce the third harmonic, we have $I_{3\omega}$ proportional to $I_\omega I_{2\omega}$ so

$$I_{3\omega}(r, t) \propto (I_{0\omega})^3 \exp\left(-2\left(\frac{2r^2}{w^2} + \frac{t^2}{T^2}\right)\right) \exp\left(-\left(\frac{2r^2}{w^2} + \frac{t^2}{T^2}\right)\right)$$

or

$$I_{3\omega}(r, t) \propto (I_{0\omega})^3 \exp\left(-3\left(\frac{2r^2}{w^2} + \frac{t^2}{T^2}\right)\right),$$

Therefore the radius at which the intensity is $1/e^2$ of the peak on axis intensity of the third harmonic is $w_{3\omega} = w_\omega/\sqrt{3}$, and the pulse duration corresponding to the $1/e$ height of the intensity in time of the third harmonic is $T_{3\omega} = T_\omega/\sqrt{3}$.

(b) Using a SHG first and then an SFG allows one to have large intensities at both w and $2w$ in the SFG so that it works efficiently. That is, it is better to sum two high-intensity beams than to triple one high-intensity beam. In each nonlinear device we deal only with a second-order nonlinearity not the much smaller third-order nonlinearity. It is also easier to find phase-matching conditions for a second-order process than it is for a third-order process.

Index

Absorption, 18, 21, 25, 61, 64, 71, 72
  coefficient, 27, 57
  cross section, 27, 57
  efficiency, 89, 91, 197, 199, 209
Acousto-optic modulator, 322–326
Acousto-optic Q-switch, 295–302
Actinide, 52
Alexandrite laser, 70–72
Amplified spontaneous emission, 94, 142–144
Amplifier
  Nd: glass, 136–142
  Nd: YAG, 128–136
Amplitude fluctuation; *see* Output fluctuation
Angular divergence; *see* Beam divergence
Aperture length, 353
Arc lamp; *see* Flashlamps
Atomic energy level; *see* Energy level
Atomic lineshapes, 21–25
Axial modes; *see* Longitudinal modes

Bandwidth; *see* Linewidth
Beam divergence definition, 155, 156, 162, 163
Beam overlap efficiency, 92, 93
Beta-barium borate, 354, 357, 370
Biaxial crystal, 348
Birefringence
  electrically induced, 289–294
  thermally induced, 248, 258–264, 267, 268
Blackbody radiation, 15, 16, 191
Bleachable dye; *see* Saturable absorber
Bohr's frequency relation, 12

Boltzmann distribution, 16, 17
Bragg scattering, 296–299, 328
Brewster angle, 269, 270, 271
Brightness, definition, 123
Brillouin scattering, 343–345, 379–383
Broadening of atomic transitions; *see* Line broadening effects

Cerium, 52
Circulating power, 86–88
Close-coupled pump cavity, 214, 215
Coherence length
  frequency doubling, 347, 348
  laser radiation, 176, 383
Collision broadening, 22
Concave-convex, resonator, 159, 160, 169
Concentric resonator, 159, 378
Confocal parameter, 156
Confocal resonator, 159, 160
Confocal unstable resonator, 180–182
Continuous arc lamp, 189, 197, 198
Conversion efficiency, 345–347, 362, 368, 377
Cooling, 219–228, 239–241, 249, 271
Cr: Forsterite laser, 53, 331
Cr: GSGG laser, 47
Cr: LiSAF laser, 53, 331
Critical phase matching, 352
Cross section, stimulated emission, 28, 56, 58, 62, 65, 66, 70, 73

Decay time, resonator, 38, 81, 82
Degeneracy, energy levels, 16, 17, 20, 28
Degenerate mode, 370

Depolarization loss, 261, 262
Depopulation losses, 142–144
Diffraction losses, resonator, 163–165
Diode arrays; *see* Laser diodes
Dipolar broadening, 23
Disk amplifier, 137–140, 221, 225–229,
    265–271; *see also* Slab laser
Dispersion compensation, 233–235
Doppler broadening, 24
Dye Q-Switch; *see* Saturable absorber
Dysprosium, 52

Efficiency factors, 89–95, 115, 117,
    126, 127, 136, 198, 211,
    217–219, 230, 240
Einstein coefficients, 17, 20
Electronic feedback loop, 329, 330
Electro-optic effect, 289–294
Electrostriction, 146, 344
Elliptical cylinder, pump cavity,
    214–219
Emissivity, 190
End-pumped lasers, 116–118, 230–239,
    271–276, 319, 323, 327, 329,
    332, 367, 377
Energy extraction, efficiency, 93, 99,
    125–127, 283
Energy level
    diagram, 29–31, 33, 55, 59, 63, 64,
        67, 69, 71
    nomenclature, 48–51
Energy storage, 125–128
Energy-transfer mechanisms, 88–95
Erbium, 50, 51
Er: glass laser, 67, 68
Er: YAG laser, 51
Etalon, 171, 176
Europium, 52
Excited state absorption, 72, 304, 305
Explosion energy, flashlamp, 192, 193

Fabry–Perot interferometer, 170–173
Fabry–Perot resonator; *see* Resonator,
    optical
Faraday effect, 178
Faraday isolator, 131, 132
Femtosecond lasers, 331–336
Fiber coupling, 236–238
Finesse, 172, 173, 176

Flashlamps, 188–197
Fluorescence, 18, 27, 32, 33, 40
Fluorides, laser host, 48
Forbidden transition, 34
Fourier transform, 313, 314
Four-level laser, definition, 33, 34
Frequency chirp, 313, 333–335
Frequency doubling; *see*
    Second-harmonic generation
Frequency stability, 179
Fresnel number, 164–166, 180, 182
Fundamental mode, 151–162; *see also*
    Gaussian beam

g parameters, resonator, 162–167, 169,
    170
GaAlAs laser diodes, 111–118, 188,
    189, 197–213
Gadolinium, 50, 52
Gain coefficient, 56, 57, 80–82, 85–96,
    98–102, 125–128, 133, 282,
    365, 366, 376, 382
Gain saturation, 84–86, 125–127
Garnet, laser host, 47; *see also* Nd:
    YAG, YB: YAG lasers
Gaussian beam, 155–157
Gaussian lineshape, 25
Gaussian temporal profile, 313, 314
Giant pulse; *see* Q-switch
Glass, laser host, 46, 60–63, 67, 68
Ground level, 31–34
GSGG, laser host, 47

Harmonic generation; *see*
    Second-harmonic generation
Heat removal; *see* Cooling
Heat-transfer coefficient, 249, 250
Hemispherical resonator, 159–161
Hermite polynomial, 151, 153
Hole burning; *see* Spatial hole buming
Holmium, 50–52
Homogeneous broadening, 22, 23
Host materials, 46–48

Idler wave, 363–366, 368, 373, 374
Indicatrix, 291, 292, 348–352
Inhomogeneous broadening, 23–25
Interferometer; *see* Fabry–Perot
    interferometer

Intracavity frequency doubling, 358–360
Inversion reduction factor, 41
Isolator, 136, 147

KDP, 291–293, 348, 350–354, 357, 361, 362
Kerr effect, 146, 147, 290, 317–322, 331, 343, 344
Kerr lens mode locking, 317–322, 331–336
Kerr lens sensitivity, 318
Krypton arc lamp, 188, 189, 197, 198
KTP, 354, 355, 357, 359, 360, 366, 367, 370, 371

Laguerre polynomial, 151, 152
Large-radius mirror resonator, 158, 160
Laser amplifier; *see* Amplifier
Laser diode pumped systems, 111–118, 131–136, 221–241
Laser diodes, 197–213; *see also* GaAlAs laser diodes
Laser threshold; *see* Oscillator, threshold condition
Lifetime broadening, 22
Light-pipe mode, 145
LiNbO$_3$, 291–294, 353, 354, 357, 372–374
Line broadening effects, 22–25
Linewidth, 21–26, 58, 62, 66, 73, 81, 173–176, 178, 179, 381
Longitudinal mode selection, 176–179
Longitudinal modes, 150, 173–176
Lorentzian lineshape, 23, 25
Loss modulation, 322–325

Metastable level, 34, 35
Mode locking, 308–336
Mode locking, active, 322–326
Mode locking, passive, 315–317
Mode matching, 92, 93, 116
Mode patterns, 152, 154
Mode radius, resonator modes, 153, 155, 156, 162, 163, 166, 167, 169, 170, 232
Mode selection
   longitudinal, 176–179
   transverse, 167–170

Monolithic laser, 177
MOPA design, 123, 137

Nd: Cr: GSGG laser, 47
Nd: glass laser, 60–63
   amplifier, 136–142
Nd: YAG laser, 57–60, 106–118, 177, 287, 288
   amplifier, 128–136
   thermal effects, 250–263
Nd: YLF laser, 48
Nd: YVO$_4$ laser, 65–67
Neodymium; *see* Nd: glass, Nd: YAG, Nd: YLF, Nd: YVO$_4$, Nd: Cr: GSGG lasers
Noncritical phase matching, 353
Nonlinear coefficient, 346, 351, 354, 366, 367
Nonlinear crystals, 354
Nonlinear optical effects, 340–345
Nonlinear refractive index, 138, 139, 147, 318, 331
Nonspherical aberration, 363

Optical parametric oscillator, 363–374
Optical phase conjugation, 379–384
Optical resonator; *see* Resonator, optical
Organic dye; *see* Saturable absorber
Oscillator, threshold condition, 80–84, 97
Oscillator loop, electronic feedback,
Output coupling, 95, 98, 99, 108, 109, 110, 111
Output fluctuation, 102–107
Output vs input calculation, 95–102

Parasitic modes, 145
Passive mode locking; *see* Mode locking, passive
Passive Q-switch, 302–305; *see also* Dye Q-switch
Periodically poled crystals, 372–374
PFN; *see* Pulse-forming network
Phase coherence, stimulated emission, 19
Phase matching, 345–356, 363, 366, 369, 370, 372–374
Phase modulation, 329–331

Picosecond lasers, 326–331
Planck's constant, 12
Planck's law, 15
Plane-parallel resonator, 159, 160
Plane wave impedance, 346
Plano-concave resonator, 160
Plastic Q-switch, 302
Pockels cell Q-switch, 290–295
Polarization, induced, 340–343
Population inversion, 13, 28–35, 82, 124
Potassium titanyl phosphate; *see* KTP
Power supplies, 194–197, 213
PPLN, 372–374
Praseodymium, 50, 52
Prelasing, 145–147
Pulse-forming network, 194–197
Pump band, 31, 33
Pump cavity, 214–221
Pump source efficiency, 89
Pumping rate, 37, 83–86, 94, 102, 104

Quasi phasematching, 372–374
Q-switch devices
    acousto-optic, 295–302
    electro-optical, 289–295
    mechanical, 288
    passive, 302–305
Q-switch theory, 280–288
Quality factor Q, 81, 279
Quantum efficiency, 37, 40
Quantum noise, 178
Quantum well, 201, 202

Radiation transfer efficiency, 89, 90
Raman laser, 374–379
Raman–Nath scattering, 296
Rare earth ions, 49–52
Rate equations, 35–40
Rectangular slab laser, 265, 268
Relaxation oscillation, 102–106
Resonant reflector, 171–173
Resonator, optical, 150–183
    configuration, 157–160
    modes, 150–154
    unstable, 179–183
Ring laser, 177, 178

Ruby laser, 54–57

Samarium, 52
Saturable absorber, 302–305
Saturation fluence, flux, 85, 86, 125–127
Second-harmonic generation, 345–360
    intracavity, 358–360
    theory, 345–353
Self-focusing, 145–148
Self mode locking; *see* Kerr lens mode locking
Semiconductor, pump source, 197–213
Sensitivity factor, thermal lensing, 257, 258
Sensitizer, 52, 67, 68
Servo loop; *see* Electronic feedback loop
Side-pumped active material, 107–116, 214–230
Simmer triggering, 195
Slab laser, 185, 221, 225–229, 265–271
Slope efficiency, 96, 97
Spatial filter, 137
Spatial hole burning, 177
Spectral characteristic, laser output, 173–176
Spiking; *see* Relaxation oscillation
Spontaneous emission, 18
Spot size, definition, 155, 156, 162, 163
Stability diagram for resonator, 161, 162, 167, 321
Stefan–Boltzmann equation, 16, 17
Stimulated emission, 19, 25
    cross section, 28
Stimulated Raman scattering, 344, 374–379
Stokes factor, 91, 92
Stokes shift, 375, 377
Storage efficiency, 94
System efficiency, 88, 115, 117, 136

TEM_{mnq}, TEM_{plq} modes, definition, 150–154
Thermal beam distortion, 245, 255–263, 267, 268, 275
Thermal broadening, 23
Thermal effect, laser rod

birefringence, 258–263
    fracture, 245, 252
    lensing, 255–258
Third-harmonic generation, 360–363
Three-level laser, 31, 32, 36–39
Threshold condition; *see* Oscillator,
    threshold condition
Threshold input, 97
Thulium, 52
Ti: sapphire laser, 72–74, 319–322,
    331
Tm: YAG, 52
Transition metals, 53, 54
Transverse mode selection; *see* Mode
    selection, transverse
Transverse modes, 150–156
Travelling wave oscillator; *see* Ring
    laser
Trigger circuit, flashlamp, 191
Tunable lasers, 70–74
Tungstate, laser host, 45
Type I, II phase matching, 351, 352,
    354

Uniaxial crystal, 348
Unstable resonator, 179–183
Upper state efficiency, 89, 92

Vanadate, laser host, 47
Variable reflectivity mirror, 183, 184
Vibronic lasers, 70–74

Waist, Gaussian beam, 155, 156
Wavefront distortion, 139, 380
Whisper modes, 145
Wien's displacement law, 16

Xenon arc, spectral data, 190, 191

YAG, laser host, 47, 57–60, 68–70; *see
    also* Nd: YAG, Er: YAG, Yb:
    YAG lasers
Yb: YAG, 68–70
YLF, laser host, 48, 63–65
Ytterbium, 52

Zig-zag slab laser, 185, 268–270